FLOW CYTOMETRY FOR BIOTECHNOLOGY

FLOW CYTOMETRY FOR BIOTECHNOLOGY

Edited by

Larry A. Sklar

UNIVERSITY PRESS

2005

OXFORD
UNIVERSITY PRESS

Oxford University Press, Inc., publishes works that further
Oxford University's objective of excellence
in research, scholarship, and education.

Oxford   New York
Auckland   Cape Town   Dar es Salaam   Hong Kong   Karachi
Kuala Lumpur   Madrid   Melbourne   Mexico City   Nairobi
New Delhi   Shanghai   Taipei   Toronto

With offices in
Argentina   Austria   Brazil   Chile   Czech Republic   France   Greece
Guatemala   Hungary   Italy   Japan   Poland   Portugal   Singapore
South Korea   Switzerland   Thailand   Turkey   Ukraine   Vietnam

Published by Oxford University Press, Inc.
198 Madison Avenue, New York, New York 10016

www.oup.com

Oxford is a registered trademark of Oxford University Press

Library of Congress Cataloging-in-Publication Data
Flow cytometry for biotechnology / edited by Larry A. Sklar.
p. ; cm.
ISBN-13 978-0-19-515234-0

1. Flow cytometry. 2. Biotechnology.
[DNLM: 1. Flow Cytometry—methods. 2. Biotechnology—methods.
QH 585.5.F56 F64407 2004]   I. Sklar, Larry A.
TP248.25.F57F57 2004
660.6—dc22   2004000655

Printed in the United States of America
on acid-free paper

Contents

Acknowledgments

This book was compiled in support of NIH Grant 1R24 GM60799/EB00264, "7TMR Drug Discovery, Microfluidics, and HT Flow Cytometry," to Larry A. Sklar, Bruce S. Edwards, Eric R. Prossnitz, Tione Buranda, and Gabriel P. Lopez.

Contributors

Frances H. Arnold
Division of Chemistry and Chemical
   Engineering
California Institute of Technology
Pasadena, CA 91126-0001

Jan Bartoš
Laboratory of Molecular Cytogenetics and
   Cytometry
Institute of Experimental Botany
Sokolovska 6, CZ-77200 Olomouc
Czech Republic

David Basiji
Amnis Corporation
2505 Third Ave., Suite 210
Seattle, WA 98121

Mark K. Bennett
Rigel, Inc.
1180 Veterans Blvd.
South San Francisco, CA 94040

Tione Buranda
Cancer Research and Treatment Center and
   Department of Pathology
University of New Mexico Health Science
   Center
MSC08-4630
Albuquerque, NM 87131-5691

Scott W. Burchiel
College of Pharmacy
University of New Mexico Health Science
   Center
MSC09-5630
Albuquerque, NM 87131-5691

Alexandre Chigaev
Cancer Research and Treatment Center and
   Department of Pathology
University of New Mexico Health Science
   Center
MSC08-4630
Albuquerque, NM 87131-5691

Alina Desphande
National Flow Cytometry Resource
   Bioscience Division
Mail Stop M-888
Los Alamos National Laboratory
Los Alamos, NM 87545

Stephen C. De Rosa
ImmunoTechnology Section
Vaccine Research Center, NIAID, NIH
40 Convent Dr., Room 5614
Bethesda, MD 20892-3015

Jaroslav Doležel
Laboratory of Molecular Cytogenetics and
   Cytometry
Institute of Experimental Botany
Sokolovska 6, CZ-77200 Olomouc
Czech Republic

Bruce S. Edwards
Cancer Research and Treatment Center and
   Department of Pathology
University of New Mexico Health Science
   Center
MSC08-4630
Albuquerque, NM 87131-5691

Anne Y. Fu
Department of Applied Physics
California Institute of Technology
Pasadena, CA 91126-0001

David W. Galbraith
303 Forbes Building
Department of Plant Sciences
University of Arizona
Tucson, AZ 85721

Duane L. Garner
P.O. Box 1939
Graeagle, CA 96103-1939

Steven W. Graves
National Flow Cytometry Resource
  Bioscience Division
Mail Stop M-888
Los Alamos National Laboratory
Los Alamos, NM 87545

George Georgiou
Department of Chemical Engineering
Department of Biomedical Engineering
Institute for Cell and Molecular Biology
University of Texas
Austin, TX 78712

Karl E. Griswold
Institute for Cell and Molecular Biology
Department of Chemistry and Biochemistry
University of Texas
Austin, TX 78712

Robert C. Habbersett
National Flow Cytometry Resource
  Bioscience Division
Los Alamos National Laboratory
Los Alamos, NM 87545

Barrett R. Harvey
Institute for Cell and Molecular Biology
University of Texas
Austin, TX 78712

Yasumichi Hitoshi
Rigel, Inc.
1180 Veterans Blvd.
South San Francisco, CA 94040

Sacha Holland
Rigel, Inc.
1180 Veterans Blvd.
South San Francisco, CA 94040

Brent L. Iverson
Institute for Cell and Molecular Biology
Department of Chemistry and
  Biochemistry
University of Texas
Austin, TX 78712

James H. Jett
National Flow Cytometry Resource
  Bioscience Division
Los Alamos National Laboratory
Los Alamos, NM 87545

Richard A. Keller
Bioscience Division
Los Alamos National Laboratory
Los Alamos, NM 87545

Alexander T. Key
Departments of Cell Biology and Physiology
  and Pathology
Cancer Research and Treatment Center
University of New Mexico Health Science
  Center
MSC08-4630
Albuquerque, NM 87131-5691

Richard S. Larson
Cancer Research and Treatment Center and
  Departments of Pathology
University of New Mexico Health Science
  Center
MSC08-4630
Albuquerque, NM 87131-5691

Gabriel P. Lopez
Department of Chemical and Nuclear
  Engineering
UNM School of Engineering
Albuquerque, NM 87131

James B. Lorens
Rigel, Inc.
1180 Veterans Blvd.
South San Francisco, CA 94040

Babetta L. Marrone
Bioscience Division
Los Alamos National Laboratory
Los Alamos, NM 87545

Esteban Masuda
Rigel, Inc.
1180 Veterans Blvd.
South San Francisco, CA 94040

Susan M. Molineaux
Rigel, Inc.
1180 Veterans Blvd.
South San Francisco, CA 94040

John P. Nolan
National Flow Cytometry Resource
  Bioscience Division
Mail Stop M-888
Los Alamos National Laboratory
Los Alamos, NM 87545

Erlina Pali
Rigel, Inc.
1180 Veterans Blvd.
South San Francisco, CA 94040

Donald G. Payan
Rigel, Inc.
1180 Veterans Blvd.
South San Francisco, CA 94040

Eric R. Prossnitz
Department of Cell Biology and Physiology
Cancer Research and Treatment Center
University of New Mexico Health Science
  Center
MSC08-4630
Albuquerque, NM 87131-5691

Ross M. Potter
Department of Cell Biology and Physiology
Cancer Research and Treatment Center
University of New Mexico Health Sciences
  Center
MSC08-4630
Albuquerque, NM 87131-5691

Mark Powell
Rigel, Inc.
240 E. Grand Ave.
South San Francisco, CA 94028

Stephen R. Quake
Department of Bioengineering
Stanford University
Clark Center
Room S166
Stanford, CA 94305-5444

Sergio A. Ramirez
Cancer Research and Treatment Center and
  Department of Pathology
University of New Mexico Health Science
  Center
MSC08-4630
Albuquerque, NM 87131-5691

Mario Roederer
Chief, ImmunoTechnology Section
Immunology Laboratory
Vaccine Research Center, NIAID, NIH
40 Convent Dr., Room 5509
Bethesda, MD 20892-3015

George E. Seidel, Jr
Animal Reproduction and Biotechnology
  Laboratory,
Colorado State University
Fort Collins, CO 80523

Howard Shapiro
283 Highland Avenue
West Newton, MA 02465-2513

Mei Shi
Departments of Cell Biology and Physiology
  and Pathology
Cancer Research and Treatment Center
University of New Mexico Health Science
  Center
MSC08-4630
Albuquerque, NM 87131-5691

Peter Simons
Departments of Cell Biology and Physiology
  and Pathology
Cancer Research and Treatment Center
University of New Mexico Health Sciences
  Center
MSC08-4630
Albuquerque, NM 87131-5691

Larry A. Sklar
Cancer Research and Treatment Center and
Department of Pathology
University of New Mexico Health Science
  Center
MSC08-4630
Albuquerque, NM 87131-5691

Adam Treister
Tree Star, Inc.
20 Winding Way
San Carlos, CA 940707

Charlotte M. Vines
Departments of Cell Biology and Physiology
Cancer Research and Treatment Center
University of New Mexico Health Sciences
  Center
MSC08-4630
Albuquerque, NM 87131-5691

James L.Weaver
US FDA
Center for Drug Evaluation and Research
Division of Applied Pharmacology Research
Laurel, MD 20708

Stuart S. Winter
Cancer Research and Treatment Center and
  Department of Pediatrics
University of New Mexico Health Science
  Center
MSC10-5590
Albuquerque, NM 87131-5691

Xiaomei Yan
Bioscience Division
Los Alamos National Laboratory
Los Alamos, NM 87545

Yohei Yokobayashiy
Division of Chemistry and Chemical
  Engineering,
California Institute of Technology,
Pasadena, CA 91126-0001

Thomas M. Yoshida
Chemistry Division
Los Alamos National Laboratory
Los Alamos, NM 87545

Feng Zhou
National Flow Cytometry ResourceBioscience
  Division
Mail Stop M-888
Los Alamos National Laboratory
Los Alamos, NM 87545

Gordon Zwartz
Cancer Research and Treatment Center and
Department of Pathology
University of New Mexico Health Science
  Center
MSC08-4630
Albuquerque, NM 87131-5691

FLOW CYTOMETRY FOR BIOTECHNOLOGY

1

The Future of Flow Cytometry in Biotechnology: The Response to Diversity and Complexity

LARRY A. SKLAR

Introduction

Flow cytometry is a mature technology: Instruments recognizable as having elements of modern flow cytometers date back at least 30 years. There are many good sources for information about the essential features of flow cytometers, how they operate, and how they have been used (5, 11, 12). For the purposes of this book, it is necessary to know that flow cytometers have fluidic, optical, electronic, computational, and mechanical features. The main function of the fluidic components is to use hydrodynamic focusing to create a stable particle stream in which particles are aligned in single file within a sheath stream, so that the particles can be analyzed and sorted. The main functions of the optical components are to allow the particles to be illuminated by one or more lasers or other light sources and to allow scattered light as well as multiple fluorescence signals to be resolved and be routed to individual detectors. The electronics coordinate these functions, from the acquisition of the signals (pulse collection, pulse analysis, triggering, time delay, data, gating, detector control) to forming and charging individual droplets, and to making sort decisions. The computational components are directed at postacquisition data display and analysis, analysis of multivariate populations and multiplexing assays, and calibration and analysis of time-dependent cell or reaction phenomena. Mechanical components are now being integrated with flow cytometers to handle plates of samples and to coordinate automation such as the movement of a cloning tray with the collection of the droplets. The reader is directed to a concise description of these processes in Robinson's article in the *Encyclopedia of Biomaterials and Biomedical Engineering* (11).

This book was conceived of to provide a perspective on the future of flow cytometry, and particularly its application to biotechnology. It attempts to answer the question I heard repeatedly, especially during my association with the National Institutes of Health–funded National Flow Cytometry Resource at Los Alamos National Laboratory: What is the potential for innovation in flow cytometer design and application? This volume brings together those approaches that identify the unique contributions of flow cytometry to the modern world of biotechnology.

Following this introductory chapter, the book is organized in two sections. The first half includes chapters that are primarily devoted to innovative tools and technology along with applications that they currently support or might support in the future. These chapters include discussions of probes, instruments, and data analysis. The second half of the book focuses on innovative applications in broad areas of biotechnology. The authors of this text include both industrial and academic researchers.

In the current state of the art, commercial flow cytometers already allow the analysis and sorting of cells or particles at rates up to 50,000/s. They offer 10 or more detectors for measurement of light scatter and fluorescence (chapter 10), and as several of the authors in this book have pointed out, flow cytometry already plays a role in all phases of drug discovery (9), which has been a major commercial driving force of biotechnology. However, with the versatility of the current generation of instrumentation and the variety of biological assays that are already part of its repertoire (3), flow cytometry is poised to play a larger role in the biotechnology revolution of the twenty-first century.

Flow cytometry benefits from the sensitivity of fluorescence and from a myriad of new probe technologies (chapter 2) that have been implemented initially in part for microscopy. It provides multiparameter analysis—in large part because of its use of multiple spatially resolved laser beams—that is not readily available to a comparable extent in other technologies. With the appropriate sample handling systems, flow cytometry is capable of time-dependent subsecond analysis of the kinetics of cell response and molecular binding interactions (chapter 9). In this book, Graves et al. provide an overview of binding interactions, resolution of free and bound ligand, rapid kinetics, and temperature control for the analysis of molecular assemblies in flow cytometry. They provide details to implement fluorescent labeling, attachment, and display strategies for biomolecules. Applications include cells, protein–protein and protein–peptide interactions on a microsphere, microsphere-based DNA-protein interactions, and lipid bilayers on microspheres.

Flow cytometry is also capable of analyzing samples that are submicroliter in volume and are repetitively sampled at rates compatible with high throughput (potentially 100,000 samples a day; chapter 3). Flow cytometry is widely used for cell-based and bead-based assays of molecular interactions that permit analysis of the affinity, stoichiometry, and kinetics of binding interactions. The sensitivity of flow cytometry is comparable to the best alternate methods, attomole to femtomole, and in special cases, single-molecule detection (chapter 7), and its applications extend to all types of molecular binding pairs (proteins, DNA/RNA, lipids, carbohydrates, toxins, etc.; 8). By combining sample handling, multiplexing, and combinatorial libraries, the assay throughput of flow cytometry could exceed hundreds of millions of distinct assays per day.

Biotechnology as a Driving Force for the Future of Flow Cytometry

A major driving force for the future of flow cytometry is nature's diversity. This diversity can be viewed in different ways: It includes the ideas that a large number of molecules can interact with living cells (or can be detected in bead-based immunoassays), that a large number of molecules in and on cells regulate cell behavior, that a large number of signaling pathways and physiological responses proceed from the initiating steps, and that a large number of distinct cells and organisms are of interest in basic research, diagnostics (chapters 4, 10, and 14), therapeutics (chapters 17 and 18), agriculture (chapter 16), reproductive biology (chapter 13), protein engineering (chapter 12), and so forth. Taken together, diversity provides a number of challenges including the complexity of cell and molecular species, the content of assays, and the ability to screen for content and complexity, as well as the potential benefit of evaluating all of these with high throughput. The chapters selected for this book show how flow cytometry can address these challenges.

It is worth noting that the science of genomics has driven much of modern biotechnology and has been defined by the sequencing of the human genome, the identification of single nucleotide polymorphisms (SNPs), the prevalence of haplotypes (SNPs that occur together) within the population, and pharmacogenomics. It is remarkable that flow cytometry played a role in initiating the genome project (1). It was specifically an outcome of the chromosome staining dyes first used in cells. As the sensitivity of detection steadily improved, individual chromosomes could be stained and sorted. A desire to sort individual chromosomes for physical mapping and sequencing was a prelude to the more highly computational analysis of chromosome fragments, which ultimately led to the completion of genomic sequencing. It should not be surprising, therefore, as described in chapter 16, that plant scientists depend on flow cytometric estimation of nuclear DNA contents and ploidy levels as well as flow cytometric analysis and sorting of isolated chromosomes, plant protoplasts, and plant organelles. Flow cytometry also now contributes in multiplex assays applied to SNP analysis and is poised for diagnostics, therapeutics, and breakpoint analysis (chapter 14).

The Challenge of Diversity

Human blood contains hundreds of different molecules, many of which respond in level to disease or infection. The molecules in blood are important for recognizing invaders, binding to cell surface receptors, and catalyzing enzymatic processes such as coagulation. Multiplex technology using immunoassays, with one assay for each particle "address," has the potential of dealing with this diversity.

Drug discovery presents a different type of diversity problem. The number of chemical species of an appropriate size, shape, and solubility to be potential drug molecules is often described as being larger than the number of atoms in the universe. Libraries of synthetic compounds, containing millions of molecular species, can be produced by combinatorial chemistry approaches than can generate millions of compounds. These libraries also arise from synthesis of single chemical species, one at a time, or from the diversity of natural products from rare or endangered species. These chemical libraries are often commercially available and could be used in activity screening by flow cytometry.

The human genome project reports that there are about 50,000 human genes that encode proteins. A significant target in diversity is characterizing the genes according to natural polymorphisms (upward of 500,000 SNPs in the human species), deletions, translocations, and other mutations. It is clear already that multiplex bead-based approaches will be an important part of these aspects of genomics. With 50,000 genes, it is anticipated that the number of protein species in human may reach 500,000 including splice variants and posttranslational modifications. There are likely to be many ways that flow cytometry will contribute to the elements of proteomics. At a minimum, flow cytometry will play roles in protein expression in cells for sorting, display of proteins in particle-based systems, detection of expressed proteins on cells, quantification of expression, and multiple means of studying protein–protein interactions.

Seemingly impossibly large sample numbers arise when there are libraries of many compounds, gene variations, and natural protein variations. Even larger numbers arise from combinatorial approaches in molecular biology. Take, for example, the number of molecules in a hexapeptide library (i.e., $\sim 20^6$ or 52,000,000) or a heptapeptide-expressing phage library ($\sim 10^8$). The Rigel group (chapter 11) has developed a drug discovery approach using retroviral vectors with combinatorial oligonculeotide inserts that create billions and billions of intracellularly expressed compounds. With one vector per cell, each cell becomes an assay for the polypeptide encoded by that insert. In this book, Pali et al. address flow cytometry–based functional genomic screening and the design and validation of specific assays.

In a conceptually similar approach, libraries of protein variants are derived from combinations and mutations of functional enzymes that are expressed in bacterial libraries. Using fluorogenic substrates that are trapped by bacteria, the group at the University of Texas, Austin (chapter 12), has developed a high-throughput means of protein engineering that uses flow cytometry to select individual bacteria expressing single-protein variants that exhibit the desired enzymatic features. Chapter 12 addresses flow cytometric isolation of binding proteins and flow cytometric screening of enzyme libraries, providing detailed technical considerations.

A related type of analysis could be used for proteomics and protein–protein interactions. This approach can be envisioned in analogy to a yeast–two hybrid assay. If one expresses a fusion protein of the bait molecule and a genomic library of potential partners, then the binding of the two components could lead to a signal, depending on the nature of the fusions. The two-hybrid approach classically leads to transcription of a detectable marker, but it could just as well lead to production of a fluorescent protein signal from two half fluorescent proteins or an energy transfer signal from two full fluorescent proteins.

Diversity at the cell, organ, and animal level leads to distinct sets of requirements for flow cytometry. The importance of multiple subsets of leukocytes in response to infection and vaccine development, for example, calls for a large number of fluorescent parameters to discriminate the subsets (chapter 10). In this text, Roederer and DeRosa provide a perspective on the value of high content analysis by flow cytometry compared to alternate approaches by other methodologies. The authors provide a context for measuring antigen-specific responses and their importance in vaccine development, and they provide details for implementing multicolor analysis including hardware, chemistry, data analysis, and presentation and optimization of reagent panels.

As living systems function in all sorts of different environments, it is likely that flow cytometers will be brought to the organisms. For marine microorganisms, the Cytobuoy literally provides a free-standing platform for automated sample uptake and analysis in the marine environment (4, 13, 14). The Cytobuoy system has had to incorporate elements of long-term stability, unattended operation, and self-diagnostics. Submicron organisms like viruses and bacteria provide challenges to optical systems as far as the sensitivity of detection. Cells infected with hazardous organisms need to be handled in isolated environments, and some activities require good manufacturing processes (www.cytopeia.com), so that flow cytometric solutions could benefit from remote operation. Multicellular organisms, such as the nematode, provide ideal tools for cytometric observation of toxicology or pharmacology, and COPAS (complex object parametric analysis and sorter) instruments are now available for millimeter-dimension objects (www.unionbio.com). The fact that environmental interactions affect animal physiology, in addition to the existence of federal regulations requiring the assessment of toxicology and pharmacology, could well lead to a major testing role for flow cytometry (chapter 15). In this book, Burchiel and Weaver address immunotoxicology screening using surface and intracellular markers, toxicity, and genomics. In addition, they identify current issues including clinical monitoring and biomarkers, animal models and reagents, and calibration of flow cytometers and assay validation

The Challenge of Throughput

The throughput rate for particle analysis, as well as the rate of sample handling, can dictate the overall assay throughput. In the examples of protein engineering or expression of combinatorial molecular libraries, the potential throughput is exhilarating. With each cell expressing a unique sequence, an analysis and sort rate of 50,000 per second translates to 4.32×10^9 or 4.32 billion separate assays every 24 hours (86,400 s). In this case, there is no particular automation required for sample input (the front end of the cytometer), as a single test tube or flask may contain the entire library of variant molecular species. Automation might be useful if the sample is not stable and needs to be replaced at regular intervals, if the cell volumes are too large, or if the process takes so long that sheath fluid needs to be replaced. Automation is, however, needed on the back end of the flow cytometer, where individual cells are sorted into plates and the plates are filled and then removed from the sorter and replaced. A simple example is sperm sorting to select the sex of the offspring (chapter 13). In this case, the raw sorting power could lead to billions of sorted sperm per day. Garner and Siedel describe functional flow cytometric analysis and sorting of spermatozoa, as well as analysis of cells from prostate, ovary, breast, uterus, milk, and fetal cells in maternal blood.

The challenge of throughput with multiple samples can be completely different than the challenge of analysis and sorting of a single sample. In one embodiment, known as multiplexing, only a modest front-end sample-handling capability can lead to impressive throughput. Consider the example of a multiplexed suspension array: If samples could be analyzed every 30 s at analysis rates of even 10,000 particles/s, approximately 250,000 particles would allow a 1000 plex assay of 250 particles for each address in the suspension array. Throughput for a completely automated system could allow for 2,880,000 assays (2000 assays/min times 1440 min/day). As a first approximation, improving the sample handling at the front end will not allow the total number of assays to increase because 10,000 particles can be evaluated per second at 5–15-psi instrument operation, and 10^7 particles per milliliter. These conditions are readily compatible with operating with conventional liquid handling and digital signal processing to exclude coincident events of the particles.

In contrast, the advantage of improved sample handling would be to allow individual samples from multiwell plates to be processed at high rates by a flow cytometer (chapter 3; see also www.hudsoncontrol.com/products/platecraneintegrations/cytek.htm). The types of applications are likely to be those such as clinical diagnostics, in which each well contains a cell sample characterized by a complex assay such as immunophenotyping. Thus, in the case in which 10,000 cells stained with 10 fluorescent dyes could be analyzed each second and leukocyte subsets identified, there would be a benefit to sampling a well each second. Another application is in drug discovery, in which each well contains a single compound along with a bead-based or cell-based assay that is sensitive to the presence of those compounds that are active. At a sample per second, a 24-hr day produces 86,400 assays acquiring up to 5000–10,000 particles per assay. Because the particle statistics could allow for 10–20 plex assays in each second, throughputs could approach 1,000,000 assays a day. There are, in fact, some interesting possibilities in such assays.

In drug discovery, it might be important to test compounds against a family of receptors for specificity. Thus, a cell mixture containing 10 sets of cells, each with a different receptor, and each with a different address tag (fluorescent intensity or color combination based on probes or quantum dots; chapter 2), could be used in testing each compound simultaneously for specificity with respect to the desired members of the receptor family. The chapter by Simons et al. provides a perspective on the G-protein coupled receptor (GPCR) family that represents the largest class of receptors in the human genome and the target of about 50% of the prescription medicines on the market. A number of cellular flow cytometric approaches are available to characterize the function of GPCR, and we have also established assemblies of molecular complexes on beads to mimic GPCR signaling events.

Chapter 18 introduces cell-adhesion biology and adhesion receptors, focusing on adhesive interactions between white blood cells with bone marrow and vascular endothelium, where measurements are required for cell activation and adhesion, quantification of receptor numbers and affinity changes, and evaluation of the effect of shear stress. These capabilities lead to emerging clinical applications in drug development, high-throughput screening, clinical prognostication, and coculture assays, as well as high-speed single-cell sorting.

Sample handling is also likely to play an important role in bead-based assays. One of the assays envisioned in chapter 17 could have both proteomic and drug discovery interactions. Our team has learned how to express G-protein-coupled receptors in soluble form as a GFP fusion protein that binds to G proteins displayed on beads. By displaying G proteins as a multiplexed array, each drug compound could be evaluated for its ability to induce association of the receptor with a particular G protein. Another drug discovery possibility is the "one compound, one bead" format currently being implemented for microscopy (7). In this scenario, appropriately labeled receptors would associate with a bead expressing a compound specific to the receptor. A modestly sized peptide array for nuclear receptors has already been reported (6). As in the case of protein variants expressed on cells as discussed above, the particles that became labeled could potentially be sorted at rates up to 4.32 billion particles per day. There are two current limitations to this idea: first, the chemical compounds on the beads may have reactivity in the region attached to the bead, which would then be lost in the target binding assay; second, unless there is a novel addressing scheme for each particle, the particle would need to be sorted so that the compound on the bead could be cleaved off and identified by mass spectrometry. In general, particles about 100 μ in diameter seem to be required for the mass spectrometry step. Sorting of large particles has been achieved with the COPAS technology.

The Challenge of Content

Biological diversity will direct flow cytometry toward new capabilities with respect to information content. As already indicated, multiplex analysis is poised to play a significant role in genetics (on the basis of SNP analysis with addressable particles), proteomics (e.g., biomarkers, protein–protein interactions), and drug discovery (e.g., specificity of receptor responses in cells and molecular assemblies in cells and beads). It is also clear that multiparameter analysis is likely to complement multiplexing by providing access to cell subsets (for immunology and vaccine development, see chapter 10) and the assessment of signal transduction pathways. In signal transduction, cascades of molecular interactions proceed from single binding events, in turn activating numerous intracellular proteins (10). The promise of multiparameter assays is in defining an entire pathway in a single experiment. In these situations, the protein cascades typically involve phosphorylated species that arise after the activating event. They can be detected by antibodies specific to both the phosphorylation and the protein backbone on which it is displayed.

Content is also likely to be a driving force in bead-based analysis. The multiplexing of 10^2 addresses has been accomplished with two fluorescent signals and by labeling particles with one of 10 different levels of brightness for each color (chapter 14). Here, Nolan et al. present a perspective on the current and future applications of flow cytometry in genomics that depends on bead-based analysis. The authors provide insight into the use of arrays, genetic variation and polymorphisms, gene expression analysis, pathogen detection, and diagnostic assays. They also provide details for implementing hybridization reactions and enzymatic modifications for SNP genotyping and gene expression analysis

The beads also permit 100 simultaneous immunoassays for blood analytes. With six colors, 10^6 simultaneous assays are conceivable, but rather than color coding each of 1 million immunoassays, the applications are likely to be in the areas of protein–protein, protein–DNA, and ligand–target interactions. In these cases, a library of molecules could be associated with a suspension array of appropriate diversity. The library might represent chemical compounds, peptides, polypeptides, or oligonucleotide sequences, as described above. These could be allowed to bind with molecules of interest one at a time or, if appropriately tagged in a manner that could be resolved from the address tags, many at a time. These farther-reaching two-dimensional multiplexing approaches, involving addresses on particles and addresses on binding elements, will require some sophistication in probe development and instrumentation to decipher the assay content. Notable in this arena is the introduction of quantum dots, fluorescent proteins, and other families of dyes that can be resolved spectrally (chapter 2).

The potential for spectral resolution has been considered time and again in flow cytometry (2). The recent development of an image flow cytometer (chapter 4) is a notable contribution to the arena of content for flow cyometry. The product specifications project an interesting perspective on the current status of the field of cytometry: simultaneous imaging in six channels, spatial resolution of 0.5-μ pixels, sensitivity of 10 mean equivalents of soluble fluorescein per pixel, self-aligning, and self-diagnostic architecture. The future technology directions for this technology include single-molecule per pixel sensitivity, three-dimensional stereo imaging, sort rates comparable to high-speed sorters, morphology-based sorting, and discrimination of very large bead libraries

The Challenge of Sensitivity

The sensitivity of current instruments is typically characterized as several hundred mean equivalents of soluble fluorescein per particle. The limitation often is not the absolute level of detection but, rather, the fact that the particles themselves exhibit varying levels of autofluorescence. The development of resolution based on spectral (wavelength) or spectroscopic (fluorescence lifetime) features will further enhance the sensitivity and contributes to the vision of a single molecule per pixel. Single-molecule detection in flow cytometry (chapter 7) has been achieved under conditions in which small sample volumes with low autofluorescence are discriminated from the single molecule contained within the illuminated volume. One of the most active areas of investigation is bacteria (biothreat) identification based on single-molecule analysis of DNA fragmentation patterns. This chapter provides extensive details on the operation of the instrument and the use of dyes and their calibration for bacterial identification.

Another aspect of sensitivity is related to sample size. With particles, several hundred events must be analyzed to statistically describe the fluorescence properties of a population. With thousands of fluorophores per particle and hundreds of particles, the number of molecules to be assayed starts as low as a million even in conventional cytometry (attomoles to femtomoles). We have found that microliter-sized sample volumes can be routinely delivered from multiwell plates (chapter 3). Moreover, integrated microfludic channel systems as described in chapter 5 hold the promise of submicroliter samples volumes and have integrated pumping, valving, detection, analysis, screening, and sorting. As a corollary, channels with stationery particles and flowing reagents (chapter 6) turns the flow cytometry paradigm on its head, converting a cytometer into a device that can be used for separation of molecules as well as their identification. This group of three chapters provides extensive detail about the design and operation of microfluidic sample delivery systems.

The Challenge of Informatics

The future of flow cytometry will be intimately involved in creating interfaces among diversity, throughput, content, and sensitivity. To practice these innovations, new means of data analysis, manipulation, archiving, and access will need to be in place (chapter 8). Triester thus places the flow cytometry standard file in the context of the file's application program interface and the flow cytometry markup language. Classically, flow cytometry is based on the concept of one file for each sample, regardless of whether the sample is simple (single population, small number of parameters) or complex (many populations and many parameters, including time). High-throughput sampling has been facilitated by acquiring data from an entire plate of samples as a single file. Innovative data analysis will be required to support this effort by identifying samples from individual wells, identifying responsive samples, and providing output suitable for cheminformatics to assess the activity of individual chemical compounds in individual wells.

References

1. Cram, L.S. 1990. Flow cytogenetics and chromosome sorting. *Hum Cell.* 3:99–106.
2. Crissman, H. and Steinkamp, J.A. 2001. Flow cytometric fluorescence lifetime measurements. *Methods Cell Biol.* 63:131–48.
3. Darzynkiewicz, Z., Crissman, H.A., and Robinson J.P. 2001. *Cytometry.* Methods in Cell Biology 64. Academic Press, San Diego, CA.
4. Dubelaar, G.B., Gerritzen, P.L., Beeker, A.E., Jonker, R.R., and Tangen, K. 1999. Design and first results of CytoBuoy: a wireless flow cytometer for in situ analysis of marine and fresh waters. *Cytometry* 37:247–54.
5. Herzenberg, L.A., Parks, D., Sahaf, B., Perez, O., Roederer, M., Herzenberg, L.A. 2002. The history and future of the fluorescence activated cell sorter and flow cytometry: a view from Stanford. *Clin Chem.* 48:1819–27.
6. Iannone, M.A., Consler, T.G., Pearce, K.H., Stimmel, J.B., Parks, D.J., and Gray, J.G. 2001. Multiplexed molecular interactions of nuclear receptors using fluorescent microspheres. *Cytometry* 44:326–37.
7. Lorthioir, O., Carr, R.A., Congreve, M.S., Geysen, M.H., Kay, C., Marshall, P., McKeown, S. C., Parr, N. J., Scicinski, J.J., and Watson, S.P. 2001. Single bead characterization using analytical constructs: application to quality control of libraries. *Anal Chem.* 73:963–70.
8. Nolan, J.P., Lauer, S. Prossnitz, E.R., and Sklar, L.A. 2000. Flow cytometry: a versatile tool for all phases of drug discovery. *Drug Discovery Today* 4:173–80.
9. Nolan, J.P. and Sklar, L.A. 1998. The emergence of flow cytometry for the sensitive, real-time analysis of molecular assembly. *Nat Biotechnol.* 16:833–8.
10. Perez, O.D. and Nolan, G.P. 2002. Simultaneous measurement of multiple active kinase states using polychromatic flow cytometry. *Nat Biotechnol.* 20:155–62.
11. Robinson, J.P. 2004. Flow cytometry. Pages 630–640 in *Encyclopedia of Biomaterials and Biomedical Engineering,* ed. G.E. Wnek and G.L. Bowlin. Marcel Dekker, New York.
12. Shapiro, H.M. 2003. *Practical Flow Cytometry,* 4th ed. Wiley-Liss, Hoboken, NJ.
13. Troussellier, M., Courties, C., and Vaquer, A. 1993. Recent applications of flow cytometry in aquatic microbial ecology. *Biol Cell.* 78:111–21.
14. Vives-Rego, J., Lebaron, P., and Nebe-von Caron, G. 2000. Current and future applications of flow cytometry in aquatic microbiology. *FEMS Microbiol. Rev.* 24:429–48.

Part I

TOOLS FOR FLOW CYTOMETRY

2

Fluorescent Probes

HOWARD M. SHAPIRO

Introduction

In the jargon of cytometry, cellular characteristics, such as size, nucleic acid content, and membrane potential, are usually referred to as *parameters*, a term that is also used for the physical characteristics, such as absorption, light scattering, and fluorescence intensity, that are measured by cytometric instrumentation. Fluorescence, as a physical parameter, plays a key role in the detection of probes on beads for multiplexed analysis.

Cellular parameters can be classed as intrinsic or extrinsic. Intrinsic cellular parameters are those that can be measured without the use of a reagent; measurement of extrinsic parameters requires the use of reagents, which are almost always referred to as *probes*, thereby occasioning confusion among molecular biologists new to cytometry.

Cellular parameters are also characterized as structural or functional; DNA and RNA content and the presence and copy number of an antigen or nucleic acid sequence are structural parameters, whereas internal pH, membrane potential, and enzyme activity are functional parameters. The distinction between structural and functional parameters blurs at the edge, but the concept has been generally useful.

Fluorescence and Flow Cytometry: Made for Each Other

Fluorescent probes allow measurement of the widest variety of extrinsic cellular parameters. For an atom or molecule to fluoresce, it must first absorb a photon, raising an electron to a higher energy level that is known as an excited state. Excitation by absorption requires only about a femtosecond. Fluorescence occurs when the electron loses all or some of the absorbed energy by emission of a photon. The fluorescence lifetime, that is, the period between excitation and emission, is typically on the order of a few nanoseconds for fluorescent organic materials but is notably longer (hundreds of microseconds) for some materials (e.g., lanthanide chelates). In almost all cases, some of the excitation energy is lost nonradiatively by transitions between different vibrational energy levels of the electronic excited state; this loss requires that the emitted energy be less than the energy absorbed, meaning that the fluorescence emission will be at a longer wavelength than the excitation. The difference between the principal excitation and emission maxima in the fluorescence spectrum is known as the Stokes shift, honoring George Stokes, who first described fluorescence in the mid-1800s. Typical Stokes shifts are no more than a few tens of nanometers.

Fluorescence is an intrinsically quantum mechanical process. The probability that a molecule will absorb is quantified in its absorption cross section and its extinction coefficient. The quantum yield and quantum efficiency of fluorescence are, respectively, the number and fraction or percentage of photons emitted per photon absorbed; they typically increase with the cross section and extinction coefficient but are also dependent on the relative probabilities of the excited molecule losing energy via fluorescence emission and nonradiative mechanisms. The quantum yields of some dyes used in cytometry are quite high, above 0.5, but it should be noted that particularly for an organic material, quantum yield is affected by the chemical environment in which the molecule finds itself. If an excited molecule that might otherwise fluoresce instead loses energy nonradiatively, for example, by collision with solvent molecules, it is said to be quenched; once returned to the electronic ground state, it can be reexcited. However, there is usually a finite probability that light absorption will be followed by a change in molecular structure, making further cycles of fluorescence excitation and emission impossible: This is called (photo)bleaching.

Although, in principle, increasing the illumination intensity can increase the intensity of light-scattering signals without limit, this is not possible for fluorescence signals because, at some level of illumination, all the available molecules will be in excited states, leaving no more to be excited if illumination intensity is further increased. This condition of photon saturation is often reached in flow cytometers that use laser powers of 100 mW or more; bleaching, which may also make the dependence of emission intensity on excitation intensity nonlinear, is noticeable at power levels of tens of milliwatts. In general, it is possible to get only a finite number of cycles of excitation and emission out of each fluorescent molecule, or fluorophore, before photobleaching occurs. Saturation and bleaching have been discussed by van den Engh and Farmer (72).

Fluorescence and flow cytometry are made for each other for several reasons, but primarily because fluorescence at least from organic materials, is a somewhat ephemeral measurement. In a flow cytometer, each cell is exposed to excitation light only for the brief period during which it passes through the illuminating beam, usually for a few microseconds, and the flow velocity is typically nearly constant for all the cells examined. These uniform conditions of measurement make it relatively easy to attain high precision, meaning that one can expect nearly equal measurement values for cells containing equal amounts of fluorescent material; this is especially desirable for such applications as DNA content analysis of tumors, in which the abnormal cells' DNA content may differ by only a few percent from that of normal stromal cells.

Another basis for the compatibility between fluorescence measurements and cytometry in general is found in the dark-field nature of fluorescence measurements. In the 1930s, unsuccessful attempts were made to detect antibody binding to cellular structures by bright-field microscopy of the absorption of various organic dyes bound to antibodies. In 1941, Coons, Creech, and Jones (12) labeled cells with antibodies containing covalently bound fluorescent organic molecules, enabling antibody-binding structures to be visualized clearly against a dark background. In general, fluorescence measurements, when compared with absorption measurements, offer higher sensitivity, meaning that they can be used to detect smaller amounts or concentrations of a relevant analyte; this is critical for demonstration of many cellular antigens and also in identifying genetic sequences or fluorescent protein products of transfected genes that are present in small copy numbers.

It is also usually easier to make simultaneous measurements of a number of different substances in cells, a process that is referred to as *multiparameter cytometry*, by fluorescence than by absorption, and the trend in recent years in both flow and static cytometry has been toward measurement of an increasingly large number of characteristics of each cell subjected to analysis. Multiparameter cytometry requires optical systems that separate fluorescence emission from cells into multiple spectral bands, as well as hardware or software fluorescence compensation to permit fluorescence contributions from probes with overlapping emission spectra to be resolved and quantified. Additional technical details about cytometry and cytometers can be found throughout this book and in my book, *Practical Flow Cytometry* (60).

If two fluorophores, A and B, are in close proximity, and the emission spectrum of A overlaps the excitation spectrum of B, nonradiative energy transfer between the two may occur, with the result that illumination at wavelengths that normally excite the donor, A, produces fluorescence from the acceptor, B. The likelihood of energy transfer increases with the degree of spectral overlap and diminishes with the distance between the fluorophores. Intramolecular energy transfer determines the fluorescence spectra of phycobiliproteins and their tandem conjugates; other fluorescent probes have been designed so that molecular rearrangement of the probe, for example, by enzyme action, results in an increase or decrease in intramolecular energy transfer (80). Intermolecular energy transfer between different probe molecules can be used to determine the proximity of the cellular structures to which the probes bind (68).

Because excitation sources used in flow cytometry are typically polarized, fluorescence emission can be expected to be as well. Measurements of fluorescence polarization can be used to determine molecular mobility; Asbury et al. (1) have discussed polarization in general.

The best single reference on fluorescent probes is Haugland's *Handbook of Fluorescent Probes and Research Products* (30). This indispensable work, now in its ninth edition and available on CD-ROM and online (www.probes.com) as well as in print form, is the catalog of Molecular Probes, Inc. (Eugene, OR), a major supplier. There is also an extensive discussion of fluorescent probes and their applications in *Practical Flow Cytometry* (60).

Probes versus Labels

It is useful to distinguish between probes and labels. If we look at stains classically used for blood cells, which date back to Paul Ehrlich's work in the 1880s, we find they are mixtures of an acid dye (typically eosin), which binds to basic elements within cells, meaning mostly proteins, and a basic dye (typically one or more of the azure dyes, all thiazines), which binds to acidic elements, notably nucleic acids but also sulfonated glycosaminoglycans. Eosin will not bind to acidic elements, and azure dyes will not bind to basic elements. However, although that is as specific as those dyes get, we can still legitimately call them probes, even if they are probes for relatively nonspecific characteristics or constituents of cells.

At the other end of the probe hierarchy are antibodies and gene probes—large, or relatively large, molecules themselves, which, respectively, are exquisitely specific for macromolecular structure and sequence when used under appropriate conditions. These probes are not, in general, intrinsically fluorescent; thus, to detect them using an optical flow cytometer, we have to attach a fluorescent label. Until the 1980s, almost all available labels were relatively low–molecular weight dyes; since then, it has become increasingly likely that the label used for a macromolecular probe will itself be a macromolecule, and specifically a phycobiliprotein, which may also have small dye molecules covalently attached to it to modify its spectral characteristics.

Some of the dye probes now in use are sensitive to changes in their chemical environment and would be useless if they were not. Measurements of intracellular pH are usually done using dyes that change their spectral characteristics as pH changes. That is an essential characteristic for a probe, but an undesirable one for a label, which we would like to behave (i.e., fluoresce) pretty much the same way regardless of the environment in which it finds itself. However, to cite one prominent real-world example, fluorescein, which is one of the all-time favorite fluorescent labels, is environmentally sensitive, and its fluorescence increases with pH enough that derivatives of fluorescein are routinely used as pH probes. Thus, sample pH needs to be controlled to minimize errors when fluorescein-labeled probes are used.

Cytometer Specifications and Probe Spectra

A fluorescent probe or label is unlikely to be useful for flow cytometry unless it can be excited by and detected by a commercially available instrument. At present, four excitation wavelength ranges are likely to be available in fluorescence flow cytometers. Blue-green light at or near 488 nm, from an argon ion or solid-state laser or an arc lamp, is available in the largest number of instruments. Many systems also provide red (633–645 nm, from a He–Ne or diode laser) or ultraviolet (UV; 325–365 nm, from an air-cooled He–Cd laser, a water-cooled ion laser, or an arc lamp) excitation. Green (532 nm from doubled YAG lasers, 543 nm from He–Ne lasers) and violet (395–415 nm, from diode lasers) excitation have been made available on some newer instruments.

In principle, a flow cytometer should be able to measure fluorescence at any wavelength longer than that used for excitation. However, the light transmission of optics used in most instruments is typically poor at wavelengths below 400 nm, and the photomultiplier tube detectors supplied as standard equipment often do not respond well at wavelengths above 700 nm. It is usually possible to replace a standard-issue photomultiplier tube with one with extended red sensitivity; it may be harder to deal with poor light transmission below 400 nm.

Multistation flow cytometry, in which multiple, spatially separated fluorescence excitation beams are used, has advantages. Even when one can choose from a number of probes to select those with desired spectral characteristics, the use of separated excitation beams generally facilitates resolution of fluorescence signals from multiple probes. When a choice of probes is not available, multistation flow cytometry may provide the only means of making correlated measurements of two or more parameters of interest.

Spectral properties of a representative sample of fluorescent dyes and labels used in flow cytometry are given in table 2.1.

Staining Mechanisms

Two mechanisms are primarily responsible for selective staining of cellular constituents by fluorescent dyes or probes. The first, which also provides a basis for staining by nonfluorescent dyes, involves the development of contrast as a result of differences in the concentration of dye from one region of the cell to another, and as a result of differences in the affinities of various cellular constituents for the dye. Thus, basic dyes are bound in relatively high concentrations to acidic materials such as nucleic acids and glycosaminoglycans, whereas dyes with high lipid solubility stain membranes and fat droplets, and so on.

The second mechanism of fluorochroming involves an increase in the quantum efficiency of a fluorescent dye when it is bound to a particular substance or in a particular environment (e.g., nonpolar vs. polar). Binding of the DNA-selective dyes Hoechst 33342 and DAPI to the minor groove of the DNA molecule results in approximately a 50-fold increase in fluorescence, as does the intercalative binding of ethidium and propidium; cyanine dyes such as thiazole orange (TO) and TO-PRO-1 increase fluorescence by more than 1000-fold on intercalative binding to nucleic acid. However, acridine orange (AO), which also intercalates into DNA, is slightly quenched when bound; the bright nuclear staining produced by this dye must therefore result almost exclusively from increased concentration of the dye in the nucleus.

Binding of acid dyes to proteins usually does not increase quantum efficiency, and background fluorescence from free dye tends to be relatively high as a result. For this reason, many investigators prefer to use reactive derivatives of dyes, for example, fluorescein isothiocyanate (FITC), for protein staining. After incubation with FITC leaves some fluorescein covalently bound to protein; the unreacted dye may be removed by washing, lowering background fluorescence.

Even "specific" stains such as the Hoechst dyes and ethidium may bind nonspecifically to some materials in cells, particularly when the dye or the interfering material is present at high concentrations. Nonspecific staining may also occur when environmental factors such as salt concentration or pH are outside the range in which specific staining has been reported.

Table 2.1. Fluorescence spectral properties of a selection of reagents usable in flow cytometry.

Excitation	UV (325–365 nm)	Violet (395–415 nm)	Blue-Green (488 nm)	Red (633–645 nm)
Fluorescent Labels	AMCA, Alexa 350 (440)	Cascade Blue (420) Cascade Yellow (520)	Fluorescein, Cy2 (520) Cy3 (565) PE (575) PE-Texas Red (610) PE-Cy5 (660) PerCP (670) PerCP-Cy5.5, PE-Cy5.5 (700) PE-Cy7 (780)	APC, Cy5 (660) APC-Cy5.5, Cy5.5 (700) APC-Cy7 (780)
DNA-Selective Dyes	Hoechst dyes (440) DAPI (455)	Hoechst Dyes (440), DAPI (455) Chromomycin, Mithramycin (560) ?7-AAD (660)	AO (520)	
Nonselective nucleic acid dyes			7-AAD (660) DRAQ5 (700) TO-PRO-1, etc. (530) Pyronin Y (575) Ethidium (600) Propidium (615) AO (650)	DRAQ5 (700) TO-PRO-3, etc. (660)
Enzyme Substrate Fluorophores	7-amino-4-chloro-methylcoumarin (470) ELF 97 (530)	3-cyano-7-hydroxycoumarin (450)	Fluorescein, rhodamine 110 (520) resorufin (585)	
Tracking Dyes			DiO, CFSE (520) DiI, PKH26 (565)	
Membrane Potential Probes			DiBAC$_n$(3), DiOC$_n$(3), JC-1, Rhodamine 123 (520) JC-1 (585) DiOC$_n$(3) (610)	DiIC$_n$(5) (660)
Ca^{++} Probes	indo-1 (405) indo-1 (480)		fluo-3 (520) Fura Red (660)	
pH Probes			BCECF (520) Carboxy SNARF-1 (580) BCECF (620) Carboxy SNARF-1 (640)	
Reporter Proteins		ECFP (470)	EGFP (510) EYFP (535) dsRED (575)	

Emission maxima are indicated next to names of probes; probes for which two maxima are listed may be usable for ratiometric measurements.

Staining, Permeancy, Permeability, and "Viability"

Applications of flow cytometry in biotechnology often involve the analysis of puta-
tively viable cells and the sorting of those with selected characteristics for biochemical analysis or for further short- or long-term observation in culture. Dyes that can enter and stain living cells have long been described as vital stains, even though many
such dyes may eventually be toxic to cells.

There is considerable disagreement about the definition of "viability" as it applies
at the cellular level. People concerned with growing cells in culture generally take
clonogenicity or reproductive viability as a criterion. This excludes fully functional differentiated cells such as nerve and muscle cells, blood granulocytes, and so forth; preservation of some specific cell function therefore seems to be a more suitable criterion
of viability for such cells. Reproductive viability is not directly measurable by flow
cytometry.

If reproductive capacity is too stringent a criterion of viability, the capacity of the
cell to exclude dyes such as trypan blue, eosin, and propidium iodide may not be stringent enough. Cells that have been so damaged as to be incapable of performing their
typical differentiated functions, for example, chemotaxis and phagocytosis in the case
of granulocytes, may still be classified as "viable" by such dye exclusion tests.

It is convenient to use the term *intact cells* to describe cells that have not been treated
with fixatives or lysing agents and that do not show obvious morphologic damage or
functional impairment. Such parameters as membrane integrity and permeability, cytoplasmic $[Ca^{++}]$ and pH, and cytoplasmic and mitochondrial membrane potential are
meaningful only as characteristics of intact cells. The validity of studies of responses
of these parameters to biologic stimuli is best established by the inclusion of controls
that demonstrate known responses to standard stimuli (e.g., changes in calcium distribution in cells treated with calcium ionophores).

To stain constituents inside intact cells, a probe must be capable of crossing the
cell membrane, either by diffusion or by some form of active transport or carrier-
mediated transport. Most vital stains are small molecules that are relatively lipid
soluble and are either positively charged or electrically neutral at physiologic pH.
High lipid solubility favors partitioning of dyes from aqueous media into the lipid
bilayer phase of the cell membrane and into membranous or lipid-containing intracellular structures. Positively charged dye molecules are attracted to negatively
charged cell constituents such as glycosaminoglycans and nucleic acids; in addition,
in living cells, positively charged molecules are concentrated from the medium into
the cytosol and from the cytosol into mitochondria because there are interior-
negative electrical potential gradients across both the cytoplasmic and the mitochondrial membranes.

Materials that can readily cross the intact cytoplasmic membranes of cells are said
to be *membrane permeant* or, more simply, *permeant*; materials that are excluded by
intact cytoplasmic membranes are described as being *membrane impermeant*, or just
impermeant. To stain living cells, that is, to act as a vital stain, a dye must be permeant. Cells that can take up a dye or other chemical are said to be permeable to the material; cells that cannot take up a material are said to be impermeable to it. The terms
permeability and *impermeability* are used to characterize cell membranes in particular,
as well as cells in general.

A number of active efflux mechanisms may partially or completely exclude permeant dyes from cells. For example, the glycoprotein efflux pump responsible for multiple drug resistance efficiently clears a wide variety of molecules from cells (see Nucleic Acid Dyes for examples).

Low–Molecular Weight Fluorescent Labels/Protein Stains

Reactive dyes such as FITC, mentioned above as a stain for total protein content, are more widely used as fluorescent labels for a variety of large and small molecules that can be bound strongly and specifically to various cellular constituents. Ligands thus labeled can be used as reagents for a number of structural and functional parameters, including surface sugars (demonstrated using tagged lectins), surface and intracellular antigens (fluorescent antibodies), surface charge (fluorescent polycations), surface and intracellular receptors (fluorescent hormones, growth factors, neurotransmitters, viruses, etc.), endocytosis (fluorescent macromolecules, microorganisms, or plastic particles), DNA synthesis (fluorescent antibody to detect BrUdR incorporation; fluorescent nucleotides), and specific nucleic acid sequences (fluorescent oligonucleotide probes).

Fluorescein, conjugated as its reactive isothiocyanate derivative (FITC), is by far the most popular fluorescent label; its excitation maximum is very close to the 488-nm argon ion laser wavelength available in almost all flow cytometers, its quantum efficiency is high, and it had been in widespread use long enough before flow cytometers became available for conjugation and staining procedures to have become well established, particularly in immunology. Fluorescein emits at approximately 520 nm.

Tetramethylrhodamine isothiocyanate, although used to provide an antibody label distinguishable from fluorescein by fluorescence microscopy, is very poorly excited at 488 nm and is not widely used in flow cytometry. It is suitable for use with instruments using 532- or 543-nm green excitation sources and emits at around 570 nm. The first label combination widely used for two-color fluorescence flow cytometry paired fluorescein with rhodamine 101, conjugated in reactive form as an isothiocyanate (XRITC) or a sulfonyl chloride (Texas Red). Rhodamine 101 emits near 615 nm and is best excited at wavelengths between 565 and 595 nm; the fluorescein–rhodamine combination thus can only be used in instruments that have a suitable source, such as a krypton or dye laser, in addition to a 488-nm excitation source.

The "Cy dyes" (43, 65) are a series of reactive derivatives of symmetric cyanine dyes that use a succinimidyl ester group to link to proteins. The oxacarbocyanine derivative Cy2 has absorption and emission spectral characteristics similar to those of fluorescein. Cy3, an indocarbocyanine, excites maximally at about 545 nm, but its absorption is high enough that it excites adequately at the 488-nm argon laser wavelength available in most cytometers. The emission peak of Cy3 is at about 565 nm; however, a substantial fraction of Cy3 emission is transmitted by the bandpass filters typically used for phycoerythrin detection. Cy5, an indodicarbocyanine, absorbs maximally near 640 nm; it is very effectively excited by 633-nm He–Ne lasers or 635–640-nm diode lasers and emits at around 660 nm. Cy5.5 is a reactive derivative of dibenzoindodicarbocyanine, with maximal absorption near 675 nm and maximal emission at 695–700 nm; its absorption at 633 nm is sufficient to make it possible to use Cy5- and Cy5.5-labeled antibodies for two-color immunofluorescence analyses in an instrument with a

633-nm He–Ne or red diode laser source. Cy7 is a reactive indotricarbocyanine dye. It absorbs in the near infrared (about 750 nm) and emits around 770 nm. The indotricarbocyanine structure itself is not as stable chemically as the indocarbocyanine and indodicarbocyanine dye fluorophores of Cy3 and Cy5 are, which makes it harder to prepare Cy7 labels and also results in a relatively short shelf life. Antibodies labeled with cyanines seem to adhere to monocytes and, to a lesser extent, to granulocytes, resulting in low levels of irrelevant staining. Antibody manufacturers have come up with various proprietary ways of minimizing such binding. The Cy dyes, particularly Cy3 and Cy5, have become popular as labels for nucleic acids and oligonucleotides in applications such as gene array scanning and fluorescence in situ hybridization.

A number of UV-excited, blue fluorescent labels have come into use. The first popular coumarin label was 7-amino-4-methylcoumarin-3-acetic acid, or AMCA, which excites maximally at about 350 nm and has an emission maximum near 455 nm. Molecular Probes's Alexa 350 has a similar spectrum but has a quantum yield almost twice as high as that of AMCA.

Cascade Blue, a reactive derivative of pyrene, was also introduced by Molecular Probes. Its excitation maximum is near 390 nm, with maximum emission at about 415 nm. This dye, and Molecular Probes's Cascade Yellow (emission maximum near 550 nm), are effectively excited by violet diode lasers.

Molecular Probes also offers the BODIPY series of dyes, which are boron dipyrromethane derivatives, and the Alexa dyes, a series of sulfonated coumarin- and rhodamine-based labels. Excitation maxima of the various BODIPY dyes cover the range from 500 to 646 nm, and emission maxima range from 506 to 660 nm. Alexa dyes have spectral characteristics similar to those of some of the more popular labels previously mentioned (e.g., AMCA, fluorescein, Texas red, Cy3, Cy5, Cy5.5, and Cy7). However, the Alexa dyes have higher quantum yields, better photostability, and better charge characteristics, allowing more dye molecules to be put on a protein molecule.

Phycobiliproteins and Tandem Conjugates as Labels

The phycobiliproteins are a family of macromolecules found in red algae and cyanobacteria (formerly called blue-green algae), in which they play critical roles in the function of the photosynthetic apparatus, participating in a chain of nonradiative energy transfers that finally makes blue-green light energy available to chlorophyll.

Phycoerythrins absorb blue-green and green light, phycocyanins green and yellow light, and allophycocyanins orange and red light. These molecules are all highly fluorescent and have been of great use in flow cytometry since Oi, Glazer, and Stryer demonstrated their utility as antibody labeling reagents (48).

The chromophores in phycobiliproteins are bilins, which are pyrrole pigments. Each phycobiliprotein molecule contains a large number of such chromophores. The extinction coefficients of phycobiliproteins are extremely high, and the quantum yields are also high. Phycobiliproteins, and phycoerythrins in particular, are characterized by broad shoulders in their excitation spectra, allowing them to be excited effectively at wavelengths substantially below their emission maxima; with excitation at 488 nm, a phycoerythrin-labeled antibody molecule will emit several times as much fluorescence as a fluorescein-labeled antibody molecule.

The peak absorption of R-phycoerythrin (R-PE) is at 565 nm, with the emission maximum at 578 nm, which allows R-PE to be used very effectively in combination with fluorescein for two-color immunofluorescence flow cytometry using only a single 488-nm excitation beam. PE-labeled antibodies have been widely available since the late 1980s.

Allophycocyanin (APC) is of particular interest as a single label because it exhibits high (about 75% of maximum) absorption in the 633–645 nm-range, in which red He–Ne and diode lasers operate. The absorption maximum of APC is 650 nm; its emission maximum is 660 nm.

The molecular weights of phycobiliproteins are high (240,000 for PE, 100,000 for APC), which makes it feasible to prepare conjugates containing defined numbers of molecules of phycobiliprotein, facilitating quantitative fluorescence measurements. BD Biosciences offers 1:1 phycoerythrin conjugates of several monoclonal antibodies.

Glazer and Stryer (24) were the first to prepare a tandem conjugate of PE and APC in which energy transfer between these proteins, with the phycoerythrin molecule acting as the donor and the allophycocyanin molecule as the acceptor, resulted in strong emission at 660 nm on excitation at wavelengths between 470 and 570 nm. The phycobiliprotein tandem conjugates that are now in widest use incorporate only a single phycobiliprotein molecule, to which are conjugated several molecules of a lower–molecular weight fluorochrome. The first conjugates prepared in this fashion incorporated phycoerythrin and Texas Red; antibodies labeled with such conjugates are available from a number of manufacturers, each using its own trade name for the conjugate. Although most of the emission from PE-Texas Red conjugates is in the 610–620-nm emission region of Texas Red, incomplete energy transfer results in some emission from the conjugates in the PE emission region around 580 nm, and PE itself has substantial emission in the 610–620-nm emission range. As a result, fluorescence compensation must be applied to separate the fluorescence signals from a PE-labeled antibody and another antibody labeled with a PE-Texas Red tandem conjugate.

PE-Cy5 tandem conjugates, introduced by Waggoner et al. (73), comprise a single phycoerythrin molecule and several molecules of the cyanine dye label Cy5 and are preferable to PE-Texas Red conjugates as a third label for immunofluorescence analyses using 488-nm excitation; they emit at the emission maximum of Cy5, near 660 nm. The list of phycoerythrin tandem conjugates also includes PE-Cy5.5 (emission maximum near 700 nm) and PE-Cy7 (emission maximum near 770 nm). There are now flow cytometers on the market that permit simultaneous measurements of the fluorescence of fluorescein, PE, PE-Texas red, PE-Cy5, PE-Cy5.5, and PE-Cy7, using a single 488-nm excitation beam.

Allophycocyanin tandem conjugates can be used in conjunction with APC itself for multicolor immunofluorescence measurements employing a 633-nm He–Ne or red diode laser source. APC-Cy7 (4, 51) emits maximally near 770 nm, and APC-Cy5.5 emits maximally near 700 nm; both are now available from a number of companies, conjugated to a variety of monoclonal antibodies.

Peridinin chlorophyll protein, or PerCP (41), is a component of the photosynthetic apparatus in a dinoflagellate; it has an absorption maximum near 490 nm and a relatively sharp emission peak at about 680 nm. The sharpness of the emission peak minimizes crosstalk between PerCP and PE and therefore also minimizes the amount of fluorescence compensation needed. However, PerCP is relatively intolerant of high illumination power levels because it may enter a long-lived triplet state (the same problem is noted to a much lesser extent with PE and APC). This problem is eliminated in a PerCP-Cy5.5 tandem conjugate (5, 19), which has maximum emission near 700 nm.

The probability that energy transfer between a donor and an acceptor species will occur varies with the extent to which the donor emission spectrum and the acceptor excitation spectrum overlap. This overlap diminishes pretty drastically as we move from PE-Texas red to PE-Cy5 to PE-Cy5.5 to PE-Cy7. As a result, energy transfer between donor and acceptor in this series is progressively less efficient, with the result that the longer-wavelength-emitting tandems also exhibit more and more emission in the spectral range in which PE normally emits. There is only a trace of PE emission in the PE-Cy5 spectrum, but there is a significant PE contribution in the PE-Cy5.5 and PE-Cy7 emission spectra.

If there is a lot of PE emission in the PE-Cy7 spectrum, it means that a substantial fraction of the PE chromophores have not donated energy to Cy7, and therefore, that there is not as much emission from Cy7 as there would be from Texas red in PE-Texas red or from Cy5 in PE-Cy5. In some instances, the number of photons an instrument actually collects from PE-Cy7 can be less than 1/100 the number it would get from PE. It is therefore inappropriate to use PE-Cy7 or other similarly inefficient labels to attempt to discriminate cells bearing small amounts of target ligands.

Semiconductor Nanocrystal Labels (Quantum Dots)

Semiconductor nanocrystals, better known as quantum dots, now appear to be becoming practical as labels for biological molecules or as tags for beads used in multiplex assays (6, 10, 11, 29, 32, 33, 76).

In semiconductors, absorption of optical or electrical energy results in one atom of the material temporarily losing an electron while another somewhere in the vicinity temporarily gains one. Some of the absorbed energy is subsequently lost in the form of a photon. In semiconductor crystals with dimensions smaller than about 10 nm, the emission wavelength becomes more dependent on the size of the crystal than on its composition. The emission wavelength of a CdSe crystal with a diameter of 2.1 nm is approximately 510 nm; that of a crystal with a 3.1-nm diameter crystal of the same material is approximately 560 nm. Larger crystals have emission wavelengths ranging into the infrared range.

The fluorescence spectrum of a nanocrystal is considerably different from that of an organic dye. Organic dyes typically have small Stokes shifts (i.e., their emission maxima are within 20 nm of their excitation maxima). The excitation spectrum and the emission spectrum of an organic dye often resemble mirror images of one another: Both are substantially skewed, with a short-wavelength "shoulder" in the excitation spectrum and a long-wavelength "tail" in the emission spectrum.

The emission spectrum of a nanocrystal is typically nearly symmetric, with a full width at half maximum of, at most, a few tens of nanometers. Moreover, the excitation spectra of nanocrystals are relatively independent of emission wavelength; progressively shorter wavelengths are increasingly effective for excitation. Excitation at 400 nm is typically at least twice that at 488 nm. This indicates that violet diode lasers will be economical and useful excitation sources for work with quantum dots in either scanning or flow cytometers.

Being inorganic, nanocrystals are much less susceptible to photobleaching than are organic dyes; a nanocrystal is likely to be emitting over 75% of its original fluorescence output after an observation time sufficient to photobleach over 95% of the fluorescence emission from a dye. Nanocrystals also have much higher absorption than do dyes: Although the quantum efficiencies of nanocrystals and dyes are about the same, the fluorescence from a nanocrystal is typically equivalent to the fluorescence from a dozen or more dye molecules.

Some practical problems that have prevented widespread use of nanocrystal labels have been solved, but some remain. Because both the emission wavelengths and the fluorescence intensities of nanocrystals depend on their size, preparative methods must yield crystals with highly homogenous size distributions to keep emission peaks confined to a small spectral range and to maintain low-emission bandwidths. Also, as nanocrystals are intolerant of aqueous media, practical labels must consist of a semiconductor core and an outer layer that allows the particle to remain dispersed in aqueous solution and that provides means to attach it to a biomolecule. Streptavidin-conjugated quantum dots with a 605-nm emission wavelength are now commercially available [from Quantum Dot Corporation, Hayward, CA], soon to be joined by nanocrystal labels emitting at other wavelengths.

Nucleic Acid Dyes

Fluorescent nucleic acid–binding dyes are used to measure cellular DNA and RNA content, to demonstrate loss of membrane integrity, and to define cell subpopulations in which active efflux pumps are present.

DNA-Selective Dyes

An ideal dye for measurement of DNA content would be DNA specific (i.e., it would form a fluorescent complex with DNA, but not with RNA or other macromolecules). It would also not exhibit any base or sequence preference; in other words, the fluorescence from a given number of dye molecules bound to a given number of base pairs' worth of DNA would be the same, regardless of the relative proportions of A-T and G-C base pairs. There are a number of dyes that are highly DNA selective, if not DNA specific.

The UV-excited (excitation maximum about 350 nm), blue fluorescent (emission maximum about 450 nm) bisbenzimidazole dyes, Hoechst 33258, 33342, 33378, and 33662, (35, 36), bind to sequences of three A-T base pairs. The dyes bind in the minor groove of the DNA helix, rather than by intercalation. Their affinity for DNA is sufficiently strong that they will displace bound molecules of a variety of intercalating dyes. The strong A-T base preference accounts for the popularity of Hoechst 33258 in combination with chromomycin A_3 (which has a G-C preference) for bivariate flow cytometric analysis of chromosomes. Hoechst 33342, which is membrane permeant, is the only compound that has been extensively used for flow cytometric determination of DNA content in living cells; sperm cells from laboratory and domestic animals and humans have been stained with the dye, sorted into X- and Y-chromosome-enriched fractions based on DNA content, and injected into eggs to produce viable offspring.

Hoechst 33342 is a substrate for several efflux pumps; a side population of cells, in which primitive stem cells are found, can be defined on the basis of relatively low UV-excited blue and red fluorescence after Hoechst 33342 staining (26, 27). The responsible transporter is from the breast cancer resistance protein family; the protein found in murine cells is Bcrp1 (79), whereas that found in human cells is ABCG2 (34, 37, 56). This should make it possible to use fluorescent antibodies to ABCG2, or probes for the gene, rather than (or in addition to) Hoechst dye fluorescence to identify, or confirm the identity of, putative side population cells.

A highly DNA-selective, impermeant dye, 4'-6-diamidino-2-phenylindole, or DAPI, has fluorescence properties similar to those of the Hoechst dyes and, like the Hoechst dyes, has a strong A-T base preference. In fixed or permeabilized cells, DAPI often yields DNA histograms with lower G_0/G_1 peak coefficients of variation than are obtained using other dyes. Both DAPI and the Hoechst dyes can be excited by violet diode lasers, albeit somewhat inefficiently.

The antitumor antibiotic mithramycin and the structurally related antibiotic chromomycin A_3 are highly DNA selective (74) and act as fluorochromes for G-C rich regions of DNA. The excitation maxima of the DNA complexes of both dyes are at approximately 440 nm (15); the emission maximum of the chromomycin A_3 complex is at about 555 nm, whereas that of the mithramycin complex is at about 575 nm. The quantum efficiency of the dye–DNA complexes is relatively low, which may limit precision in work with chromosomes or bacteria due to photon statistics. The dyes are impermeant: Chromomycin A_3 is used primarily as a chromosome stain in combination with Hoechst 33258.

7-Aminoactinomycin D (7-AAD; 77), a fluorescent analog of the antitumor antibiotic actinomycin D, is impermeant and highly DNA selective, exhibiting a G-C base preference. The complex of this dye with DNA has an absorption maximum at about 550 nm and an emission maximum at about 660 nm; it can be excited effectively at 488 nm.

DRAQ5 (63, 64) is a permeant anthraquinone dye that, when bound to DNA, has an excitation maximum near 650 nm and an emission maximum near 700 nm. DRAQ5 appears to be reasonably DNA selective, and the combination of DNA selectivity and permeancy allows it to be used to produce reasonably good DNA content histograms when it is used as a vital stain, making it the only dye other than Hoechst 33342 usable for that purpose. The fluorescence of DRAQ5 is not significantly increased when the dye is bound to DNA, and some manipulation of the relative concentrations of dye and cells may be necessary to obtain the best-quality DNA content measurements. On the plus side, it is possible to excite the dye at 488 nm as well as with red He–Ne or diode lasers.

Dyes that Stain Both DNA and RNA

Ethidium bromide (20) and propidium iodide (14) were among the first dyes used for DNA content determination by flow cytometry. Both are excited at 488 nm; the original rationale for the use of propidium lay in the fact that its emission maximum (about 615 nm) is 10–15 nm farther into the red region of the spectrum than that of ethidium, making it easier to separate red and green fluorescence signals from propidium and fluorescein, using optical filters. Neither ethidium nor propidium is DNA selective—both form complexes with double-stranded DNA and RNA by intercalating between base pairs, increasing fluorescence 20 to 30 times on binding. Neither ethidium nor propidium exhibits a strong base preference. When either dye is used to stain DNA in fixed or permeabilized cells, specimens must be treated with RNAse to eliminate artifactual broadening of DNA content distributions that will otherwise result from the fluorescence of dye bound to double-stranded RNA; the same consideration applies to DNA staining with the other nucleic acid dyes, discussed below, which are not DNA selective.

Because ethidium does not rapidly cross the membranes of intact cells and is likely to be pumped out when it does, it has widely, but erroneously, been regarded as impermeant and has been used in dye exclusion tests; this is not a good idea. Propidium, which by virtue of its double positive charge is impermeant, is a more suitable alternative. The double charge also gives propidium a higher binding affinity for double-stranded nucleic acid than ethidium; the former dye will displace the latter from cells that are permeable to both.

Asymmetric cyanine dyes have been widely used for nucleic acid staining since the mid-1980s; when Lee et al. (38) developed thiazole orange (TO), which chemically is 1,3'-dimethyl-4,2'-quinothiacyanine, for use in blood reticulocyte analysis in instruments using 488-nm light sources. Thiazole orange, when bound to RNA, has an absorption maximum at 509 nm and an emission maximum at 533 nm, and its fluorescence quantum efficiency is increased approximately 3000 times over that of the free dye. The dye also behaves as a DNA fluorochrome. It and related dyes are useful for reticulocyte counting only because reticulocytes normally contain RNA but no DNA. Lee et al. also synthesized 1,3'-dimethyl-4,2'-quinothiacarbocyanine, which they named thiazole blue. This dye also behaves as a DNA and RNA fluorochrome and, when bound to nucleic acid, has an excitation maximum at about 640 nm, making it useful with red He–Ne or diode laser sources.

Both thiazole orange and thiazole blue are permeant; Molecular Probes has developed both permeant and impermeant derivatives of both dyes for a variety of applications. TO-PRO-1 has the same ring structure and fluorescence spectrum as TO but has a quaternary ammonium side chain, making it impermeant. TO-PRO-3 shares the thiazole blue ring structure and spectrum but is also made impermeant by a quaternary ammonium group. Other dyes in the TO-PRO series have shorter and longer excitation and emission wavelengths than those of TO-PRO-1 and TO-PRO-3.

Molecular Probes's SYTO series of dyes are cell-permeant relatives of the TO-PRO dyes. Their permeancy indicates that, similar to their parent TO (the TO in TO-PRO, TOTO, and SYTO), they lack the quaternary ammonium groups present in TO-PRO and TOTO dyes. Their structures have not been revealed, but these dyes presumably have extra side chains that improve their permeancy. The SYTO dyes stain both DNA and RNA, and some of them also stain other cellular structures (e.g., mitochondria). The SYTOX series of dyes are highly impermeant, indicating that they bear at least three positive charges (structures have not been published). They are promoted as indicators of "nonviability" for dye exclusion tests. Other Molecular Probes cyanines include PicoGreen, SYBR Green I and II, and SYBR Gold.

TOTO-1 and YOYO-1 (60) are dimeric cyanine nucleic acid dyes in which two molecules of TO, in the first case, and oxazole yellow (YO), its quinooxacyanine analog, in the second, are joined by a diazaundecamethylene linker. The binding affinity of these dyes for DNA is sufficiently high that fragments labeled with TOTO-1 or

YOYO-1 and with ethidium can be mixed and separated by electrophoresis. TOTO-1 has the spectral characteristics of thiazole orange; YOYO-1 has those of oxazole yellow (excitation maximum for nucleic acid–bound dye 489 nm; emission maximum 509 nm), and both are over 1000 times more fluorescent in the nucleic acid–bound than in the free form. TOTO-3, which combines two thiazole blue chromophores using the same linker as is used in TOTO-1 and YOYO-1, is red excited. All three dyes and some related dimers are now available from Molecular Probes.

DNA/RNA Stains: AO and Hoechst 33342/Pyronin Y

Darzynkiewicz and his coworkers (16–18, 70) established that, under carefully controlled conditions, the blue-excited green fluorescence of AO molecules intercalated into DNA can be used to provide accurate and precise estimates of cellular DNA content, whereas the red metachromatic fluorescence of complexes of the dye with single-stranded RNA or denatured DNA, respectively, can provide indications of RNA content or of chromatin structure.

Measurements of RNA and DNA content define subcompartments of the cell cycle (18). The quiescent state, described as G_0 or G_{1Q}, in which cells such as peripheral blood lymphocytes normally remain, is characterized by a "diploid" (2C) DNA content and a low RNA content. Within 12 hours or so following exposure to mitogens, lymphocytes enter the G_1 phase and begin to synthesize RNA. RNA content continues to increase during the S phase, beginning about 30 hours after stimulation, in which DNA synthesis occurs.

Analysis of DNA content alone cannot discriminate cells in G_0 (G_{1Q}) from cells in the proliferative G_1 state, because the DNA content remains at 2C until the S phase begins. Measurements of RNA content can be used to make this distinction and, in addition, to define different stages within G_1. Cells pass from G_{1Q} through a brief transitional phase called G_{1T}, in which RNA content is slightly increased, and then into G_{1A}, during which RNA content increases further but remains lower than the RNA content of any S-phase cell. The cells then enter G_{1B}, in which RNA content is at or above the lowest value seen in S-phase cells. RNA content increases approximately linearly during S and G_2.

In exponentially growing cultures, which lack cells in G_{1Q}, cells appear to pass from S through G_2 and M back into G_{1A}. Normal cells, such as stimulated lymphocytes, when maintained in long-term culture, tend to revert back to a G_{1Q} state, although quiescent, low-RNA "S_Q" and "G_{2Q}" populations can appear transiently in cells deprived of nutrients or exposed to cold or to inhibitors of protein synthesis. Transition to quiescent (Q) states during S and G_2 appears to be somewhat more common in transformed and malignant cells.

Patterns of cellular DNA and RNA content observed using AO staining can also be demonstrated when other dyes, notably Hoechst 33342 and pyronin Y (58), are employed to stain DNA and RNA, respectively. Although this approach has the disadvantage of requiring the use of dual-wavelength excitation (UV for the Hoechst dye and blue-green or green for pyronin Y, with pyronin emission measured at 575 nm), the Hoechst/pyronin technique produces comparable results and can be used in combination with surface immunofluorescence measurements. The RNAse-sensitive fluorescence of pyronin Y in cells comes predominantly from dye bound to double-stranded

RNA in polyribosomes (69). Viability can be maintained after Hoechst/pyronin staining; Srour and Jordan (66) routinely sort and culture human hematopoietic stem cells after staining with Hoechst 33342 at a concentration of 1.6 μM (1 μg/mL) and pyronin Y at a concentration of 3.3 μM (also 1 μg/mL), adding 50–100 μM verapamil to block efflux of the dyes.

Fluorescent Indicators of Enzyme Activity

Enzyme activity in single cells can be demonstrated and quantified by flow cytometry following incubation of the cells with fluorogenic substrates, which yield fluorescent products. The majority of fluorogenic substrates are derivatives of fluorescein or of coumarins. Substrates derived from fluorescein, as would be expected, yield blue-excited, green fluorescent products, while coumarin-based substrates form products that require UV or violet excitation and emit in the blue and green regions of the spectrum. Derivatives of resorufin, which form green- to yellow-excited, orange-red fluorescent products, can be used to demonstrate esterases and oxidative enzymes. The excitation maximum of resorufin, however, is at about 570 nm, making 488 nm excitation suboptimal at best. Both the Molecular Probes *Handbook* (30) and *Practical Flow Cytometry* (60) contain extensive information on fluorogenic substrates.

The best-known fluorogenic substrate is fluorescein diacetate (FDA), more properly called diacetylfluorescein. This nonfluorescent ester of fluorescein is taken up by living cells; once inside, it is hydrolyzed by nonspecific esterases to fluorescein, which, because of its anionic character, is retained in cells for minutes to hours. Retention of fluorescein produced by FDA hydrolysis is considered a criterion of "viability", but actually only indicates that the cytoplasmic membrane is not damaged enough to permit the dye to leak out more rapidly. Dye retention tests are the "flip side" of dye exclusion tests, which detect membrane damage in terms of uptake by cells of normally nonpermeant dyes such as propidium and Trypan blue. Esters of carboxyfluorescein and calcein are preferable to FDA for dye retention tests because cells without membrane damage retain their fluorescent hydrolysis products much longer than fluorescein is retained.

Oxidative metabolism of cells, also used as an indicator of functional viability, can be studied, using a variety of fluorogenic substrates. Cyanoditolyl tetrazolium chloride (57, 67) is enzymatically reduced by actively metabolizing cells and bacteria to a water-insoluble fluorescent formazan that has an excitation maximum at 450 nm and emits in the range from 580 to 660 nm. 2,7-dichlorodihydrofluorescein diacetate (3), also known as dichlorofluorescin diacetate (DCFH-DA or H_2DCFDA), is, like FDA, hydrolyzed by nonspecific intracellular esterases, which transform the ester into a non-fluorescent intermediate, DCFH. In the presence of peroxidase, and of H_2O_2 formed during the respiratory burst in activated granulocytes, DCFH is converted to the fluorescent 2,7-dichlorofluorescein (DCF). DCFH-DA and other dyes are therefore used for quantitative assessment of neutrophil function, sometimes in conjunction with fluorescent plastic particles or stained bacteria, which can be used to quantify phagocytosis. Dihydrorhodamine 123 (52), which is oxidized to the green fluorescent cationic dye rhodamine 123, is also used to detect the respiratory burst in neutrophils; the product is produced only in cells containing peroxidase, meaning that the dye, like DCFH-DA, is relatively insensitive to superoxide production.

Hydroethidine (HE), or dihydroethidium (22), is the product of reduction of ethidium by sodium borohydride. The UV-excited, blue fluorescent HE, unlike ethidium, readily enters live cells, where it may be oxidized to ethidium. Ethidium then exhibits its characteristic blue- or green-excited red fluorescence, enhanced by binding to double-stranded nucleic acids. Although DCFH-DA and dihydrorhodamine 123 respond primarily to hydrogen peroxide generation, HE is more affected by superoxide, and the dyes, used separately, allow discrimination of these two types of reactive oxygen species (8, 53).

Caspases (cysteine–aspartic acid–specific proteases) are activated by cell death–inducing stimuli, and several reagents have been described for detection of caspase activity in apoptosis. Caspases can also be detected using fluorescent inhibitors (60).

By making appropriate kinetic flow cytometric measurements, it is possible to characterize intracellular enzyme reactions in considerable detail, determining relevant rates of reaction and inhibition kinetics.

Tracking Dyes

It is often desirable to attach permanent or relatively permanent fluorescent labels to cells for purposes such as detecting cell aggregation or hybridization, determining the localization and fate of cells isolated from and reinjected into an animal, and establishing the rate of turnover of various components of the membrane itself. A number of fluorescent compounds have been used in this way as "tracking dyes" (60).

One class of labels is incorporated into the membrane lipid bilayer itself. Among the most widely studied labels are derivatives of cyanine dyes, including *DiI* (dioctadecylindocarbocyanine) and *DiO* (dioctadecyloxacarbocyanine), both of which can be excited at 488 nm, with DiO emitting green fluorescence at about 500 nm and DiI emitting yellow fluorescence at about 565 nm. Fluorescein and phycoerythrin emission filters are, respectively, well suited for measurement of DiO and DiI. The octadecyl (C_{18}) side chains of these probes reside in the lipid bilayer, whereas the cyanine chromophores remain at the surface.

If a cell labeled with a tracking dye subsequently proliferates, each daughter cell will get approximately half the label; in the next generation, each daughter cell will carry one-quarter of the label, and so on. Analysis of the labeling intensities of cells grown in vivo or in vitro after labeling will, therefore, allow determination of the number of division cycles through which each cell has progressed since the label was applied. The dye first widely used for such studies was PKH26, a yellow fluorescent indocarbocyanine dye with long alkyl side chains. Estimation of the numbers of cells in various daughter generations after PKH26 labeling requires application of a mathematical model; the dye and software are available from Sigma (St. Louis, MO) and Verity Software House (Topsham, ME).

An alternative to PKH26, carboxyfluorescein diacetate succinimidyl ester (CFSE; 39), is a nonfluorescent fluorescein ester that enters cells and is hydrolyzed to a reactive dye by nonspecific esters; the end result is that fluorescein molecules are bound covalently to intracellular protein. Distributions of CFSE fluorescence in proliferating populations usually show peaks indicating the positions of cells in different daughter generations. These distributions can be analyzed with mathematical models, but it is also possible to combine sorting with CFSE labeling to isolate cells from different generations (46), which cannot be done reliably when PKH26 is used as a tracking dye. CFSE labeling has been used extensively for studies of lymphocyte activation and differentiation (23, 40, 75).

Probes of Membrane Potential, pH, and Calcium

Changes in cytoplasmic and mitochondrial membrane potential, intracellular pH, and calcium ion concentration are associated with cell activation processes mediated by ligand-receptor interactions. These associations stimulated interest in probing these functional responses to elucidate the effects of natural products and drugs on cell signaling.

Cytoplasmic and Mitochondrial Membrane Potential Indicators

Electrical potential differences are present across the cytoplasmic membranes of most living prokaryotic and eukaryotic cells and also between the cytosol and the interior of organelles such as chloroplasts and mitochondria. Membrane potential ($\Delta\Psi$) is generated and maintained by transmembrane concentration gradients of ions such as sodium, potassium, chloride, and hydrogen.

Changes in cytoplasmic $\Delta\Psi$ play a role in transmembrane signaling in the course of surface receptor–mediated processes related to the development, function, and pathology of many cell types. Cytoplasmic $\Delta\Psi$ is reduced to zero when the membrane is ruptured by chemical or physical agents; mitochondrial $\Delta\Psi$ is reduced when energy metabolism is disrupted, notably in apoptosis. In bacteria, $\Delta\Psi$ reflects both the state of energy metabolism and the physical integrity of the cytoplasmic membrane.

Flow cytometry can be used to estimate membrane potential in eukaryotic cells, mitochondria in situ, isolated mitochondria, and bacteria (59, 61). Older methods, using lipophilic cationic dyes such as the symmetric cyanines dihexyloxacarbocyanine [$DiOC_6(3)$] and hexamethylindodicarbocyanine [$DiIC_1(5)$] or rhodamine 123, or lipophilic anionic dyes such as bis (1,3-dibutyl-barbituric acid) trimethine oxonol, [$DiBAC_4(3)$, often (and incorrectly) referred to as bis-oxonol], can detect relatively large changes in $\Delta\Psi$ and identify heterogeneity of response in subpopulations comprising substantial fractions of a cell population. All of the dyes just mentioned can be excited at 488 nm and emit green fluorescence, with the exception of $DiIC_1(5)$, which is red excited and emits near 670 nm. Newer techniques that use energy transfer or ratios of fluorescence emission at different wavelengths allow precise measurement of $\Delta\Psi$ to within 10 mV or less (25,47).

Because in most eukaryotic cells $\Delta\Psi$ across mitochondrial membranes is larger than $\Delta\Psi$ across cytoplasmic membranes, exposure of cells to lipophilic cationic dyes results in higher concentrations of dye in the cells than in the suspending medium and in higher concentrations in mitochondria than in the cytosol. If cells are washed after being loaded with dye, then staining of the cytosol may be minimized while mitochondrial staining persists. This is the basis for the use of $DiOC_6(3)$, $DiIC_1(5)$, rhodamine 123, and other cationic dyes to estimate mitochondrial $\Delta\Psi$. The procedure has become commonplace for studies of apoptosis, in which early increases in mitochondrial membrane permeability result in loss of $\Delta\Psi$. JC-1, a cyanine, exhibits green fluorescence in monomeric form and red fluorescence when aggregated at higher concentrations (49, 62) and has become popular for demonstrating mitochondrial deenergization in apoptosis (13). However, the most accurate measurements of mitochondrial potentials are made using very low (~1 nM) concentrations of $DiOC_6(3)$ (54).

Action of efflux pumps, changes in membrane structure, and changes in protein or lipid concentration in the medium in which cells are suspended, among other factors, can produce changes in cellular fluorescence that may be interpreted erroneously as changes in $\Delta\Psi$. For example, it was observed in the 1980s that hematopoietic stem cells were not stained by rhodamine 123, and some people concluded that this reflected low mitochondrial $\Delta\Psi$; it was later found that the dye was being actively extruded. Getting good results from cytometric techniques for estimation and measurement of $\Delta\Psi$ demands both careful control of cell and reagent concentrations and incubation times and good selection of appropriate controls.

Indicators of Cytoplasmic [Ca^{++}]: Advantages of Ratiometric Measurements

The importance of calcium fluxes in cell signaling was appreciated when flow cytometry was in a relatively early stage of development, but it was not until some years later that suitable probes became available (7). The first probes exhibited differences in the intensity of fluorescence in the presence of low and high intracellular [Ca^{++}] but did not change either their fluorescence excitation or their emission spectral characteristics to a significant degree. The distribution of fluorescence intensity from cells loaded with the probes was typically quite broad (a problem also associated with membrane potential probes). Thus, it was possible to appreciate how large changes in cytoplasmic [Ca^{++}] affected all or most of the cells in a population, which would shift the entire distribution substantially, but not to detect even a large change in cytoplasmic [Ca^{++}] involving only a small subpopulation of cells. This came as a disappointment to immunobiologists, who had hoped to use flow cytometry to detect calcium responses associated with activation of lymphocytes by specific antigens.

Tsien's group (28) made a substantial improvement in 1985 with the development of Indo-1. Like other probes, Indo-1 is a selective calcium chelator, but it does not significantly perturb cellular calcium metabolism. Its fluorescence is excited by UV light; wavelengths between 325 and about 365 nm are suitable. There are substantial differences in emission spectra between the free dye, which shows maximum emission at about 480 nm, and the calcium chelate, which emits maximally at about 405 nm. The ratio of emission intensities at 405 and 480 nm in cells loaded with Indo-1 (it is introduced as an acetoxymethyl [AM] ester) therefore, provides an indication of cytoplasmic [Ca^{++}]. The ratiometric measurement cancels out many extraneous factors, most notably including the effect of cell-to-cell variations in dye content, which plague older techniques for calcium measurement and for measurement of $\Delta\Psi$. Effects of uneven illumination and of light source noise also are eliminated by virtue of their equal influences on the numerator and denominator of the ratio. This advantage, it should be noted, is common to other ratiometric measurements (e.g., of $\Delta\Psi$ and of intracellular pH) in which both parameters used in the ratio are measured at the same time in the same beam. If aliquots of loaded cells are placed in solutions with various known Ca^{++} concentrations and treated with a calcium ionophore such as A23187 or ionomycin, the fluorescence ratio measurement can be calibrated to yield accurate molar values of cytoplasmic [Ca^{++}].

There are other calcium indicators suitable for use with 488-nm excitation. The most widely used of these is Fluo-3, which has the spectral characteristics of fluorescein but is almost nonfluorescent unless bound to calcium. Unlike Indo-1, Fluo-3 does not exhibit a spectral shift with changes in calcium concentration. However, another dye, Fura red, which is also suitable for 488-nm excitation, exhibits high fluorescence when free in solution (or cytosol) and low fluorescence when bound to calcium; the ratio of fluo-3 to Fura red fluorescence provides a precise indicator of cytoplasmic $[Ca^{++}]$ that can be calibrated and can be used in the majority of fluorescence flow cytometers (7). Both Fluo-3 and Fura red are, similar to Indo-1, loaded into cells as AM esters.

Indicators of Intracellular pH

Both the Molecular Probes *Handbook* (30) and *Practical Flow Cytometry* (60) discuss this subject at some length. Two dyes are currently favored for flow cytometric pH measurement. 2′,7′-bis-(carboxyethyl)-5,6-carboxyfluorescein (44, 50) is excited at 488 nm, and pH is estimated from the ratio of emission intensities at 520 and 620 nm. Carboxy SNARF-1 can be excited at wavelengths from 488 to 532 nm, and the ratio of emission intensities at 640 and 580 nm provides a measure of pH. Both dyes are introduced as AM esters.

Reporter Genes and Fluorescent Proteins

In this era of molecular biology, flow cytometry has come into increasing use as a method for detecting and selecting cells expressing reporter genes. Reporter genes encode protein products that are relatively readily detectable; they are linked by standard recombinant DNA technology to a gene or genes of interest to an investigator, and the resulting construct is introduced into cells. Those cells in which the gene of interest is expressed can then be identified by the presence of the reporter gene product.

In the 1980s, genes encoding surface antigens not normally present on the cells of interest were used as reporter genes; expression of these genes was detected by immunofluorescence. This was unsatisfactory, as what was generally observed were very broad distributions of relatively weak signals, and successful isolation of cells bearing the genes of real interest to investigators generally required multiple cycles of cell sorting, in which the brightest cells were sorted and cloned, the brightest progeny were resorted, and so froth.

In 1988, the Herzenberg lab (45) reported the development of a new method, called *FACS-Gal*, in which the *lacZ* gene from *Escherichia coli* was used as the reporter gene; this gene encodes the enzyme β-D-galactosidase. The activity of this enzyme is detected by flow cytometry, using the fluorogenic substrate fluorescein di-β-D-galactoside. The FACS-Gal technique made it much simpler to detect and sort cells expressing the reporter gene; the stronger signals also made it possible to distinguish cells expressing different amounts of the reporter gene and, therefore, of the genes of interest, and to sort these separately for further analyses (21, 55).

Enzymes are still being used as reporter genes. Tsien's group (80) developed a β-lactamase reporter gene; the fluorogenic substrate used to detect it is called CCF2. This gene is loaded into cells as the AM ester. CCF2 contains a coumarin donor and a fluorescein acceptor attached to a cephalosporin β-lactam ring. When the ring is intact, the coumarin and fluorescein moieties of the probe are close enough together for energy transfer to occur, and excitation (the excitation maximum is 409 nm) produces green (530-nm) fluorescence. When the ring is cleaved by β-lactamase, the distance between the coumarin and the fluorescein is increased sufficiently to greatly reduce the efficiency of energy transfer, and the probe emits blue (450-nm) fluorescence. The intensity of blue emission from the hydrolyzed probe is about twice the intensity of green emission from the intact probe.

In early 1994, Chalfie et al. (9) described the use of a gene encoding an intrinsically fluorescent protein from the bioluminescent jellyfish *Aequorea victoria* as a reporter gene. The native green fluorescent protein (GFP) absorbs maximally at 395 nm but can be excited moderately effectively at 488 nm, where its absorption is about one-third maximum; the emission spectrum has a sharp peak at 510 nm with a shoulder at 540 nm.

Matz et al. (42) note that members of the GFP family are unique among pigment proteins in that they act as enzymes and synthesize their fluorophores from amino acids in their own polypeptide chains. Other pigment proteins, such as the phycobiliproteins, make their chromophores or fluorophores from small molecules and usually require several enzymes to get the job done. Thus, it would be a much harder job to make a phycobiliprotein reporter than it is to make a GFP reporter. The self-contained palette of GFP proteins also makes it much easier to modify their spectra by site-directed mutagenesis than it would be to change the spectrum of another type of pigment protein, and a number of mutants with spectral characteristics suitable for flow cytometry have been developed (71).

ECFP (cyan; excitation maximum 434 nm, emission maximum 477 nm), EGFP (green; excitation maximum 489 nm, emission maximum 508 nm), and EYFP (yellow-green, excitation maximum 514 nm, emission maximum 527 nm) are all produced by mutants of the *Aequorea* GFP gene. DsRed (orange; excitation maximum 558 nm, emission maximum 583 nm; 2) is derived from a coral of the species *Discosoma*. The constructs needed to introduce the appropriate genes into cells are available from Clontech, a division of BD Biosciences. Hawley et al. (31) recently described flow cytometric detection of all four fluorescent proteins at once using 458-nm excitation from an argon laser for ECFP, EGFP, and EYFP, and 568-nm excitation from a krypton laser for DsRed.

GFPs themselves have, with help, "evolved" far beyond the point of merely marking transfection (60): There are fluorescent protein variants that are sensitive to pH, calcium ion concentration, and membrane potential, and others that monitor cell surface receptor interactions and kinase activities. Applications involved molecules with single chromophores as well as composites. There is also a "timer" protein that gradually changes color from green to red. We can expect flow cytometric applications of fluorescent proteins to increase in variety and significance over the next few years (78).

References

1. Asbury, C.L., Uy, J.L., and van den Engh, G. 2002. Polarization of scatter and fluorescence signals in flow cytometry. *Cytometry* 40:88–101.

2. Baird, G.S., Zacharias, D.A., and Tsien, R.Y. 2000. Biochemistry, mutagenesis, and oligomerization of DsRed, a red fluorescent protein from coral. *Proc. Natl. Acad. Sci. USA* 97:11984–9.

3. Bass, D.A., Parce, J.W., DeChatelet, L.R., et al. 1983. Flow cytometric studies of oxidative product formation by neutrophils: a graded response to membrane stimulation. *J. Immunol.* 130:1910–7.

4. Beavis, A.J. and Pennline, K.J. 1996. Allo-7: a new fluorescent tandem dye for use in flow cytometry. *Cytometry* 24:390–4; also see Publisher's Notice, *Cytometry* 24:390–5.

5. Bishop, J.E., Davis, K.A., Abrams, B., Houck, D.W., Recktenwald, D.J., and Hoffman, R.A. 2000. Mechanism of higher brightness of PerCP-Cy5.5. *Cytometry Supp.* 10:162–3.

6. Bruchez Jr., M., Moronne, M., Gin, P., Weiss, S., and Alivisatos, A.P. 1998. Semiconductor nanocrystals as fluorescent biological labels. *Science* 281:2013–6.

7. Burchiel, S.W., Edwards, B.S., Kuckuck, F.W., Lauer, F.T., Prossnitz, E.R., Ransom, J.T., and Sklar, L.A. 2000. Analysis of free intracellular calcium by flow cytometry: multiparameter and pharmacologic applications. *Methods* 21:221–30.

8. Carter, W.O., Narayanan, P.K., and Robinson, J.P. 1994. Intracellular hydrogen peroxide and superoxide anion detection in endothelial cells. *J. Leukoc. Biol.* 55:253–8.

9. Chalfie, M., Tu, Y., Euskirchen, G., Ward, W.W., and Prasher, D.C. 1994. Green fluorescent protein as a marker for gene expression. *Science* 263:802–5.

10. Chan, W.C., Maxwell, D.J., Gao, X., Bailey, R.E., Han, M., and Nie, S. 2002. Luminescent quantum dots for multiplexed biological detection and imaging. *Curr. Opin. Biotechnol.* 13:40–6.

11. Chan, W.C. and Nie, S. 1998. Quantum dot bioconjugates for ultrasensitive nonisotopic detection. *Science* 281:2016–8.

12. Coons, A.H., Creech, H.J., and Jones, R.N. 1941. Immunological properties of an antibody containing a fluorescent group. *Proc. Soc. Exp. Biol. Med.* 47:200.

13. Cossarizza, A., Baccarani-Contri, M., Kalashnikova, G., and Franceschi, C. 1993. A new method for the cytofluorimetric analysis of mitochondrial membrane potential using the J-aggregate forming lipophilic cation 5,5′,6,6′-tetra-chloro-1,1′,3,3′-tetraethylbenzimidazolcarbocyanine iodide (JC-1). *Biochem. Biophys. Res. Commun.* 197:40–5.

14. Crissman, H.A. and Steinkamp, J.A. 1973. Rapid simultaneous measurement of DNA, protein and cell volume in single cells from large mammalian cell populations. *J. Cell Biol.* 59:766–71.

15. Crissman, H.A., Stevenson, A.P., Orlicky, D.J., et al.1978. Detailed studies on the application of three fluorescent antibiotics for DNA staining in flow cytometry. *Stain Technol.* 53:321–30.

16. Darzynkiewicz, Z. 1990. Differential staining of DNA and RNA in intact cells and isolated cell nuclei with acridine orange. *Methods Cell Biol.* 33:285–98.

17. Darzynkiewicz, Z. 1994. Simultaneous analysis of cellular RNA and DNA content. *Methods Cell Biol.* 41:402–20.

18. Darzynkiewicz, Z., Traganos, F., and Melamed, M.R. 1980. New cell cycle compartments identified by multiparameter flow cytometry. *Cytometry* 1:98–108.

19. Davis, K.A. and Houck, D.W. 1998. A novel red dye for flow cytometry. *Cytometry Supp.* 9:141.

20. Dittrich, W. and Göhde, W. 1969. Impulsfluorometrie bei Einzelzellen in Suspensionen. *Z. Naturforsch.* 24b:360–1.

21. Fiering, S.N., Roederer, M., Nolan, G.P., Micklem, D.R., Parks, D.R., and Herzenberg, L.A. 1991. Improved FACS-Gal: flow cytometric analysis and sorting of viable eukaryotic cells expressing reporter gene constructs. *Cytometry* 12:291–301.

22. Gallop, P.M., Paz, M.A., Henson, E., and Latt, S.A. 1984. Dynamic approaches to the delivery of reporter reagents into living cells. *Biotechniques* 1:32.

23. Gett, A.V. and Hodgkin, P.D. 1998. Cell division regulates the T cell cytokine repertoire, revealing a mechanism underlying immune class regulation. *Proc. Natl. Acad. Sci. USA* 95:9488–93.

24. Glazer, A.N. and Stryer, L. 1983. Fluorescent tandem phycobiliprotein conjugates. Emission wavelength shifting by energy transfer. *Biophys J.* 43:383–86.

25. Gonzalez, J.E. and Tsien, R.Y. 1997. Improved indicators of cell membrane potential that use fluorescence resonance energy transfer. *Chem. Biol.* 4:269–77.
26. Goodell, M.A. 2002. Stem cell identification and sorting using the Hoechst 33342 side population (SP). Unit 9.18. In Robinson, J.P., Darzynkiewicz, Z., Dean, P., Hibbs, A.R., Orfao, A., Rabinovitch, P., and Wheeless, L., eds. *Current Protocols in Cytometry*, Wiley, New York, pp. 9.18.1–9.18.11.
27. Goodell, M.A., Brose, K., Paradis, G., Conner, A.S., and Mulligan, R.C. 1996. Isolation and functional properties of murine hematopoietic stem cells that are replicating in vivo. *J. Exp. Med.* 183:1797–806.
28. Grynkiewicz, G., Poenie, M., Tsien, and R.Y. 1985. A new generation of Ca^{2+} indicators with greatly improved fluorescence properties. *J. Biol. Chem.* 260:3440.
29. Han, M., Gao, X., Su, J.Z., and Nie, S. 2001. Quantum-dot-tagged microbeads for multiplexed optical coding of biomolecules. *Nat. Biotechnol.* 19:631–5.
30. Haugland, R.P. 2002. *Handbook of Fluorescent Probes and Research Products*, 9th ed. Molecular Probes, Eugene, OR. Available at: www.probes.com.
31. Hawley, T.S., Telford, W.G., Ramezani, A., and Hawley, R.G. 2001. Four-color flow cytometric detection of retrovirally expressed red, yellow, green, and cyan fluorescent proteins. *BioTechniques* 30:1028–34.
32. Jaiswal, J.K., Mattoussi, H., Mauro, J.M., and Simon, S.M. 2003. Long-term multiple color imaging of live cells using quantum dot bioconjugates. *Natl. Biotechnol.* 21:47–51.
33. Jovin, T.M. 2003. Quantum dots finally come of age. *Natl. Biotechnol.* 21:32–3.
34. Kim, M., Turnquist, H., Jackson, J., Sgagias, M., Yan, Y., Gong, M., Dean, M., Sharp, J.G., Cowan, K. 2002. The multidrug resistance transporter ABCG2 (breast cancer resistance protein 1) effluxes Hoechst 33342 and is overexpressed in hematopoietic stem cells. *Clin. Cancer Res.* 8:22–8.
35. Latt, S.A. 1973. Microfluorimetric detection of deoxyribonucleic acid replication in human metaphase chromosomes. *Proc. Natl. Acad. Sci. USA* 70:3395–99.
36. Latt, S.A. and Stetten, G. 1976. Spectral studies on 33258 Hoechst and related bisbenzimidazole dyes useful for fluorescent detection of deoxyribonucleic acid synthesis. *J. Histochem. Cytochem.* 24:24–33.
37. Lechner, A., Leech, C.A., Abraham, E.J., Nolan, A.L., and Habener, J.F. 2002. Nestin-positive progenitor cells derived from adult human pancreatic islets of Langerhans contain side population (SP) cells defined by expression of the ABCG2 (BCRP1) ATP-binding cassette transporter. *Biochem. Biophys. Res. Commun.* 293:670–4.
38. Lee, L.G., Chen, C-H., and Chiu, L.A. 1986. Thiazole orange: a new dye for reticulocyte analysis. *Cytometry* 7:508–17.
39. Lyons, A.B., Hasbold, J., and Hodgkin, P.D. 2001. Flow cytometric analysis of cell division history using dilution of carboxyfluorescein diacetate succinimidyl ester, a stably integrated fluorescent probe. *Methods Cell Biol.* 63:375–98.
40. Lyons, A.B. and Parish, C.R. 1994. Determination of lymphocyte division by flow cytometry. *J. Immunol. Methods* 171:131–7.
41. Mandy, F.F., Bergeron, M., Recktenwald, D., and Izaguirre, C.A. 1992. A simultaneous three-color T cell subsets analysis with single laser flow cytometers using T cell gating protocol. Comparison with conventional two-color immunophenotyping method. *J. Immunol. Methods* 156:151–62.
42. Matz, M.V., Lukyanov, K.A., and Lukyanov, S.A. 2002. Family of the green fluorescent protein: journey to the end of the rainbow. *Bioessays* 24:953–9.
43. Mujumdar, R.B., Ernst, L.A., Mujumdar, S.R., Lewis, C.J., and Waggoner, A.S. 1993. Cyanine dye labeling reagents: sulfoindocyanine succinimidyl esters. *Bioconjugate Chem.* 4:105–11.
44. Musgrove, E., Rugg, C., and Hedley, D. 1986. Flow cytometric measurement of cytoplasmic pH: a critical evaluation of available fluorochromes. *Cytometry* 7:347–55.
45. Nolan, G.P., Fiering, S., Nicolas, J.F., and Herzenberg, L.A. 1988. Fluorescence-activated cell analysis and sorting of viable mammalian cells based on β-D-galactosidase activity after transduction of *Escherichia coli lacZ*. *Proc. Natl. Acad. Sci. USA* 85:2603–7.
46. Nordon, R.E., Ginsberg, S.S., and Eaves, C.J. 1997. High resolution cell division tracking

demonstrates the Flt3 ligand dependence of human marrow CD34$^+$CD38$^-$ cell production in vitro. *Br. J. Haematol.* 98:528–39.

47. Novo, D., Perlmutter, N.G., Hunt, R.H., and Shapiro, H.M. 1999. Accurate flow cytometric membrane potential measurement in bacteria using diethyloxacarbocyanine and a ratiometric technique. *Cytometry* 35:55–63.

48. Oi, V.T., Glazer, A.N., and Stryer, L. 1982. Fluorescent phycobiliprotein conjugates for analyses of cells and molecules. *J. Cell Biol.* 93:981–6.

49. Reers, M., Smith, T.W., and Chen, L.B. 1991. J-aggregate formation of a carbocyanine as a quantitative fluorescent indicator of membrane potential. *Biochemistry* 30:4480–6.

50. Rink, T.J., Tsien, R.Y., and Pozzan, T. 1982. Cytoplasmic pH and free Mg^{2+} in lymphocytes. *J. Cell Biol.* 95:189–96.

51. Roederer, M., Kantor, A.B., Parks, D.R., and Herzenberg, L.A. 1996. Cy7PE and Cy7APC: bright new probes for immunofluorescence. *Cytometry* 24:191–7.

52. Rothe, G., Oser, A., and Valet, G. 1988. Dihydrorhodamine 123: a new flow cytometric indicator for respiratory burst activity in neutrophil granulocytes. *Naturwissen-schaften* 75:354–5.

53. Rothe, G. and Valet, G. 1990. Flow cytometric analysis of respiratory burst activity in phagocytes with hydroethidine and 2′,7′-dichlorofluorescin. *J. Leukoc. Biol.* 47:440–8.

54. Rottenberg, H. and Wu, S. 1998. Quantitative assay by flow cytometry of the mitochondrial membrane potential in intact cells. *Biochim. Biophys. Acta* 1404:393–404.

55. Saalmuller, A. and Mettenleiter, T.C. 1993. Rapid identification and quantitation of cells infected by recombinant herpesvirus (pseudorabies virus) using a fluorescence-based beta-galactosidase assay and flow cytometry. *J. Virol. Methods* 44:99–108.

56. Scharenberg, C.W., Harkey, M.A., Torok-Storb, B. 2002. The ABCG2 transporter is an efficient Hoechst 33342 efflux pump and is preferentially expressed by immature human hematopoietic progenitors. *Blood* 99:507–12.

57. Severin, E. and Stellmach, J. 1984. Flow cytometry of redox activity of single cells using a newly synthesized fluorescent formazan. *Acta Histochem.* 75:101–5.

58. Shapiro, H.M. 1981. Flow cytometric estimation of DNA and RNA content in intact cells stained with Hoechst 33342 and pyronin Y. *Cytometry* 2:143–50.

59. Shapiro, H.M. 2000. Membrane potential estimation by flow cytometry. *Methods* 21:271–9.

60. Shapiro, H.M. 2003. *Practical Flow Cytometry*, 4th ed. Wiley-Liss, Hoboken, NJ.

61. Shapiro, H.M., Natale, P.J., and Kamentsky, L.A. 1979. Estimation of membrane potentials of individual lymphocytes by flow cytometry. *Proc. Natl. Acad. Sci. USA* 76:5728–30.

62. Smiley, S.T., Reers, M., Mottola-Hartshorn, C., et al. 1991. Intracellular heterogeneity in mitochondrial membrane potentials revealed by a J-aggregate-forming lipophilic cation JC-1. *Proc Natl. Acad. Sci. USA* 88:3671–5.

63. Smith, P.J., Blunt, N., Wiltshire, M., Hoy, T., Teesdale-Spittle, P., Craven, M.R., Watson, J.V., Amos, W.B., Errington, R.J., Patterson, L.H. 2000. Characteristics of a novel deep red/infrared fluorescent cell-permeant DNA probe, DRAQ5, in intact human cells analyzed by flow cytometry, confocal and multiphoton microscopy. *Cytometry* 40:280–91.

64. Smith, P.J., Wiltshire, M., Davies, S., Patterson, L.H., and Hoy, T. 1999. A novel cell permeant and far red-fluorescing DNA probe, DRAQ5, for blood cell discrimination by flow cytometry. *J. Immunol. Methods* 229:131–9.

65. Southwick, P.L., Ernst, L.A., Tauriello, E.W., Parker, S.R., Majumdar, R.B., Majumdar, S.R., Clever, H.A., and Waggoner, A.S. 1990. Cyanine dye labeling reagents—carboxymethyl-indocyanine succinimidyl esters. *Cytometry* 11:418–30.

66. Srour, E.F., Jordan, C.T. 2001. Isolation and characterization of primitive hematopoietic cells based on their position in cell cycle. In Klug, C.A., Jordan, C.T., eds. *Hematopoietic Stem Cell Protocols (Methods in Molecular Medicine, Vol. 63)*, Humana Press, Totowa, NJ, pp. 93–111.

67. Stellmach, J. and Severin, E. 1987. A fluorescent redox dye. Influence of several substrates and electron carriers on the tetrazolium salt-formazan reaction of Ehrlich ascites tumor cells. *Histochem. J.* 19:21–6.

68. Szöllösi, J., Damjanovich, S., and Mátyus, L. 1998. Application of fluorescence resonance energy transfer in the clinical laboratory: routine and research. *Cytometry (Comm. Clin. Cytometry)* 34:159–79.

69. Traganos, F., Crissman, H.A., and Darzynkiewicz, Z. 1988. Staining with pyronin Y detects changes in conformation of RNA during mitosis and hyperthermia of CHO cells. *Exp. Cell Res.* 179:535–44.
70. Traganos, F., Darzynkiewicz, Z., and Sharpless, T., et al. 1977. Simultaneous staining of ribonucleic and deoxyribonucleic acids in unfixed cells using acridine orange in a flow cytofluorometric system. *J. Histochem. Cytochem.* 25:46–56.
71. Tsien, R.Y. 1998. The green fluorescent protein. *Annu. Rev. Biochem.* 67:509–44.
72. van den Engh, G. and Farmer, C. 1992. Photo-bleaching and photon saturation in flow cytometry. *Cytometry* 13:669–77.
73. Waggoner, A.S., Ernst, L.A., Chen, C.H., and Rechtenwald, D.J. 1993. PE-CY5. A new fluorescent antibody label for three-color flow cytometry with a single laser. *Ann. N. Y. Acad. Sci.* 677:185–93.
74. Ward, D.C., Reich, E., and Goldberg, I.H. 1965. Base specificity in the interaction of polynucleotides with antibiotic drugs. *Science* 149:1259–63.
75. Wells, A.D., Gudmundsdottir, H., and Turka, L.A. 1997. Following the fate of individual T cells throughout activation and clonal expansion. Signals from T cell receptor and CD28 differentially regulate the induction and duration of a proliferative response. *J. Clin. Invest.* 100:3173–83.
76. Wu, X., Liu, H., Liu, J., Haley, K.N., Treadway, J.A., Larson, J.P., Ge, N., Peale, F., and Bruchez, M.P. 2003. Immunofluorescent labeling of cancer marker Her2 and other cellular targets with semiconductor quantum dots. *Natl. Biotechnol.* 21:41–6.
77. Zelenin, A.V., Poletaev, A.I., Stepanova, N.G., Barsky, V.E., Kolesnikov, V.A., Nikitin, S.M., Zhuze, A.L., and Gnutchev, N.V. 1984. 7-amino-actinomycin D as a specific fluorophore for DNA content analysis by laser flow cytometry. *Cytometry* 5:348–54.
78. Zhang, J., Campbell, R.E., Ting, A.Y., and Tsien, R.Y. 2002. Creating new fluorescent probes for cell biology. *Natl. Rev. Mol. Cell Biol.* 3:906–18.
79. Zhou, S., Morris, J.J., Barnes, Y., Lan, L., Schuetz, J.D., and Sorrentino, B.P. 2002. Bcrp1 gene expression is required for normal numbers of side population stem cells in mice, and confers relative protection to mitoxantrone in hematopoietic cells in vivo. *Proc. Natl. Acad. Sci. USA* 99:12339–44.
80. Zlokarnik, G., Negulescu, P.A., Knapp, T.E., Mere, L., Burres, N., Feng, L., Whitney, M., Roemer, K., and Tsien, R.Y. 1998. Quantitation of transcription and clonal selection of single living cells with β-lactamase as reporter. *Science* 279:84–8.

3

Automation and High-Throughput Flow Cytometry

BRUCE S. EDWARDS AND LARRY A. SKLAR

Introduction

The flow cytometer is unique among biomedical analysis instruments in its ability to make multiple correlated optical measurements on individual cells or particles at high rates. Moreover, an ever-expanding arsenal of fluorescent probes enables the modern flow cytometer to quantify a large and growing diversity of cell-associated macromolecules and physiological processes. Modern flow cytometers have achieved such a level of sophistication and reliability that unattended operation by automated systems is a practical reality.

From its inception, flow cytometry has been in the vanguard of automation in cytological analysis (26). One of the most powerful automated features is cell sorting, an operation in which highly purified subsets of cells or particles are isolated from heterogeneous source populations on the basis of a targeted, multiparameter phenotype. The method most widely used for sorting today, which is based on electrostatic deflection of charged droplets, was developed over 30 years ago (11) and led to commercial flow cytometers that were capable of sorting cells at rates of hundreds of cells per second. Influenced by the need of the Human Genome Project for efficient isolation of purified chromosomes, a high-speed chromosome flow sorter was developed and patented in 1982 that increased sort rates to tens of thousands of events per second (13). Commercial systems subsequently became available in the 1990s that permitted sorting of cells at such high rates (www.bdbiosciences.com; www.dakocytomation.com). Thus, since the initial development of the technology, the throughput of automated cell sorting has increased by nearly two orders of magnitude.

In single cell analysis and sorting, throughput is determined by the rate at which the flow cytometer can process individual cells as they pass single file through the point of detection. Another aspect of flow cytometer throughput concerns the rate at which the flow cytometer can sequentially process multiple discreet collections of cells. This component of throughput will be important, for example, in the screening of collections of test compounds for their effects on bulk populations of cells. This is of particular relevance for modern drug discovery, in which there is a need to test cellular targets against millions of potentially valuable compounds that may bind cellular receptors to effect clinically therapeutic cellular responses (20). Although flow cytometry represents one of the more powerful analytical technologies by which to accomplish this objective, multiple sample processing is at present a rate-limiting bottleneck in commercial instruments. Recently, commercial systems have become available that enable automated processing by flow cytometers of multiple samples presented in tube carousels and 96-well plates (e.g., www.bdbiosciences.com, www.cytekdev.com, www.dakocytomation.com, www.luminex.com). However, as of this writing, sample throughput by these systems is typically in the range of two samples per minute, a rate comparable to manual sample introduction methods that have been used since the advent of commercial flow cytometers. The aim of this review is to describe some recent innovations in sample and data handling that promise to alleviate this shortcoming and promote what may be truly considered to be high-throughput flow cytometry.

Throughput Rate-Limiting Factors

Before discussing new methodological approaches in any detail, it is worth first considering what are aspects of conventional flow cytometry that limit sample throughput.

One Sample–One File Data Storage

In conventional flow cytometry, there are two periods of significant "dead time" encountered during sample analysis that are related to how data from the individual samples are partitioned for storage. Because data from each sample are typically saved in separate data files, there is first a delay at the onset of data acquisition, during which the hardware and software are initialized for data collection, and a second delay at the end of data collection, during which the data are saved to a file on a hard disk or other storage media. The magnitude of the total delay is determined by innate characteristics of the hardware and software and by the number of recorded events. Optimizing hardware and software performance is a feasible solution to minimize such dead time but is not usually a practical one for most flow practitioners. An alternate solution is to process multiple samples during a single round of data collection and to store the resulting data in a single, time-resolved data file (7). This requires methods to ensure that data from each sample are correctly partitioned and identified during subsequent post hoc analyses.

Optical Alignment

High rates of sample throughput often require high volumetric sample stream flow rates. This results in the expansion of the diameter of the sample stream within the enclosing sheath stream and in a greater probability that some flowing sample particles will stray from the optical alignment axis during transit through the detection point. There are certain types of assays in which only the very best possible optical alignment is acceptable. For example, even modest instrument misalignment can result in failure to detect subtle abnormalities of cell DNA content that are indicative of neoplasia. However, there are many types of assays in the conventional repertoire in which high-quality quantitative data can be obtained even under suboptimal conditions of optical alignment (7, 8, 24). Appropriate positive and negative control samples are necessary to ascertain data quality under such circumstances.

Particle and Fluid Carryover between Samples

As with optical alignment, a small amount of cell or particle carryover can be prohibitive in some types of assays and inconsequential in many others. Washing of the sample uptake probe between samples may minimize or eliminate such carryover, but it does so at the expense of a significant increase in sample processing time. With careful assay selection and appropriate control samples, high-quality data may be obtained under conditions in which probe-washing steps are abbreviated or eliminated using the high-throughput methodologies described below.

Number of Events Analyzed Per Sample

Conventional wisdom is that more is better when determining how many cells should be analyzed per sample to get statistically valid results. However, there is always a point of diminishing returns, when accumulation of more measurements has only a proportionately small and incremental effect on measurement precision. A practical number of total cells to analyze will vary with the type of measurement being made and the frequency of the cell subpopulation of interest. Moreover, although the making of a large number of measurements on a single sample will provide a good assessment of the measurement variance for that sample, it may be misleading with respect to the true variance associated with the biological system under investigation. Such an approach fails to provide any information about other important components of variance such as that associated with bioassay sample preparation. In many types of assays, preparation of two or three replicate samples and analysis of 1000 cells from each sample will take less time and probably achieve a more biologically relevant mean and variance estimate than analysis of 10,000 cells from a single sample. In high-throughput flow cytometry, this is a practical and conveniently achieved objective.

Automation in Flow Cytometry

Automated sample handling for flow cytometry has been commercially available for many years in a sample tube carousel format. Only recently have systems capable of automated sampling from 96-well microplates become available. These systems recapitulate most of the features associated with conventional, manually performed analysis of end-point assays. For example, in a typical end-point assay, samples are individually resuspended just before sample uptake, the sampling probe may be thoroughly washed between samples, and the rate of sample analysis is typically about two samples per minute at top speed. Thus, these systems are extremely versatile and permit end-point assay analysis of multiple samples from 96-well microplates without operator intervention, but they have throughput rates no faster than can be typically achieved by manual flow cytometer operation.

Another important automation concept in flow cytometry is the online addition of reagents to cells to allow faster and more complex dynamic processes of cells to be investigated. The earliest incarnations of this concept involved direct syringe injection of reagent into the sample head of the flow cytometer (9), construction of specialized sample introduction compartments for online addition of up to three reagents (3, 16, 17, 21, 22), and a sample introduction system based on continuous pumping of cells and reagents with a peristaltic pump (23).

Recently, computer-controlled sample mixing and delivery systems have been developed to provide flow cytometers with advanced and more fully automated sample handling capabilities [reviewed in Nolan et al. (20)]. Lindberg et al. (19) were the first to adapt flow injection analysis (FIA)-based techniques to flow cytometry in 1993. In this prototype system, a six-port, two-position injection valve was connected on the sample flow line of the flow cytometer. Sample and reagent were first loaded into a valve loop that contained a 70-μL mixing chamber and stir bar, and then the valve was switched to its alternate position to enable injection of the mixed chamber contents into the flow cytometer by a syringe pump. This system was successfully used to demonstrate online staining of cells with different concentrations of DNA-reactive dyes. A similar system was later developed to perform unattended online monitoring of yeast and bacterial growth dynamics, protein production, and cell cycle in bioreactors (28). Other experimental systems have been developed to enable execution of subsecond resolution kinetics experiments. A coaxial mixing device with a specially designed flow nozzle permitted continuous mixing of cells and reagents to allow measurements at a fixed time point within 100 ms of mixing (1). Another mixing approach employed a stepper-motor driven mixing device (Bio-Logic, Claix, France) that provided mixing and sample delivery to a flow cytometer with a 300-msec dead time (27). The dead time was subsequently reduced to 55 ms by optimizing flow nozzle design parameters and establishing electronic gates to exclude misaligned particles (12). A less expensive FIA-based system was also developed that provided nearly as good performance (500 ms dead time) at a lower cost (25). Another interesting recent development has been the commercial production of a first-generation flow cytometer specially designed for the in situ and autonomous unattended analysis and monitoring of composition and abundance of phytoplankton in natural water (www.cytobuoy.com). Although they represent significant advances in flow cytometer automation, these various sample-handling approaches were not designed to facilitate the high rates of throughput required for large-scale sample screening operations.

Plug Flow Cytometry

We have recently developed an alternative FIA-based approach to enable automated high-throughput sample handling for flow cytometry (5, 8). As with earlier FIA approaches, a multiport valve was inserted into the sample delivery pathway to enable sample to be injected from a valve holding loop into the flow cytometer. However, the design of the valve interface was modified in significant ways so as to optimize sample throughput. First, an eight-port, two-position valve was used, a geometry that enabled two sample holding loops to be positioned in two discrete pathways conducting continuously flowing fluid streams through the valve (figure 3.1).

One of the streams, the sample uptake pathway, conducts sample particles into the valve from a source such as a multiwell plate. The other stream conducts saline buffer from a pressurized source vessel into the flow cytometer. When the valve switches positions, the holding loop from the sample uptake pathway is repositioned into the flowing buffer pathway. Sample particles in the loop are thus eluted into the buffer stream as a bolus or "plug" of precisely defined initial volume and are transported by the flowing stream to the point of analysis. The valve switch simultaneously repositions the other holding loop from the buffer stream pathway into the sample uptake pathway. Thus, as sample particles are eluted from the first loop, the second loop is loaded with a subsequent set of sample particles. In practice, the valve continuously reciprocates between the two positions and thereby delivers a series of sample plugs separated by empty volumes of fluid (the buffer of which the stream is composed) for analysis (figure 3.1).

A second important design feature of the plug flow system valve interface is the use of holding loops of relatively small volume (typically 5 μL). This minimizes the time required for sample to be completely eluted (i.e., injected) into the buffer stream as compared to earlier FIA systems in which much larger injection volumes were used (19, 28). The rate at which sample is injected is determined by the flow rate of the buffer stream. At the upper range of acceptable buffer stream flow rates ($\sim$2 μL/s), the plug flow system can deliver up to 10 clearly resolved sample plugs per minute (7) but at the cost of suboptimal optical alignment conditions, as discussed earlier and elsewhere (8).

Applications

To date, we have used two different methods to control fluid flow in the sample uptake pathway. In the first (figure 3.2A), an automated stepper motor-driven syringe is connected to the outlet port of the valve via a length of Teflon tubing. Another length of Teflon tubing connects the valve inlet port to the sample source. The sample is aspirated into one of the valve holding loops by applied negative pressure from the syringe. This configuration has been used for online analysis of adhesion between cells sampled from the controlled fluid shear environment of a cone-plate viscometer (4, 6). While the viscometer is continuously rotating to generate appropriate fluid shear conditions, samples of a cell mixture are aspirated at sequential time intervals and transported to the flow cytometer for quantification of adherent cell frequencies. The lag time between cell removal from the viscometer and analysis in the flow cytometer is typically less than 10 s (4, 6).

In the second method (figure 3.2B), a peristaltic pump controls the flow of sample particles to the valve. Sample is aspirated under negative pressure and then pushed into one of the valve holding loops under positive pressure. Flexible tubing compatible with peristaltic pumping requirements is used to transport sample from source to valve. This configuration was used to interface the flow cytometer with an automated fluidics-based commercial pharmacology platform (High-Throughput Pharmacology System [HTPS], Axiom Biotechnologies, San Diego, CA [acquired by Sequenom]). Applications of this configuration included end-point fluorescence assays of cells sampled from 96-well microplates at analysis rates of 9–10 cell samples per minute and assays of cellular cal-

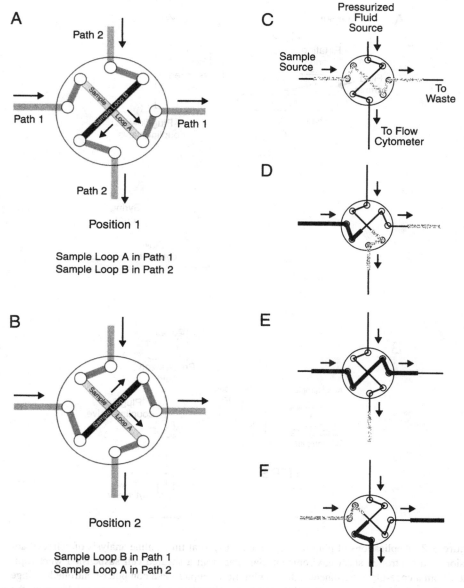

Figure 3.1. Plug flow cytometry. A two-position, eight-port valve reciprocates between two positions. The entry and exit ports for each of the two fluid pathways that pass through the valve (Path 1 and Path 2) remain constant. (*A*) In valve position 1, fluid in Path 1 passes through sample loop A while fluid in Path 2 passes through sample loop B. (*B*) In valve position 2, the sample loops are switched so that each is in the alternate fluid pathway. In this fashion, a plug of fluid from each pathway is inserted into the alternate pathway. (*C–F*) Hypothetical sequence of events illustrating sequential sample plug formation and delivery to the flow cytometer. Gray and black bands represent discrete boluses of particles that sequentially pass through the valve and from which 5-μL sample plugs are trapped in a sample loop and eluted into the common fluidic pathway leading to the flow cytometer. [Reprinted from (8) with permission from Wiley & Sons, Inc.]

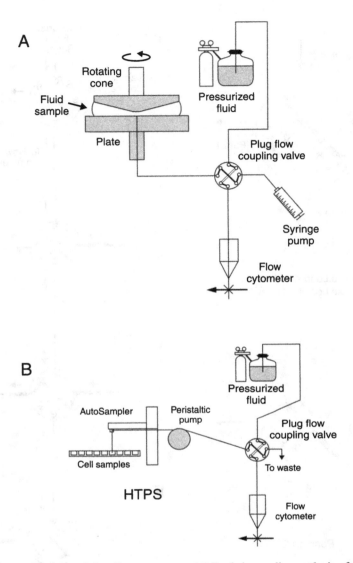

Figure 3.2. Applications of plug flow cytometry. (*A*) Real-time online analysis of cell–cell adhesion in a controlled shear environment. Samples from a sheared cell suspension were aspirated from a cone-plate viscometer into the plug flow coupler via a computer-controlled syringe pump. The sample plugs were then injected into a pressurized fluid stream for delivery to the flow cytometer for analysis. [Reprinted from (6) with permission from Wiley & Sons, Inc.] (*B*) Automated analysis of cell samples in end-point assay format by plug flow cytometry as adapted to the Axiom Biosciences High-Throughput Pharmacology System (HTPS). Configuration of the HTPS and its plug flow interface with the flow cytometer: The peristaltic pump was adjusted to deliver sample from microplate wells to the plug flow coupling device through peristaltic tubing (inner diameter = 0.02 in) at a flow rate of 3.5 μL/s. [Reprinted from (7) with permission from Sage Publications.]

cium responses to soluble receptor ligands sampled from microplate wells at analysis rates of three to four samples per minute (7). The HTPS interface also enabled rapid (2–3-min) automated characterization of the dose–response profile of cells to ligand gradients spanning three or more logs of concentration. We exploited the multiparametric measurement capabilities of the flow cytometer in these gradient assays to make novel measurements of the quantitative relationship between receptor occupancy and cell response (7). Particle carryover from one sample to the next is typically between 1% and 2% at the highest rates of sample throughput (7).

Multisample File Storage and Analysis

In all applications of the plug flow system, a critical design element to facilitate high-sample throughput is the processing of multiple samples during a single round of data collection and the storage of the resulting data in a single, time-resolved data file. As discussed above, this minimizes the number of hardware/software initialization and file storage operations required per sample. Individual samples are represented as discrete, time-resolved clusters of events (figure 3.3). A specialized software analysis program (FCSQuery) enables automatic detection and independent analysis of the individual sample clusters.

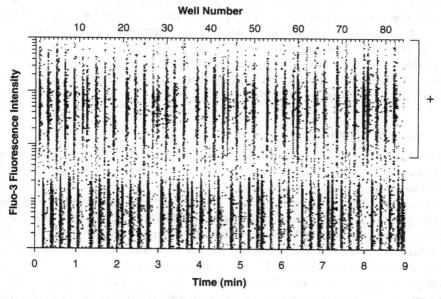

Figure 3.3. Fluorescence intensity profile of Fluo-3 labeled and unlabeled U937 cells sampled from 84 wells over a 9-min interval using High-Throughput Pharmacology System plug flow cytometry. Labeled and unlabeled cells were added to wells at concentrations of 10^5 and 2×10^5 cells/mL, respectively. Each discrete cluster of dots represents cells sampled from a separate well. Labeled and unlabeled cells were predominantly placed in alternating wells, but periodically, cells of the same labeled or unlabeled phenotype were added to adjacent pairs of wells. Cells falling in the region of Fluo-3 fluorescence intensity indicated by the bracket were considered to be fluorescence positive (+). [Reprinted from (7) with permission from Sage Publications.]

Particle Counting for Concentration Determinations

As well as enabling increased sample throughput, the plug flow system offers the additional feature that the sample plug volume is well defined and consistently the same throughout the sampling period. Thus, determination of the number of cells or particles in each sample plug provides a direct readout of particle concentration in the source suspension. We have exploited this particle-counting capability to extend the accuracy of flow cytometric cell–cell adhesion assays (4, 6) and described it in chapter 18. Examples of other potential applications include assays of the death or survival of cell subsets (as dyes such as propidium iodide fail to account for dead cells that have disintegrated) and cell subset proliferation.

HyperCyt: The Next Generation

More recently, we developed a simple method to rapidly process multiple samples that does not require a sample injection valve (18). This involves connecting the sampling probe of a commercial autosampler directly to the sample input port of a flow cytometer with commercial peristaltic tubing. A peristaltic pump is used to sequentially aspirate sample particle suspensions from multiwell microplates and to insert air bubbles between individual samples. This results in the generation of a tandem series of bubble-separated samples that are delivered directly to the flow cytometer for serial analysis (figure 3.4). The sample size and air bubble size are determined by the time that the autosampler probe is in a microplate well or is above a well, taking in air. As in plug flow cytometry (PFC), multiple samples are processed during each round of data collection, and the resulting data are stored in time-resolved data files. This high-throughput flow cytometry system has been designated HyperCyt.

Sample Volume

In HyperCyt, the sample volume is determined by the length of time the sampling probe is immersed in the sample source well. Thus, the sample volume may be easily varied over a wide range by simple manipulation of autosampler controlling software. Moreover, the entire volume of sample aspirated from the source well is passed through the detection point of the flow cytometer. In contrast, in PFC it is necessary to aspirate a volume of sample particles at least two to three times larger than the sample holding loop volume to ensure that a representative sample concentration will be captured in the loop when the valve switches. The unused sample is discarded to waste. Thus, the HyperCyt system is much more efficient in sample usage than PFC. We have recently shown that several types of assays can be efficiently performed with the HyperCyt system in 10-μL volumes, using aspirated volumes ranging from 1 to 5 μL per assay (24).

Sample Analysis Rate

In PFC the sample analysis rate is limited primarily by the time required to fully elute the captured sample particles from the sample holding loop. This is typically the time required to pass three loop volumes of buffer through the sample holding loop (i.e.,

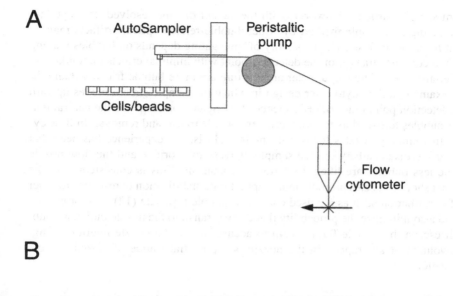

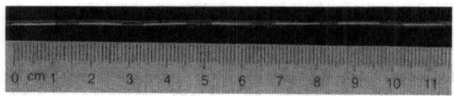

Figure 3.4. High-throughput flow cytometry with HyperCyt. (*A*) Schematic view of the flow cytometer, autosampler, and peristaltic pump. (*B*) Adjacent samples of latex microspheres separated by air in the 0.01-in (254-μm) inner diameter tubing between the peristaltic pump and the flow cytometer. [Reprinted from (18) with permission from Wiley & Sons, Inc.]

15 μL buffer for a 5-μL sample loop). At a buffer flow rate of 2.5 μL/s, which is the maximal rate compatible with a reasonable degree of optical alignment, this corresponds to about 6 s per sample for an analysis rate of approximately 10 samples/min, as previously reported (7). In HyperCyt, the air bubbles that separate the samples also restrict particle dispersion as the samples are transported to the flow cytometer. If a 5-μL volume of sample is initially aspirated and delivered to the flow cytometer at 2.5 μL/s, the analysis time is approximately 2 s (30 samples/min). Thus, a comparable sample volume may be analyzed nearly three times faster in HyperCyt as compared to PFC. By reducing the aspirated sample volume to 2 μL or less, it has been possible to achieve effective analysis rates of up to 100 sample/min (18).

Effects of Air Bubbles

Conventional wisdom in flow cytometry has been that air bubbles in the sample stream are to always be avoided as being disruptive of analysis. Clearly, a bubble passing through the point of detection introduces optical aberrations unrelated to optical signals from sample particles. However, with the advent of time-resolved flow cytometry, time-gating is a simple method by which bubble-related optical artifacts may be excluded from the analysis. The feasibility of time gating depends on bubbles making smooth and complete transits of the detection point with minimal effects on fluidic stability. Certain types of flow cytometer architecture are more bubble friendly than others. For example, a flow cytometer design in which the sample stream moves upward past the detection point (e.g., BD Biosciences FACScan) complements the natural tendency of bubbles to rise, thus favoring efficient bubble transit and removal. In flow cytometers that employ nozzles for stream-in-air analysis, our experience has been that the shorter the distance between the sample delivery tube orifice and the flow nozzle outlet, the less bubbles are likely to perturb the analysis. This is consistent with the finding that shorter nozzles promote more rapid fluidic stabilization in response to other types of perturbation such as a pulsed change in sample flow rate (12). A shorter nozzle should also minimize the probability that during transit to the nozzle outlet, air bubbles will escape the sample fluid stream to accumulate in the nozzle interior. A significant volume of air trapped in the nozzle is a potential source of chronic fluidic destabilization.

Quantitative End-Point Assay Applications with HyperCyt

HyperCyt has been successfully adapted to a number of cell-based end-point assay applications (24). In an example of an immunophenotyping assay (figure 3.5), human mononuclear cells were stained in wells of a 96-well plate with fluorescein- and phycoerythrin-conjugated antibodies directed against T, B, and NK cell antigens. Triplicate samples were taken from each of three wells with an aspiration time of 2 s per sample and no washing of the sampling probe between samples. The nine total samples were resolved as temporally discrete data clusters when fluorescein fluorescence intensity was plotted as a function of time (figure 3.5, top right panel). Analysis of two-color fluorescence intensity dot plots for each of the nine data clusters indicated that lymphocyte subset percentage measurements were generally reproducible within each triplicate group (figure 3.5, bottom panel) and were in good agreement with data obtained by the conventional manual sampling method (24). In addition to this and other types of quantitative immunofluorescence measurements, HyperCyt has also been successfully adapted to assays measuring cell–cell adhesion (chapter 18) and fluorescent ligand binding to cellular receptors (Ramirez, S., Edward, B., and Sklar, L.A., unpublished results).

Online Mixing

HyperCyt has also been successfully used in conjunction with high-throughput online mixing protocols (15). For example, if compounds are contained in microplate wells, drawn by one sample line, bubbles can be inserted to separate the compounds. A separate sample line draws cells or particles from a reservoir. The two lines are brought together in a Y, using the bubbles associated with the compounds to separate the samples in the mixed sample output line. Sample input and output lines have an inner diameter of 254 μm, and fluid flow is laminar at a low Reynolds number. Under such circumstances, it is expected that mixing will be limited to a diffusion-based mechanism. However, we determined that the use of a peristaltic pump to drive fluid trans-

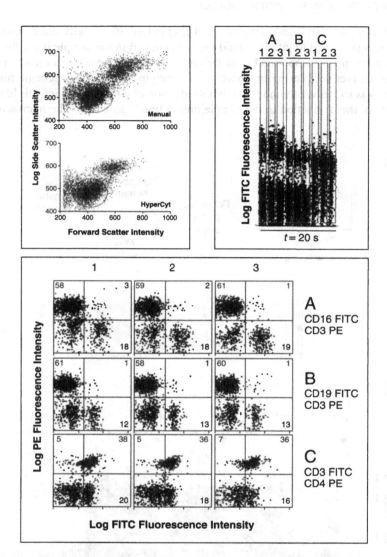

Figure 3.5. Immunophenotype analysis with HyperCyt. Peripheral mononuclear cells from a healthy donor were stained with fluorescent monoclonal antibodies to distinguish B, T, and NK lymphocyte subpopulations. Top, left panel: Light scatter analysis of cells under manual (top) and HyperCyt (bottom) analysis conditions with circles representing lymphocyte scatter gates. Top, right panel: Three wells containing immunostained mononuclear cells were sampled in triplicate from a 96-well microplate by HyperCyt with an aspiration time of 2 s per sample. In post–data acquisition analysis with FCSQuery software, the resulting nine data clusters were automatically resolved on the horizontal time axis into separate groups for fluorescence analysis. Bottom panel: Dot plots of fluorescein isothiocyanate and phycoerythrin fluorescence intensity data for each of the time domain-gated data clusters A1-3 (top row), B1-3 (middle row), and C1-3 (bottom row). Numbers in dot plot quadrants represent the percentage of total events detected in the quadrant. [Reprinted from (24) with permission from Wiley & Sons, Inc.]

port in both sample input lines promoted rapid efficient mixing of cells and compounds (15). This was attributed to pulsatile fluid motions elicited in the sample input lines by the peristaltic pump, as supported by the failure to detect mixing when a syringe pump was used to deliver samples more smoothly. The originally described scheme for on-line mixing was subsequently modified by introduction of a magnetic microstir bar just downstream of the Y junction in the sample output line (figure 3.6). The microstir bar

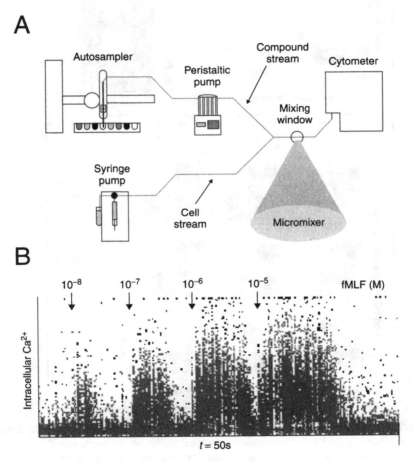

Figure 3.6. HyperCyt mixing system. (*A*) The system includes an autosampler with peristaltic tubing connected to one bifurcation of a Y fitting. The other segment of the Y is a continuous stream of cells or particles delivered with a syringe. These two streams exit through a common outlet and are mixed with an inline microstirbar before reaching the flow cytometer. The stirbar is controlled by a magnetic stirrer located in a plane below the horizontal tubing. [Reprinted from (14) with permission from Eaton Publishing.] (*B*) A HyperCyt mixing experiment to determine the intracellular Ca^{2+} response of U937 cells transfected with the formyl peptide receptor to a 1000-fold range of fMLF peptide concentration. Arrows indicate sampling of the indicated fMLF concentration from one well of a 96-well plate. Eleven wells containing rinse buffer were sequentially sampled after each fMLF well to remove residual fMLF. [Modified from (14) with permission from Eaton Publishing.]

significantly enhanced the efficiency of mixing of soluble compounds with cells, as compared to the mixing that resulted from peristaltic pumping alone or from a dimensionally smaller magnetic stirring wire (14).

In online mixing experiments with HyperCyt, sample throughput rates of up to nine samples per minute have been achieved with good resolution of the response following in-line mixing (15). This more than doubles the maximum rate of sample throughput previously achieved with PFC in combination with HTPS (7). The throughput-limiting feature of HyperCyt mixing is cross contamination (carryover) of sequential samples. Typically, several rinse steps are required to adequately reduce carryover of fluid between a soluble compound from one sample to the next. This is in distinct contrast to results obtained with sequential samples of cells or particles, in which only 1%–2% of particles in a first sample typically carry over into the next sample even in the absence of an intervening wash step (15, 18). Unlike particle carryover, fluid compound carryover increases proportionally with the length and inner diameter of the sample tubing (1A, unpublished data). This indicates that interior tube wall surface area is an important element in fluid carryover and that fluid compounds are less efficiently swept from tube walls by air bubbles than are particles.

Prospects

HyperCyt versus Plug Flow Cytometry

The HyperCyt system offers several advantages over PFC with respect to features such as sample volumes required, efficiency of sample usage, and sample analysis rates. However, PFC may be preferable in circumstances such as sorting, where air bubbles are problematic. Alternatively, with appropriate valve fluidic pathway modifications and bubble detection methodology, PFC could be used in combination with HyperCyt as a means to extract bubbles just before sample introduction into the flow cytometer. In HyperCyt sampling, there are also variations in the volumes aspirated for each sample because of factors such as fluid level differences from well to well and minor variations in volumetric fluid flow rates characteristic of peristaltic pumping. If an assay requires consistent delivery of uniform sample volumes to the flow cytometer, PFC will be the method of choice. We have successfully implemented these technologies on instruments from each of the three major flow cytometer manufacturers: Becton Dickinson, Beckman-Coulter, and Dako Cytomation.

Multiplexing

As described in detail elsewhere in chapters 9 and 14, the multiparameter detection capabilities of flow cytometry have begun to be exploited in microsphere-based assays in a process designated *multiplexed analysis*. Multiple sets of microspheres are combined in a single sample, in which each set detects and reports a different analyte. In one approach, commercialized by Luminex Corporation (www.luminex.com), this is accomplished by uniquely color coding each microsphere set with varying amounts of two spectrally distinct fluorophores. Each color-coded set is engineered to mediate a reaction with a unique analyte. Detection agents, tagged with a third fluorophore, are used to quantitatively report analytes bound to the individual microsphere sets. With this technology, it has been possible to measure dozens (10), and potentially hundreds (2), of analytes in a single sample. Other microsphere color-coding schemes have more recently been commercialized that use multiple levels of a single fluorophore (www. bdbiosciences.com, www.dukescientific.com) and distinctive combinations of fluorophore and light-scatter intensity (28). Likewise, cells in a heterogeneous mixture can be color coded on the basis of membrane protein profiles or other phenotypic characteristics to enable, for example, multiplexed analysis of cell responses, reported by a distinctly colored fluorescent probe, to a library of experimental compounds. The use of such multiplexed assays in combination with HyperCyt offers the promise of extending flow cytometry analysis throughput rates by an additional one to two orders of magnitude, a scale that is compatible with requirements of drug discovery applications.

Microtechnology for the Clinical Research Lab

Because of the reduced sample volume requirement with HyperCyt, it is feasible to perform flow cytometry assays on cells seeded in volumes of 10 μL or less (24). It is expected that this microtechnology capability will be of importance for enabling multiple quantitative multiparameter flow cytometry measurements to be made when cell numbers are limiting, a circumstance often encountered with tissue specimens of clinical origin. Likewise, this microtechnology approach will enable optimal concentrations of quantity-limited or expensive reagents to be achieved with smaller total quantities of reagents. Another attractive feature is that the use of small volume samples in Terasaki microplates permits a simple robust solution to the problem of cell settling, a circumstance encountered, for example, when there is the requirement of an extended bioassay incubation period before analysis. Because of the strong influence of liquid surface tension forces at this volumetric scale, Terasaki microplates can be completely inverted without spilling of well contents. We have shown that uniform cell suspensions suitable for flow cytometry can be maintained for at least 30–60 min by periodic inversion of Terasaki plates on a continuously rotating drum (24). Taken together, these features offer the promise of new opportunities for low-cost, efficient, automated, high-throughput molecular target assays to benefit the clinical research enterprise.

Integrated Systems to Ensure Proper Assignment of Event Clusters to Source Wells

Routine practical application of HyperCyt and PFC high-throughput flow cytometric analysis technologies will require additional considerations that are now beginning to be addressed. One such issue is synchronization between the autosampler and the flow cytometer so that analyzed event clusters can be unambiguously assigned to the source wells from which they were sampled, and hence properly identified. Misregistration between the sequence of sampled source wells and the resulting sequence of time-resolved event clusters might occur, for example, if a sample is missed (e.g., empty well), merges with another sample, or subdivides during transit. A general solution to detecting such occurrences is to develop coding systems by which proper registration is reestablished at fixed sampling intervals. As a simple example, one might add a uniquely identifiable sample of "tracer" cells or beads to the last well of each six-well row of a Terasaki plate. Off-line analysis software could automatically detect tracer-containing event clusters, count the number of event clusters in between, and flag for manual reanalysis any row in which the count differs from five.

More sophisticated coding schemes can easily be envisioned involving markers such as multilevel fluorescent beads mixed in varying proportions to denote multiple discrete levels of source well/event cluster registration. We have recently developed a system involving fluorescent marker beads and custom bead detection circuitry with which a first level of online synchronization between autosampler and flow cytometer has been achieved (8A, manuscript in preparation).

Data Management Systems

A second issue in the routine practical application of HyperCyt and PFC will be the implementation of a data management system to facilitate the storage, retrieval, and quality control of large volumes of rapidly accumulating multiparameter data. This will especially be of concern in extended screening activities involving the analysis of thousands to tens of thousands of samples per day. Fortunately, there exist sophisticated commercial data management software packages that are routinely used for managing high-throughput screening data from a diversity of analytical instruments. Of particular interest are packages capable of importing data from spreadsheet programs (e.g., IDBS ActivityBase), since we have already implemented the output of formatted time-resolved data to Microsoft Excel spreadsheets as a standard feature of our in-house FC-SQuery software, a computer program that automatically detects and analyzes time-resolved event clusters. Discussed in another chapter are additional approaches to list-mode data storage that could further facilitate the management of time-resolved data such as is generated by the HyperCyt system (see chapter 8).

References

1A. Bartsch, J.W., Tran, H.D., Waller, A., Mammoli, A., Baranda, T., Sklar, L.A., Edwards, B.S. 2004. An investigation of liquid carryover and sample residual for a high-throughput flow cytometer sample delivery system. *System. Anal. Chem.* 76:3810–3817.

 1. Blankenstein, G., Scampavia, L.D., Ruzicka, J., Christian, G.D. 1996. Coaxial flow mixer for real-time monitoring of cellular responses in flow injection cytometry. *Cytometry* 25:200–204.

 2. Chandler, V.S., Denton, D., Pempsell, P. 1998. Biomolecular multiplexing of up to 512 assays on a new solid-state 4 color flow analyzer. *Cytometry* (Suppl. 9):40

 3. Dunne, J.F. 1991. Time window analysis and sorting. *Cytometry* 12:597–601.

 4. Edwards, B.S., Curry, M.S., Tsuji, H., Brown, D., Larson, R.S., Sklar, L.A. 2000. Expression of P-selectin at low site density promotes selective attachment of eosinophils over neutrophils. *J. Immunol.* 165:404–410.

 5. Edwards, B.S., Kuckuck, F., Sklar, L.A. 1999. Plug flow cytometry: an automated coupling device for rapid sequential flow cytometric sample analysis. *Cytometry* 37:156–159.

 6. Edwards, B.S., Kuckuck, F.W., Prossnitz, E.R., Okun, A., Ransom, J.T., Sklar, L.A. 2001. Plug flow cytometry extends analytical capabilities in cell adhesion and receptor pharmacology. *Cytometry* 43:211–216.

 7. Edwards, B.S., Kuckuck, F.W., Prossnitz, E.R., Ransom, J.T., Sklar, L.A. 2001. HTPS flow cytometry: a novel platform for automated high throughput drug discovery and characterization. *J. Biomol. Screen* 6:83–90.

8A. Edwards, B.S., Andrzejewski, Ramirez, S., Sklar, L.A. 2002. Multi-threaded integration of high throughput flow cytometry autosampling and analysis. *Cytometry* S11–120.

8. Edwards, B.S., Sklar, LA. 2001. Plug Flow Cytometry. *Current Protocols Cytometry*: 1.17.1–1.17.10.

9. Finney, D.A., Sklar, L.A. 1983. Ligand/receptor internalization: a kinetic, flow cytometric analysis of the internalization of N-formyl peptides by human neutrophils. *Cytometry* 4:54–60.

10. Fulton, R.J., McDade, R.L., Smith, P.L., Kienker, L.J., Kettman, J.R. Jr. 1997. Advanced multiplexed analysis with the FlowMetrix system. *Clin. Chem.* 43:1749–1756.

11. Fulwyler, M.J. 1965. Electronic separation of biological cells by volume. *Science* 150:910–911.

12. Graves, S.W., Nolan, J.P., Jett, J.H., Martin, J.C., Sklar, L.A. 2002. Nozzle design parameters and their effects on rapid sample delivery in flow cytometry. *Cytometry* 47:127–137.

13. Gray, J.Y., Alger, T.W., Lord, D.E. 1982. U.S. Patent No. 4361400.

14. Jackson, W.C., Bennett, T.A., Edwards, B.S., Prossnitz, E., Lopez, G.P., Sklar, L.A. 2002. Performance of in-line microfluidic mixers in laminar flow for high-throughput flow cytometry. *Biotechniques* 33:220–226.

15. Jackson, W.C., Kuckuck, F., Edwards, B.S., Mammoli, A., Gallegos, C.M., Lopez, G.P., Buranda, T., Sklar, L.A. 2002. Mixing small volumes for continuous high-throughput flow cytometry: performance of a mixing Y and peristaltic sample delivery. *Cytometry* 47: 183–191.

16. Kelley, K.A. 1989. Sample station modification providing on-line reagent addition and reduced sample transit time for flow cytometers. *Cytometry* 10:796–800.

17. Kelley, K.A. 1991. Very early detection of changes associated with cellular activation using a modified flow cytometer. *Cytometry* 12:464–468.

18. Kuckuck, F.W., Edwards, B.S., Sklar, L.A. 2001. High throughput flow cytometry. *Cytometry* 44:83–90.

19. Lindberg, W., Ruzicka, J., Christian, G.D. 1993. Flow injection cytometry: a new approach for sample and solution handling in flow cytometry. *Cytometry* 14:230–236.

20. Nolan, J.P., Lauer, S., Prossnitz, E.R., Sklar, L.A. 1999. Flow cytometry: a versatile tool for all phases of drug discovery. *Drug Discov. Today* 4:173–180.

21. Nooter, K., Herweijer, H., Jonker, R., van den, E.G. 1990. On-line flow cytometry: a versatile method for kinetic measurements. *Methods Cell Biol.* 33:631–645.

22. Omann, G.M., Coppersmith, W., Finney, D.A., Sklar, L.A. 1985. A convenient on-line device for reagent addition, sample mixing, and temperature control of cell suspensions in flow cytometry. *Cytometry* 6:69–73.

23. Pennings, A., Speth, P., Wessels, H., Haanen, C. 1987. Improved flow cytometry of cellular DNA and RNA by on-line reagent addition. *Cytometry* 8:335–338.

24. Ramirez, S., Aiken, C.T., Andrzejewski, B., Sklar, L.A., Edwards, B.S. 2003. High throughput flow cytometry: validation in microvolume bioassays. Submitted for publication. 53A:55–65.

25. Seamer, L.C., Kuckuck, F., Sklar, L.A. 1999. Sheath fluid control to permit stable flow in rapid mix flow cytometry. *Cytometry* 35:75–79.

26. Shapiro, H.M. 1988. *Practical Flow Cytometry*. New York: Alan R. Liss

27. Sklar, L.A., Seamer, L.C., Kuckuck, F., Posner, R.G., Prossnitz, E.R., Edwards, B., Nolan, J.P. 1998. Sample handling for kinetics and molecular assembly in flow cytometry. *Proc. SPIE* 3256:144–153.

28. Zao, R., Natarajan, A., Srienc, S. 1999. A flow injection flow cytometry system for on-line monitoring of bioreactors. *Biotechnol. Bioeng.* 62:609–617.

4

Multispectral Imaging in Flow: A Technique for Advanced Cellular Studies

DAVID BASIJI

Introduction

Flow cytometry is one of the most sophisticated technologies for the study of cellular biology, with the unique ability to analyze cell populations numbering in the millions. In addition to their great speed, flow cytometers can be configured with 10 or more detectors, so each cell can be characterized by multiple parameters corresponding to light scattered or emitted as they pass through the system.[20] These capabilities have led to the development of a wide variety of applications for flow cytometry in cell biology, hematology, immunology, oncology, cell culture, and other fields.[1,7,13,16] Nevertheless, a wide range of applications are not suited to flow cytometric analysis because the resulting data fail to capture cell morphology or the spatial distribution of signals within a cell.

In contrast to conventional flow cytometry, microscopy can elucidate cell morphology by a number of means, including absorbed light imaging (brightfield), scattered light imaging (darkfield), and fluorescence imaging.[10] An image of a typical mammalian cell measuring approximately 10 μm in diameter will cover over 300 pixels, assuming a detector resolution of 0.5 μm per pixel. Hence, even a single cell image represents orders of magnitude more data than is acquired by a flow cytometric analysis of that same cell and therefore may represent far more information about that cell.[17,19] In theory, a set of brightfield, darkfield, and fluorescence images of each cell would provide all the information of flow cytometry plus morphological features, such as the nuclear to cytoplasmic ratio, membrane texture, the distribution of a fluorescent probe, and so forth. However, as a practical matter, it is difficult to acquire and compare more than a few images of a cell because of the need to change fluorescence filter combinations, reconfigure the microscope for brightfield and darkfield imaging, and register the various images to compensate for shifts and distortion from the different optical arrangements. As a result, flow cytometry and microscopy have remained complementary techniques. The goal of imaging cells in flow has been elusive, despite several approaches that have achieved various levels of success over the last several decades.

The challenge in flow imaging is to combine good image fidelity and high sensitivity with high throughput. The long signal integration times employed in static microscopy for increased sensitivity conspire against image quality in flow. To acquire images of high quality, the cell must be positioned and maintained in the plane of focus with minimal uncontrolled rotation or translation relative to the detector during the imaging process. To freeze cell motion on the detector, the signal integration period of the imaging system must be short enough and the relative velocity between the cell and the detector low enough that the distance the cell travels during integration falls below the resolution of the imaging system.

Consider a flow-imaging system configured much like a flow cytometer, with a sample velocity of 5 m/s and a signal integration time of 5 μs, with illumination provided by a pulsed Xenon source. In such a system, a cell will move 25 μm during the illumination pulse, and the image will be hopelessly streaked. To maintain an image resolution of 0.5 μm, either the velocity of the sample stream or the duration of the pulse must be reduced 50-fold. Because the Xenon pulse in this example is largely fixed, the velocity must be reduced, and the throughput will suffer accordingly. This could be acceptable trade-off, as even a system running at 100 cells/s can analyze over 100,000 cells in less than 20 min. However, in comparison to a flow cytometer, such a system would have low fluorescence sensitivity because of the lower brightness of the pulsed Xenon source relative to even a modest, 20-mW laser, and because of the fact that only one pulse could be employed before a cell flowed out of the field of view. Attempting to increase sensitivity by increasing illumination power helps to an extent, particularly with brightfield and darkfield imagery. However, pulsed arc lamps of higher power tend to have longer pulses and bigger, not brighter, arcs so most of the additional light cannot be projected on the field of view.

In contrast to arcs, pulsed lasers have high pulse energies, are readily projected on the field of view, and can have nanosecond or shorter pulse durations. The last property is seemingly ideal for flow imaging, as by the analysis presented above, cell motion would be completely frozen over the course of a single pulse. Unfortunately, short-pulsed lasers can be inefficient excitation sources because of the nature of fluorescence excitation. A fluorescent dye molecule enters an excited state following absorbance of a photon and typically does not emit a fluorescent photon for several nanoseconds.[11] During the excited state, the molecule is "blind" to further excitation, so if the laser pulse is sufficiently powerful, it can cause a saturation effect, where a large fraction of molecules are excited and the remaining pulse energy is ineffective at producing additional fluorescence.[5] Further, by the time the molecules have relaxed, the laser pulse has passed and cannot stimulate additional excitation cycles.

Despite the challenges of imaging in flow, a number of techniques have been developed for the task. These approaches include strobed illumination techniques,[9,12] flying spot scanning,[4] mirror tracking,[8] and slit scanning flow cytometry.[23] Of the various techniques, slit scanning, which was originally developed in the 1970s, is the most mature. Slit scanning involves measuring the intensity profile of a cell as it passes across a narrow field of view, corresponding to a one-dimensional fluorescence image. Though the technique was demonstrably superior for several common cytometric assays, it did not achieve commercial success because of a combination of factors that included the lack of a clearly defined "killer" application.[22] In contrast, today there are a wealth of applications for flow imaging in basic research, drug discovery, and clinical diagnostics thanks to our understanding of the human genome, the development of a huge array of fluorescent probes, and advances in molecular biology and combinatorial chemistry.

The remainder of this chapter describes a new technology being developed by the Amnis Corporation for multispectral imaging and analysis of cells in flow. The technology, called ImageStream®, employs a number of novel subsystems that greatly increase signal collection efficiency and the information content of the data.[2,3] The ultimate goal is to combine the speed, sample-handling, and cell-sorting capabilities of flow cytometry with the imagery, sensitivity, and resolution of brightfield, darkfield, and multicolor fluorescence microscopy.

The ImageStream Platform

Figure 4.1 illustrates the layout and key components of the platform. As in a flow cytometer, cells are hydrodynamically focused into a single-file core stream approximately 10 μm in diameter. Sheath and sample flow are controlled by precision stepper motor-driven pumps that have been optimized to reduce pulsatility. In the initial commercial embodiments of the platform, which are designed to operate at approximately 100 cells/s, cells flow at approximately 30 mm/s. As shown in the figure, cells are illuminated from the side for scatter and fluorescence imaging, as well as from behind for brightfield.

Light is collected from the cells with an imaging objective lens and is ultimately projected on a charge-coupled detector. Before hitting the detector, a fraction of the scattered light is reflected out of the imaging path and transmitted to a system for the determination of cell velocity and autofocus. The velocity information is used to maintain proper synchronization of the charge-coupled detector, which is operated in time-delay integration mode (TDI), as described later.

Before projection on the detector, the majority of the light is passed to a spectral decomposition element. The decomposition element is a fan arrangement of dichroic mirrors that direct different spectral bands of an image at different angles. When the spectral decomposition element is placed before the detector in aperture space, it causes the different spectral bands to be focused to different positions across the TDI detector. With this technique, a single cell image is optically decomposed into a set of subimages, each corresponding to a different color component. The arrangement of cell image features across a five-channel detector is shown in figure 4.1, which depicts five subimages consisting of violet laser light scattered from the cytoplasm; blue light from a fluorescent nuclear stain; and green, orange, and red fluorescent probes, all located in their respective spectral image channels. Spectral decomposition can facilitate the location, identification, and quantitation of signals within the cell by physically separating on the detector signals that may originate from overlapping regions of the cell,

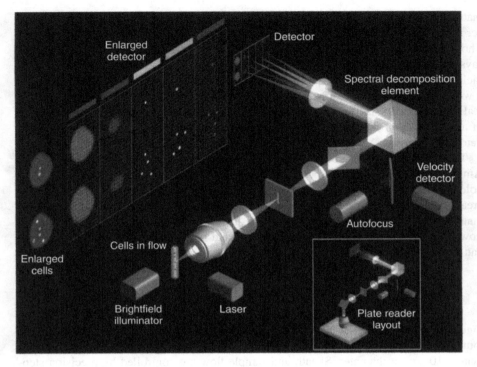

Figure 4.1. ImageStream layout and key components.

as depicted in figure 4.1. This image processing occurs in real time during the image formation process, rather than via digital image processing of a conventional composite image. Spectral decomposition also allows multimode imaging, the simultaneous detection of brightfield, darkfield, and multiple colors of fluorescence, by choosing distinct spectral bands for the different illumination modes. For example, 488-nm laser light can be used to excite fluorescein and phycoerythrin, which are detected in 500–550-nm and 550–600-nm channels, whereas laser light scattered from the cells can be used for darkfield imaging in a 480–505-nm channel, and 620-nm backlighting is used for brightfield imaging in a 600–630-nm channel.

Optical image processing and multimode imaging can greatly increase the information obtained from a cell. To complement the high information content of multimode imaging with increased sensitivity of detection, ImageStream employs TDI, a specialized detector readout mode that is commonly used in machine vision applications, such as semiconductor wafer inspection and assembly line quality assurance, in which there is fast relative linear movement between the camera and the object being imaged. In TDI, the image on the detector is read out continuously, one row of pixels at a time, from the bottom of the detector chip. As each row is read out, the signals in the remaining detector pixels are shifted down by one row, causing the latent image to translate down the detector during readout.[14] If the readout rate of the detector is matched to the velocity of the object being imaged, the image will not blur.

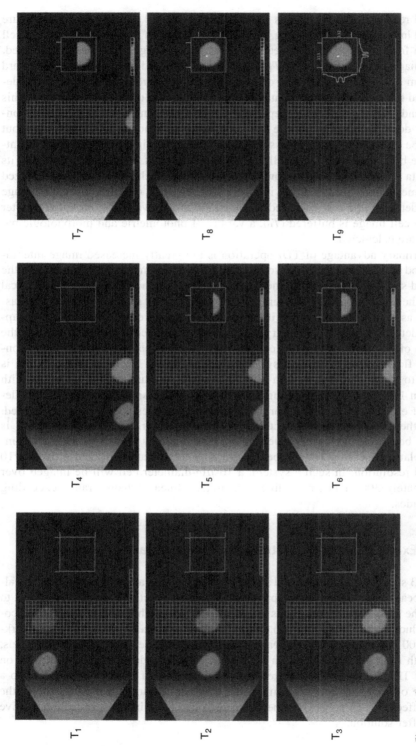

Figure 4.2. Time-delay integration readout time sequence.

A TDI readout time sequence is illustrated in figure 4.2. Each frame in the figure is divided into four parts, consisting, from left to right, of an illumination source, a cell in motion from top to bottom, a detector on which an image of the cell is projected, and an image buffer. Just before T_1, a cell enters the field of view, moving toward the bottom of the detector. A row of image data is read from the bottom of the detector, and the remaining image data are shifted down the detector by one row. This readout-and-shift process occurs repeatedly, causing the image data to translate continuously down the detector. If the speed of the cell is known, the detector readout rate can be matched to the progress of the cell, as shown in T_1 through T_3, preventing image blur. Eventually, the cell reaches the bottom of the field of view, and its image data are read out row by row, as shown in T_4. Each row of data is analyzed in real time to determine whether it contains background or cell imagery. If image data are detected, the row is placed in the buffer, as shown in T_5 through T_8. After the entire cell image is buffered (T_9), a variety of photometric and morphologic parameters are calculated.

The primary advantage of TDI operation is the greatly increased image integration period it affords. In comparison to standard video imaging, TDI increases the integrated signal proportional to the number or rows on the detector.[6] The practical limit on the number of rows is determined by the accuracy of the cell velocity measurement, as velocity errors result in cumulative tracking errors. Past efforts to apply TDI detection to flow cytometry met with limited success in part because of the difficulty of accurately measuring cell velocity to better than 1% accuracy.[15] In contrast, the flow velocity detection system employed in the ImageStream platform is accurate to better than one part in 1000, permitting the use of TDI detectors with more than 1000 rows. The resulting 1000-fold increased signal levels allow the detection of even faint fluorescent probes within cell images acquired at high speed. Though the charge-coupled detector is a linear detector, high dynamic range is achieved because each image generally covers numerous pixels. In the first commercial platform, each pixel will be 0.5 μm at the object and will be digitized at 10 bits/pixel resolution. In such a system, a 10-μm-diameter cell will be imaged over approximately 300 pixels, providing a theoretical linear dynamic range exceeding four decades.

Example of Multispectral Flow Image Data

Figure 4.3 shows typical image and statistical data from an analysis of fluorescent calibration beads in an ImageStream prototype with four spectral channels. From left to right in the image data the channels are configured for brightfield (>595 nm), phycoerythrin fluorescence (560–595 nm), fluorescein fluorescence (500–560 nm), and darkfield (<500 nm). Images of each bead appear in the brightfield and darkfield channels, along with a fluorescence image in the channel corresponding to the dye present on the bead. This imagery was gathered at a magnification of 20×, corresponding to a pixel size of approximately 0.65 μm at the object. The darkfield imagery shows the lensing effect of each polystyrene bead because of its large index of refraction relative to the buffer solution.

Figure 4.3. Image and statistical data from an analysis of fluorescent calibration beads in an ImageStream four-channel prototype.

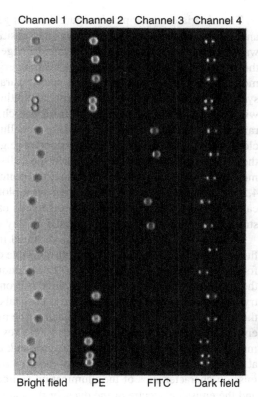

Channel 1   Channel 2   Channel 3   Channel 4

Bright field   PE   FITC   Dark field

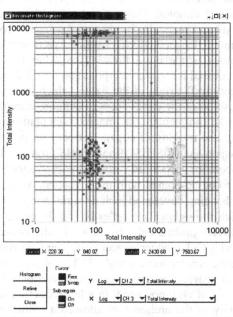

The bivariate scatter plot in figure 4.3 was generated by Amnis's analysis software and shows the discrimination power and sensitivity of the system. The analysis software is designed to facilitate the detailed image analysis of individual cells as well as the statistics of large populations. Cell imagery in each channel is used to calculate more than 35 morphologic and photometric parameters, so a six-channel version of the system will characterize each cell by calculating over 200 parameters. In addition, it will be possible to use parameters from one channel to modify the calculation of parameters in another (e.g., total intensity of cellular fluorescence not located in the nucleus) or to derive multichannel parameters (e.g., nuclear-to cytoplasmic area ratio, or the relative position of the nucleus in the cell). In addition to representing the parametric data derived from a cell image, the points in the bivariate scatter plot of figure 4.3 are active controls, so by clicking on a plotted point, the associated cell imagery can be inspected. Conversely, groups of cells can be selected from the image file and statistically analyzed and plotted based on any calculated parameter.

Figure 4.4 shows three sets of four-channel image data (brightfield, two channels of fluorescence, and darkfield) taken from a sample of K562 cells labeled with a Cy3 marker for CD71 and stained with SYTO-16 DNA-binding dye. The imagery to the left side of the figure represents raw data taken directly from the image buffer. As in flow cytometry and other detection systems, there is crosstalk between the spectral channels because the illumination and detection filters cannot be perfectly matched to each other or to the emission spectra of the fluorescent dyes. However, the linearity of the detection system makes it possible to perform effective crosstalk compensation, as shown in the imagery at the center of the figure. A compensation matrix is calculated on the basis of the transmission characteristics of the illumination sources, the spectral decomposition element, and the emission spectra of the dyes, or the matrix can be derived directly using a set of control samples. The image data are then spatially registered to within a fraction of a pixel and spectrally compensated using the compensation matrix. As the center imagery shows, the method effectively corrects any unsaturated signal to within a few digital counts of the noise floor. Once it has been crosstalk-corrected, the imagery can be processed for visual interpretation by false color composition. This is shown in the imagery to the right of the figure, which juxtaposes the brightfield imagery with a superposition of the darkfield, SYTO fluorescence, and Cy3 fluorescence signals.

To complement the ImageStream platform, Amnis developed protocols for the staining and probing of cells in suspension including fluorescence in-situ hybridization in suspension (FISHIS). The FISHIS protocol differs somewhat from previous FISH protocols for suspended cells in that it preserves both cytoplasm and membrane, allowing it to be combined with immunophenotyping and cytoplasmic probing.[21] Other protocols being developed include simultaneous fluorescence and chromogenic staining and high-throughput multiplexing of fluorescent probes within a cell. An example of FISHIS-probed cells is shown in the two-channel ImageStream data of figure 4.5. UM-UC-3 bladder cancer cells were probed in suspension, using a fluorescently labeled chromosome 9 probe.[18] The brightfield image defines the boundaries of the cells and provides morphological information, whereas the fluorescence image shows trisomy of chromosome 9 in a large fraction of the cells. For this assay, the high sensitivity of the architecture allowed a reduction in collection numeric aperture, thereby trading sensitivity and resolution to increase the depth of field and increasing the number of in-focus FISH spots.

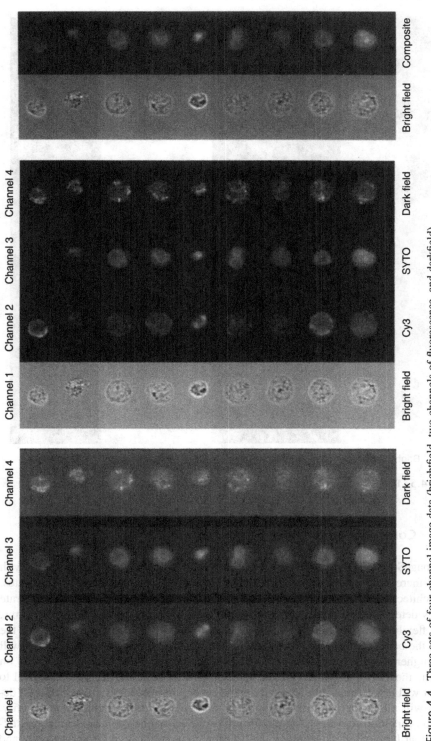

Figure 4.4. Three sets of four-channel image data (brightfield, two channels of fluorescence, and darkfield).

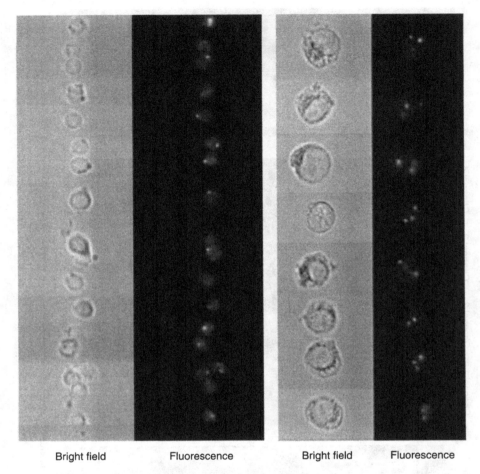

| Bright field | Fluorescence | | Bright field | Fluorescence |

Figure 4.5. FISHIS-probed cells in a two-channel ImageStream.

Conclusion and Future Directions

The ImageStream platform has been conceived as a method of imaging cells with greatly increased sensitivity and information content, both in flow and on substrates. The architecture is reliant on several key technology components, including accurate velocity detection, optimized pumping and fluidics, novel optics, real-time algorithms, highly flexible analysis software, and suspension-based cell probing protocols. In the future, the basic platform will be enhanced in a number of ways. Faster detectors will allow higher throughput. Increased computing power will allow cell sorting based on real-time morphological analysis. The addition of a second imaging leg orthogonal to the first will deliver real-time, three-dimensional cell morphology. Intensified TDI detectors will allow the platform to approach single-molecule fluorescence sensitivity. With the capabilities of the basic platform plus the envisioned enhancements, it is expected that ImageStream technology will find wide application in drug discovery and development, in instruments for basic research, as point-of-care devices, in analytical systems for optically encoded microbeads, and as a platform for rare-cell clinical applications of diagnostic and therapeutic importance.

References

1. Al-Rubeai, M. and Emery, A.N., editors. *Flow Cytometry Applications in Cell Culture*. New York: Marcel Dekker, 1996.
2. Basiji, D.A. and Ortyn, W.E. U.S. Patent 6,211,955; 2001.
3. Basiji, D.A. and Ortyn, W.E. U.S. Patent 6,249,341; 2001.
4. Ehrlich, M.P., Stolle, M., Grand, S., and DeCote, R. U.S. Patent 3,699,336; 1972.
5. van den Engh, G. and Farmer, C. Photo-bleaching and photon saturation in flow cytometry. *Cytometry* 1992;13:669.
6. Holst, G.C. *CCD Arrays, Cameras, and Displays*, 2nd ed. Winter Park, FL: JCD Publishing, and Bellingham, WA: SPIE; 1998.
7. Jaroszeski, M.J. and Heller, R., editors. *Flow Cytometry Protocols*. New Jersey: Humana Press; 1998.
8. Kay, D.B., Cambier, J.L., and Wheeless, L.L. Imaging in flow. *J Histochem Cytochem* 1979;27:329.
9. Kay, D.B. and Wheeless, L.L. Laser stroboscopic photography: technique for cell orientation studies in flow. *J Histochem Cytochem* 1976;24:265.
10. Kohen, E. and Hirschberg, J.G., editors. *Cell Structure and Function by Microspectrofluorometry*. San Diego, CA: Academic Press; 1989.
11. Lakowicz, J.R. *Principles of Fluorescence Spectroscopy*. New York: Plenum Press, 1983.
12. Maekawa, Y. and Kosaka, T. U.S. Patent 5,159,398; 1992.
13. Melamed, M.R., Lindmo, T., and Mendelsohn, M.L. editors. *Flow Cytometry and Sorting*, 2nd ed. New York: Wiley; 1990.
14. Ong, S.H. *Development of a System for Imaging and Classifying Biological Cells in a Flow Cytometer*. Ph.D. Thesis, 1985.
15. Ong, S.H., Horne, D., Yeung, C.K., Nickolls, P., and Cole, T. Development of an imaging flow cytometer. *Cytometry* 1987;9(5):375.
16. Owens, M.A. and Loken, M.R. *Flow Cytometry Principles for Clinical Laboratory Practice*. New York: Wiley-Liss; 1995.
17. Pawley, J.B., editor. *Handbook of Biological Confocal Microscopy*, 2nd ed. New York: Plenum; 1995.
18. Reeder, J.E., O'Connell, M.J., Yang, Z., Moreeeale, J.F., Collins, L., Frank, I.N., Messing, E.M., Cockett, A.T., Cox, C., Robinson, R.D., et al. DNA cytometry and chromosome 9 aberrations by fluorescence in situ hybridization of irrigation specimens from bladder cancer patients. *Urology* 1998;51(5):58.
19. Shannon, C.E. and Weaver, W. *The Mathematical Theory of Communication*. Chicago: Illinois; 1963.
20. Shapiro, H.M. *Practical Flow Cytometry*, 3rd ed. New York: Wiley-Liss; 1995.
21. Trask, B., van den Engh, G., Landegart, J., in de Wal, N.J., and van der Ploeg, M. Detection of DNA sequences in nuclei in suspension by in situ hybridization and dual beam flow cytometry. *Science* 1985;230:1401.
22. Wheeless, L.L. personal communication 2002.
23. Wheeless, L.L. Slit scanning. In: Melamed M.R., Lindmo T., and Mendelsohn M.L., editors. *Flow Cytometry and Sorting*, 2nd ed. New York: Wiley; 1990, p. 109.

5

Elastomeric Microfabricated Fluorescence-Activated Cell Sorters

ANNE Y. FU, YOHEI YOKOBAYASHI, FRANCES H. ARNOLD,
AND STEPHEN R. QUAKE

Introduction

This chapter describes the development of elastomeric microfabricated cell sorters that allow for high sensitivity, no cross contamination, and lower cost than any conventional fluorescence-activated cell sorting. The course of this development depends heavily on two key technologies that have advanced rapidly within the past decade: microfluidics (19, 32, 38) and soft lithography (46, 47). Sorting in the microfabricated cell sorter is accomplished via different means of microfluidic control. This confers several advantages over the conventional sorting of aerosol droplets: novel algorithms of sorting or cell manipulation can be accomplished, dispensing of reagents and biochemical reactions can occur immediately before or after the sorting event, completely enclosed fluidic devices allow for studies of biohazardous/infectious cells or particles in a safer environment, and integration of other technologies can be implemented into the cell sorter. In addition, because of the easy fabrication process and inexpensive materials used in soft lithography, this elastomeric microfabricated cell sorter is affordable to every research laboratory and can be disposable just as a gel in gel electrophoresis, which eliminates any cross contamination from previous runs.

Because of the advent of soft lithography, many inexpensive, flexible, and microfabricated devices could be designed to replace flow chambers in conventional flow cytometers. Soft lithography is a micromachining technique that uses the process of rapid prototyping and replica molding to fabricate inexpensive elastomeric microfluidic devices with materials such as plastics and polymers (47). The elastomeric properties of plastics and polymers allow for an easy fabrication process and for cleaning for reuse or disposal. A variety of biological assays can also be carried out as a result of the chemical compatibilities of different plastic materials with different solvents. More accurate sorting of cells can be accomplished because the sorting region is at or immediately after the interrogation point. On-chip chemical processing of cells has been accomplished and can be observed at any spot on the chip before or after sorting. Time-course measurements of a single cell for kinetic studies can be implemented using novel sorting schemes. Furthermore, linear arrays of channels on a single chip, the multiplex system, may be simultaneously detected by an array of photomultiplier tubes (PMT) for multiple analysis of different channels. Multiplexing in these microsystems increases the through-put rate and allows for synchronous measurements that cannot be done using a conventional FACS (although linear flow velocities of these chips may not attain the high linear flow velocity of a conventional FACS, 10 m/sec). Furthermore, other sophisticated biological assays can also be implemented on-chip because of the simplicity and flexibility of sample handling, mixing, incubation, and massive parallelization, such as cell lysis (44), polymerase chain reaction (PCR) (24), optical tweezer/cell trapping (1), and even transformation of cells by electroporation (27) or optoporation (41). In terms of optical light collection efficiency, these microchips allow for a minimal volume of cell suspension, approximately 100 fL, at the interrogation point. The minimal volume of optical interrogation greatly reduces the amount of background light scatter from the suspending medium and sheath flow and from the materials of the flow chamber. Because of the planar configuration of these microchannels, higher numerical aperture (NA) oil immersion objectives can be used to collect more fluorescence instead of the conventional NA of 0.6 dry lens used in both fluidic and aerosol flow chambers. One additional feature for these soft-lithographed microchips is that most of them are disposable because of their inexpensive material. These properties relieved many of the concerns for sterilizing and permanent adsorption of particles onto the flow chambers.

This chapter presents the construction of microfabricated FACS devices (μFACS) using soft lithography. Sorting of cells was accomplished using electrokinetic flow (16) or with pneumatically actuated microvalves and micropumps (15). Many future potential applications of μFACS lie in its unique capability to integrate into other technologies for measurements of electrical (40), magnetic (17), or other physical cellular properties. At present, μFACS serves as an inexpensive, robust, and powerful tool to perform high-throughput screening in various fields, such as directed evolution, digital genetic circuits, microbiology, and cell biology of gene expression and regulation. Ultimately, the development of μFACS lays down the foundation of future work in cell sorting and single-cell analysis. The vision of a complete integrated lab-on-chip system, in which cell sorting is just one of the steps, is now being realized.

An Electrokinetic μFACS

Within the last decade, high-throughput analysis has become an essential part of genetic and biotechnology research. These studies of gene expression and gene evolution often need to analyze up to 10^6 different species to acquire substantial information concerning the genetic or evolutionary pathways. Thus, new and improved high-throughput screening technologies have to be implemented in parallel. Conventional FACS, laser scanning image analysis, and silicon microfluidics were all developed to perform high-throughput data acquisition in these studies.

The oldest technology, flow cytometry, analyzes and sorts cells in a single profile as they pass through a point of detection in a jet stream. Despite their extensive capabilities of analysis and sorting, these conventional flow cytometers also have major problems of low sensitivity for bacterial cells and DNA and cross contamination between runs, and they are mechanically inflexible to work with different cell types and sizes. Hence, to this date, most work in flow cytometry is primarily focused on mammalian cells.

High-throughput image analyses of DNA (7, 29, 39), protein (13, 20, 30) and cell microarrays (35, 48) using laser scanning fluorescence microscopy or an ultrasensitive charge-coupled device camera allow up to 10^6 samples to be detected in a short period of time. Micropipettes or laser tweezers can be used to manipulate a single cell or molecule for further analysis or recovery. Yet recovery of a population of cells or particles from the arrays is difficult. If further analysis needs to be done, the identities of the individual spots on the arrays must be known a priori.

Many silicon microfluidic devices were also developed to perform high-throughput analysis. Electrokinetic forces can be used within microchannels to separate ionic species, such as DNAs and peptides (19, 21). These DNAs and peptides are stained with fluorescent dyes either before or after separation and are identified as they pass through the detection region. Multiplexing these capillaries allows arrays of samples to be analyzed simultaneously and increases the throughput rate. Large-scale DNA or protein sequencing can be done in this way. This kind of passive separation, however, depends on the inherent differences in the electrophoretic mobility of samples being interrogated. More recently, separation of cells through a microfabricated lattice (6) or dielectrophoresis (1, 34, 36) was also done, using the differences in their inherent morphology and anatomy. Detection of these species occurs only after separation.

Microfabricated silicon devices also created valveless switches for pressure switching (4) and dielectrophoresis (14) to perform active sorting. Yet the delicate and tedious fabrication process along with the complicated electronic and buffer requirements in dielectrophoresis have inhibited the silicon microfluidic technologies from gaining popularity among biochemists and biologists.

In conjunction with these advancements, an elastomeric μFACS was developed to address these needs for high-throughput screening tools. Active sorting occurs via electrokinetic flow when the fluorescence of the cell passes a preset threshold. Sorted cells can be recovered at the output wells of the sorter, and thus the identities of the cells do not have to be known a priori. Soft lithography offers an easy way to fabricate inexpensive elastomeric microfluidic devices with plastics and polymers instead of silicon. Unlike conventional FACS, this μFACS is easy to operate and highly sensitive for bacteria and DNA, and it can be disposable. The flexibility in the design of the flow cell also allows different cell types and sizes to be analyzed.

Design of an Electrokinetic μFACS

Adopting from a pioneering work in DNA sizing using an elastomeric microfabricated device (8), a microfluidic flow cell, an optical detection system, and electronics were designed and constructed to perform cell sorting. The disposable, soft-lithographed mi-

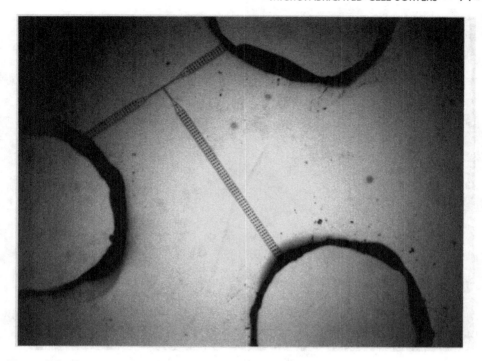

Figure 5.1. Optical micrograph of the microfabricated fluorescence-activated cell sorters de-vice made from GE RTV 615.

crofabricated flow cell is a silicone elastomer impression of an etched silicon wafer, with three channels joined at a T-shaped junction (see Figures 5.1 and 5.2) (8, 16). The channels are 100 μm wide at the wells, narrowing down to about 5–10 μm at the T junction. Channel depth is 4 μm. The channels are sealed with a glass coverslip. A buffer solution is introduced at the input channel and fills the device by capillary ac-tion. The pressure is equalized by adding buffer to the two output ports and then adding a sample containing the cells to the input port. The fluid within the channels is ma-nipulated via electrokinetic flow, which is controlled by three platinum electrodes at the input and output wells. The whole cell-sorting device is mounted on an inverted microscope with an oil immersion objective for fluorescence excitation and detection, as shown in Figure 5.3.

Optical Setup

A schematic of the optical setup for excitation and detection of the microfabricated cell sorter is as follows: The laser beam was collimated to achieve uniform illumination of the samples. A 5-W Argon laser (Coherent Innova 70) was used as an excitation source. For cell sorting, the plasma tube current was set between 10 and 25 amps, with an out-put power of 100–500 milliwatts at 488 nm. A half waveplate is placed in front of the laser to rotate the polarization of the beam at an angle of 20°–60°. This beam is then split by a polarizing cube beam splitter into its p-plane and s-plane polarized compo-

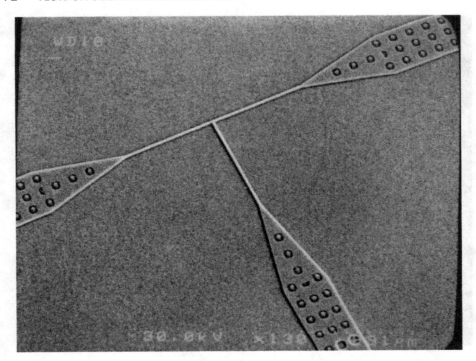

Figure 5.2. Scanning electron microscope image of an etched silicon mold. The etched silicon wafer is fabricated as follows (11). A <100> lightly doped silicon wafer with an oxide layer of 200–300 nm thickness was used. After using standard contact photolithography techniques to pattern the photoresist (Shipley SJR 1813) onto the silicon wafer, a mixture of C_2F_6 and CHF_3 gases is used to etch the wafer by reactive ion etch (RIE). RIE etches away the photoresist and the oxide layer, exposing the silicon layer underneath. A wet etch using KOH anisotropically etches deeper into the silicon layer. The final product is an etched silicon wafer that becomes a mold for the silicone microfluidic chip.

nents. By adjusting the angle of the half waveplate, the angle of polarization of the laser beam with respect to the optical axis will differ, thus varying the difference in the intensity of its p-plane and s-plane components. The p-polarized beam, about half the output of the laser, is expanded and collimated by a homemade "telescope." This beam expander is composed of two lenses; the first lens has a focal length of 12 mm, and the second lens has a focal length of 85 mm. The expanded beam is then directed by a pair of 2-in steering mirrors into the back of an inverted microscope (Zeiss Axiovert 35). Within the Zeiss microscope, the collimated beam is focused by a pair of lenses into the back focal plane of the objective (Olympus Plan Apo 60×, 1.4 NA, oil immersion). The objective again collimates the beam to achieve uniform illumination of the sample.

The fluorescence detection scheme also uses the principle of collimating the emitted light. The fluorescence is collected by the objective at infinity focus. The image of the sample is then focused onto an adjustable slit at the sideport of the microscope. This adjustable slit controls the field of detection and the amount of light entering into

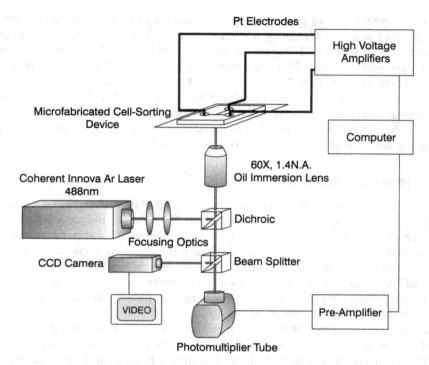

Figure 5.3. A schematic of the setup for microfabricated fluorescence-activated cell sorters. The cell sorting device is mounted on an inverted microscope (Zeiss Axiovert 35) with an oil immersion objective (Olympus Plan Apo 60×, 1.4 NA). Epi-fluorescent excitation was provided by an Argon ion laser (Coherent Innova 70) for cells and a 100-W mercury lamp for beads. Fluorescence was collected with the same objective and projected onto the cathode of a Hamamatsu H957-08 photomultiplier tube with custom current-to-voltage amplifier. Part of the light can be directed onto a charge-coupled device camera for imaging. The detection region is 5–10 μm below the T junction and has a window of 15 × 5 μm dimension. The window is implemented with a Zeiss adjustable slit. Cells or particles can be directed to either side of the T channels, depending upon the voltage potential settings. The voltages on the electrodes are provided by a pair of Apex PA42 HV op amps powered by Acopian power supplies. The third electrode is ground. The PMT signal is digitized by the PC, which also controls the high-voltage settings via a National Instruments Lab PC1200 card.

the PMT. The fluorescence reaches infinity focus again by a lens with a focal length of 75 mm. A 20/80 beamsplitter images 20% of the collected light into the charge-coupled device camera for observation, and the rest is directed onto the PMTs. It is straightforward to extend the system to include multiple-color fluorescence and light scattering detection, as in conventional FACS machines.

Electronics

A side-on photomultiplier tube (H957-08; Hamamatsu, Bridgewater, NJ) was used as the detector for cell sorting. A transimpedence amplifier (OPA128; Burr-Brown, Tucson, AZ) converts the photocurrent to voltage at a gain of 107 V/A. A 15-V direct current power supply was built for both the detector and the current-to-voltage preampli-fier. The signal was then filtered by a resistor-capacitor low-pass analog filter at 1.6 kHz and then digitized by a NI-Daq card (LAB-PC-1200; National Instruments, Austin, TX) on a personal computer. Running a sorting algorithm in C, the signals were ana-lyzed, and appropriate voltages were set on the PC1200 board analog outputs (+5 V to −5 V). These voltages were then amplified by a pair of APEX PA42 (30 V/V) to two platinum electrodes that were inserted, one each, into the collection and waste wells. The third electrode was ground at the input well. These voltages were set to ma-nipulate the direction of the fluid flow inside the channels.

Theory of Electrokinetic Flow

When a potential difference is applied across a microfluidic channel, an electro-osmotic flow is induced. Beads, cells, or other particles in the electrolytic solutions are being carried along with electro-osmotic flow. Manipulation of these particles can be accomplished by adjusting the potential difference between two platinum electrodes inserted at the ends of the microchannels. The switching of fluid flow at the T junc-tion of μFACS is almost as instantaneous as the switching of the potentials. Electro-osmotic flow is thus a good and easy way to manipulate particles and create valveless switches within any microfluidic device.

According to (3), electrokinetic phenomena arise from forces occurring at mobile electrified interfaces. For example, a potential difference is applied across a glass cap-illary filled with electrolytic solution. Instead of current flowing through the capillary, the electrolytic solution begins to move within the capillary, resulting in electro-osmotic flow. This simply means that a potential difference has the same effect as a pressure difference within the capillary. Thus, the flow velocity, V, of electrolytes within a capillary depends on two components; a pressure gradient, ΔP, and an elec-tric field, X.

$$V = a_1 \Delta P + a_2 X \qquad (1)$$

So if there is no pressure gradient $\Delta P = 0$, there is still flow, $V = a_2 X$, resulting from the electric field, where a_2 is the electrophoretic mobility.

Vice versa, one can predict that a current, i, can also result from an electric field and a pressure difference within the capillary.

$$i = a_3 \Delta P + a_4 X \qquad (2)$$

So in the absence of an electric field, $X = 0$, a streaming current occurs from the pres-sure gradient,

$$i = a_3 \Delta P \qquad (3)$$

where a_3 = streaming current density. If equation 3 is divided by the specific conduc-tivity of the electrolytic solution, σ, then

$$\frac{i}{\sigma} = \frac{a_3 \Delta P}{\sigma} = X_i \qquad (4)$$

where X_i is the electric field of the streaming potential resulting from the application of a pressure difference. Thus, as equation 5 shows, one can predict that the flow velocity of fluid resulting from an applied electric field within the capillary is the same as the current resulting from a pressure difference.

$$\left(\frac{i}{\Delta P}\right)_{x=0} = a_3 = a_2 = \left(\frac{V}{X}\right)_{p=0} \tag{5}$$

From an atomistic view on electrokinetic phenomena, the electrolytic solution consists of many diffuse layers acting as planar electrodes. Each layer is a few angstroms thick and is at a distance, χ, from the wall of the capillary. If a potential difference is applied across the capillary, a layer of charge, q, at a distance χ from the wall of the capillary will experience an electric force q_X, where X is the electric field. This force will cause the layer to move across the capillary. Yet this motion will be opposed by the viscous force. Thus, when the electrolytic solution reaches a steady gradient, the electric force is exactly equal to the viscous force. The same phenomenon will occur if a pressure gradient occurs within the capillary.

In addition to electrokinetic phenomena, another motion occurs within the electrified capillary, called ion migration. Instead of seeing the electrolytic solution as layers of diffuse charges, individual ions within the solution will move in a specific way in the presence of an electric field. In the absence of any field, ions perform a diffusive, random walk with equal probabilities in all directions. Yet, in the presence of an electric field, these ions will migrate toward either the positive or negative electrode according to their individual charge and experience collisions with other ions, shielding effects, and viscous forces from the medium, depending on their distances from the electrodes. Redox reactions inevitably will occur at the interface of electrodes and electrolytic solutions. As a result, ions are being depleted constantly, and thus the electro-osmotic mobility of the solution changes over time (3).

In a more complex system, such as the microfabricated cell sorter, all these electrokinetic phenomena occur simultaneously and interact with each other. Many experiments were carried out to search for the best conditions for sorting beads and cells. Erratic behaviors and clogging of the beads and cells were often observed as a result of ion depletion and other unfavorable conditions. Surface treatments of the elastomeric chip and glass coverslips to produce hydrophilicity are described in detail in the next section. Different buffer conditions for beads and cells and the fabrication of microelectrodes are also described.

Surface Chemistry

In microfabricated chips, where the dimensions of microchannels are comparable to the size of the particles flowing within them, surface chemistry becomes very important. Cells and proteins can nonspecifically adsorb onto any hydrophobic or hydrophilic surface, which may result in clogs or reduce throughput rates. Electro-kinetic phenomena are highly dependent on the ionic strength of the fluid and the surface charge of the capillary. Poly(dimethylsiloxane) is inherently hydrophobic, consisting of a Si–O backbone with methyl and vinyl substituents. Its hydrophobicity prevents aqueous solutions, cells, and other bioparticles from flowing smoothly inside these microchannels. Clogging of channels as a result of adsorption of cells was often observed. Thus, several methods were developed to render the surfaces of the elastomeric chip and the glass coverslips more hydrophilic.

Three effective methods were developed to make the elastomeric devices more hydrophilic (11). Coating the surface with polyurethane (Hydrogel RL#153-87, Tyndale; 3% w/v in 95% ethanol and diluted 10× in ethanol) and curing at 90°C for 60 min deposits a hydrophilic layer on the surface of the microchannels. Yet, such a coating would seal up channels with depths of 2–3 μm. In addition, the polyurethane coating deteriorates after each use, reducing the channels' hydrophilicity and giving inconsisent flow results. A different method exploits the addition of a surfactant, MAKON 6 (Stephan Canada, Longford Mills, Canada; 0.2% v/v) into the mixture of General Electric RTV 615 components, followed by curing in an oven as before. MAKON 6 effectively renders the surface of the elastomer hydrophilic, but unexpectedly, it also increases the background fluorescence of the device. As a consequence, the third method is chemically modifying the surface of the elastomer by an acid treatment. Immersing the chip in dilute HCl (pH 2.7, 0.0074% in water) for 40 min at 60°C will break up the Si–O backbone of the elastomer, modifying the surface to become hydrophilic, consisting of Si–H and Si–OH substituents. This HCl treatment avoids clogging caused by excessive coating, has negligible fluorescence background, and does not deteriorate with use.

To increase the hydrophilicity of the whole-cell sorting device, methods were also sought to lessen the adsorption of beads/cells onto the glass coverslip. Coating the coverslips with various chemicals and treatments had not proven to be successful. Rainex, bovine serum albumin (BSA), and successive multiple ionic polymer layer (SMIL) (22) did not reduce the amount of adsorption of particles onto glass, and SMIL even caused cell death. However, cleaning dusts and etching metallic and organic residues off the glass coverslips renders the surface more hydrophilic by exposing its polar silanol surface. Two wash formulas seem to work best: a base wash called RCA and an acid wash for glass surfaces, called Chromerge. RCA treatment is a base wash that consists of six parts H_2O, four parts NH_4OH, and one part H_2O_2. Glass coverslips were immersed in a stirred RCA solution at 60°C for 1 hr. Then the coverslips were rinsed and stored in high-purity water for later use. Chromerge (Manostat Corporation, New York) is a chromic acid solution derived from chromium trioxide. Coverslips were immersed in a stirred Chromerge solution for 1 hr without any heat. Cell adsorption was minimized the most when Chromerge-washed coverslips were used, although hydrophilicity of both RCA-washed and chromerge-washed coverslips were very comparable.

Buffers and Microelectrodes

A good buffer system is critical for the success of electro-osmotic sorting within the chip. There were two major problems encountered in the search for an ideal buffering system: ion depletion and adsorption of cells and beads onto surfaces of glass and PDMS. Ion depletion results from migration of ions to the electrodes, exhausting the amount of ions remaining inside the solution. Erratic movements of beads and cells within the channels were often observed within 30 min in the presence of an electric field. Thus, two criteria were used to select the best buffer system: the buffer has to have a run time of up to 2 hr without experiencing heavy ion depletion, and this buffer also minimizes the amount of cell adsorption and maximizes the viability of the cells within an electric field. Various buffers of different ionic strengths such as phosphate-buffered saline (PBS), piperazine-N′,N′-bis(2-ethane-sulfonic acid) (PIPES), N-2-

hydroxyethylpiperazine-N'-2-ethane sulfonic acid (HEPES), and distilled water were examined. Each buffer was also tested with different salts of various concentrations (CaCl$_2$, MgCl$_2$, NaCl, and KCl). In dealing with adsorption, neutral surfactants, such as Triton X-100, Tween 20, and MAKON 6; positively charged surfactant, cetyltrimethyl ammonium bromide; and negatively charged surfactants, BSA, and sodium dodecyl sulfate (SDS), were investigated at various concentrations in the buffers to alleviate adsorption. Two buffer systems were discovered that were optimal for these two conditions, one for the carboxylate-modified (CML) beads and the other for HB101 *Escherichia coli* cells. For beads, extra reservoir wells have to be incorporated in addition to the input and the collection wells to prevent ion depletion. Although these two buffer systems may not be the absolute best systems for sorting these particles, they are sufficient for carrying out the experiments within a few hours of run time.

Microfabricated electrode pads were sputtered or evaporated onto the surface of glass coverslips to prevent electrolysis at the electrode/buffer interface. Metals, such as Au, Ni, Au/Ni, and Pt/Ni, were all evaporated on the surface of the coverslip in a specific pattern using photolithography. Two microelectrodes, placed at either side of the T junction, were about 200–400 μm apart and about 50–100 nm thick. The cell sorting device was adhered to the surface of the coverslip containing microelectrodes. Yet at an electric field of 6 V/cm, electrolysis (detected as bubbles from the microelectrodes) and metal plating occurred within the channels. In addition, except for Pt/Ni electrodes, all of the other metals did not adhere well to glass. They can be easily peeled off as the chip detaches from the glass surface. In the end, platinum wires instead were used as electrodes for the electrokinetic sorting experiment.

Performance of μFACS

The performance of any flow cytometer depends on three factors: sensitivity, resolution, and measuring rate. Sensitivity, in terms of fluorescence detection, is defined as the minimum number of dye molecules per cell that can be resolved. Resolution is expressed as a coefficient of variation for precision in fluorescence detection. The measuring rate for flow cytometers is the number of cells per given time passing through the analysis point without any coincidences. In the case of μFACS, the measuring rate is the fastest time the sorter can respond on the detection of a desired cell. Thus, three parameters were measured to verify the performance of the microfabricated cell sorter: the coefficient of variance in measurements, the sensitivity of the detection system, and the response time in a sorting event.

The coefficient of variance (CV, defined as the standard deviation of the peak divided by the mean for an uniform population of particles) is routinely used as a measure of the system's precision in fluorescence detection. The CV for this microfabricated cell sorter system ranges from 2% to 5%, measured with LinearFlow Green Flow Cytometry Intensity Calibration Beads (Molecular Probes, Eugene, OR). This discrepancy may depend on several factors, such as the relative intensity of different calibration beads and the bias settings on the PMT. Specifically, the depth of the channel also affects the CV of the system. A higher CV is obtained when a deeper channel is used because of a greater variation in the position of the beads. Interestingly, different methods of modifying the elastomer to be more hydrophilic also influence the CV. Polyurethane coating of the elastomer yields the best CVs, but the CVs deteriorate with each subsequent use. Although HCl-treated channels produce CVs that are 0.5%–1% higher than those of the polyurethane coating, their CVs remain consistent with each usage. The MAKON 6 channels give the highest CVs among the three different hydrophilic treatments, about 1%–2% higher than the CVs obtained from the HCl-treated channels. Regardless, these CVs are already comparable to those obtained from a conventional FACS.

The sensitivity of this system is about 200 dye molecules at 100 events/s of YO-YO 1 (quantum yield = 0.52), a DNA intercalating fluorescent dye (8). This is at least twice as sensitive as the most sensitive flow cytometers commerically available, which can detect about 1000 fluorescein molecules (quantum yield = 0.9) at 1000 events/s. In addition, an advantage of μFACS is the small detection volume, which in this case is approximately 100 fL, which greatly reduces background fluorescence in the suspension.

The response time of μFACS consists of two parts: time from a signal on the PMT to the actual change in voltage settings of the Pt electrodes, and time from the switching of the voltage settings to the actual change of fluid flow at the T junction. Measurements were taken using wild-type GFP HB101 cells flowing through the detection window to observe the response times of the hardware in μFACS. The time from a signal on the PMT to the actual change in the voltage settings was determined to be between 1.25 and 1.50 ms, and the actual switching in fluids occurs in less than 1 ms. The whole response time of the device was less than 3 ms for sorting particles electroosmotically.

The linear flow velocities of GFP expressing *E. coli* cells in μFACS in response to the applied electric field strengths were measured and plotted in figure 5.4. A PMT was used to observe the fluorescence from each cell as it passed through the interrogation region. The signal pulses from the cells were recorded by an oscilloscope, and the widths of the pulses were calculated as the time the cells took to travel through the

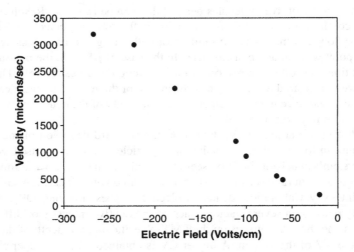

Figure 5.4. A graph of the linear flow velocities of cells in response to the applied electric field strengths inside microfabricated fluorescence-activated cell sorters. These data were taken on the same device but on two separate days. The width of the detection volume was 12 μm.

interrogation point. The linear flow velocities of the cells were derived and plotted against the applied electric field strengths. As figure 5.4 shows, cell velocities escalate with increasing electric field strengths. However, at above a certain electric field strength (greater than the absolute value of -600 V/cm), these cells began to lose viability, as indicated by their loss of fluorescence inside the microchannels and at the wells. Thus, the fastest velocity that could be attained by *E. coli* cells inside μFACS is ~3–5 mm/s. Cell viability would be greatly compromised if higher electric field strength is applied.

Sorting Schemes

Different algorithms for sorting in the microfluidic device can be implemented by the computer. The standard "forward" sorting algorithm consists of running the cells from the input channel to the waste channel until a cell's fluorescence is above a preset threshold, at which point the voltages are temporarily changed to divert the cell to the collection channel. With electrokinetic flow, switching is virtually instantaneous, and throughput is limited by the highest voltage that can be applied to the sorter (which also affects the run time through ion depletion effects). In contrast, a pressure-switched scheme does not require high voltages and is more robust for longer runs. However, mechanical compliance in the system is likely to cause the fluid switching speed to become rate-limiting with the "forward" sorting program. Because the fluid is at low Reynolds number and is completely reversible, when trying to separate rare cells, one can implement a sorting algorithm that is not limited by the intrinsic switching speed of the device. The cells flow at the highest possible static (nonswitching) speed from the input to the waste. When an interesting cell is detected, the flow is stopped. By the time the flow stops, the cell will be past the junction and part way down the waste channel. The system is then run backward at a slow (switchable) speed from waste to input, and the cell is switched to the collection channel when it passes through the detection region. At that point, the cell is saved and the device can be run at high speed in the forward direction again. This "reverse" sorting method is not possible with standard FACS machines and should be particularly useful for identifying rare cells or making multiple time-course measurements of a single cell. Higher throughput rates could be achieved with this algorithm. We first demonstrated reverse sorting with beads, using electrokinetic flow, and later with cells in the second-generation cell sorter, which was with integrated valves and pumps (see figure 5.5.)

Decision Making and Sorting Algorithms

Methods of sorting in microfluidic devices are essentially different from the conventional aerosol droplet sorters. In any sorting logic, the detection of each cell first determines whether or not the criteria are met; when these criteria are met, a logic signal is generated to trigger sorting. For conventional aerosol droplet sorters, the detection of the cell usually occurs in a jet stream, and the sorting occurs after the stream has been broken up into droplets that contain zero to two cells. Depending on the distance between the interrogation point and the breakoff point, the time lag between these two points may be tens to hundreds of microseconds. For μFACS, detection occurs ~5 μm before the T junction, and sorting is immediately performed in

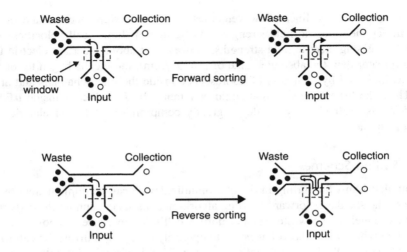

Figure 5.5. A schematic of forward and reverse sorting.

continuous flow. This may allow for more accurate and synchronous sorting (within 10 ms time frame), which can be critical for time-course measurements or any downstream analysis. In addition, with the reverse sorting algorithm, the detection of the cell actually occurs two to three times more during sorting before the cell is finally shuttled into the collection channel.

For the electrokinetic cell sorter, both forward and reverse sorting algorithms were written in C++. One electrode at the input was at ground, and voltage potentials were applied to the other two electrodes at the collection and waste wells. As figure 5.6 shows, the arms of the T channel can be considered as three wires with identical resistance (because they have the same dimensions), so the currents from the collection and the waste wells combine at the T junction and go into the input well.

$$I_c R = V_c - V_T; \; I_w R = V_w - V_T; \; I_T R = V_T \tag{6}$$

$$I_c + I_w = I_T \tag{7}$$

$$(V_c - V_T) + (V_w - V_T) = V_T \text{ since } I_c R + I_w R = I_T R \tag{8}$$

$$V_c + V_w = 3V_T \tag{9}$$

If the collection channel is set to be "floating," then $V_w = V_T$. Ideally, there should not be any current flowing into the "floating" collection channel, and hence no electrokinetic flow. This also means that $V_c = 2V_w$.

Forward Sorting

For forward sorting, the voltage potentials are initially set as $V_c = 2V_w$, so all the electrokinetic flow goes into the waste channel; that is, $I_c = 0$. When a cell's fluorescence reaches above a set threshold voltage, these two voltage potentials switch; that is, $V_w = 2V_c$. The cell is directed into the collection channel for a certain period of time before the voltage potentials switch back.

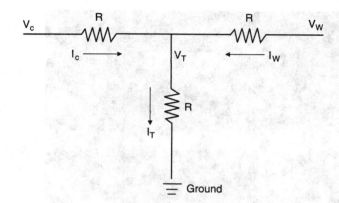

Figure 5.6. T-channel wire diagram. The arms of the T channel can be considered as three wires with identical resistance because they have the same dimensions.

Reverse Sorting

Similarly, the reverse sorting algorithm is such that the default voltage setting is $V_c = 2V_w$. When the sorting signal triggers, the cell is slowly reverted back into the detection region again at a tenth of the original flow rate; that is, $-\frac{1}{10} V_c = -\frac{1}{10} 2V_w$. After the second detection, the cell is slowly directed into the collection channel, $\frac{1}{10} V_w = \frac{1}{10} 2V_c$, before switching back to the default potentials. Unfortunately, because of evaporation and uneven pressure buildup in the two output wells, the voltage potentials have to be readjusted from time to time to maintain accurate sorting for both forward and reverse sorting.

Sorting with Beads and Bacterial Cells

The use of μFACS for forward and reverse sorting with electrokinetic flow was demonstrated with fluorescent beads of different emission wavelengths in different ratios and up to 33,000 beads per hour throughput. Extra reservoir wells were incorporated on the outer side of the three wells to avoid ion depletion, and platinum electrodes (with the ground electrode in the input well) were inserted into the reservoir wells. The collection wells were filled with buffer, and a mixture of red and blue fluorescent beads was injected into the input well in 10–30-μL aliquots. The optical filter in front of the PMT passed only red fluorescence, allowing selective sorting of red beads. Figure 5.7 shows a snapshot of a sorting event. Sorting can be performed for as long as 3 hr with occasional readjustment of the voltage settings. The coefficient of variation in bead intensity was measured to be 1%–3%, depending on the depth of the channel and surface treatment of the elastomer.

Sorting of Beads

Table 5.1 shows that a single pass through the μFACS produced a highly enriched sample of red beads. Whereas the initial concentration of red beads was 7.4%, the collection well held 84% red beads, whereas the waste well had less than 1%. Similar results were obtained when running in reversible sorting mode when the initial concentration of red beads was lowered to 1% (Table 5.1). Run times varied from 10 min up

Figure 5.7. A snapshot of a sorting event. A bead (oval-shaped) is being sorted into the right, collection, channel. The beads (round-shaped) in the left, waste, channel are stagnant because of a voltage change that directs the electrokinetic flow into the collection channel. Preparation of beads is as follows. Fluorescent beads (1 μm diameter, Interfacial Dynamics Corporation, Portland, OR) were suspended in phosphate-buffered saline (137 mM NaCl, 2.7 mM KCl, 4.3 mM $Na_2HPO_4 \cdot 7H_2O$, 1.4 mM KH_2PO_4) with 10% bovien serum albumin (1 mg/mL) and 0.5% Tween 20 and overall concentration of 1.5%. Fluorescence of the beads was excited by a 100-W mercury lamp with 488DF20 optical filter. A 630DF30 optical filter (Chroma Technology Corp., Brattleboro, VT) was used to select the red fluorescent emission. The μFACS device had 3 × 4-μm channels.

to 3 hr. With both forward and reverse sorting, enrichments of 80×–97× were obtained in single runs, where the enrichment is defined by the increase in the fractional concentration of red beads.

Sorting of E. coli Cells

Living *E. coli* cells can also be sorted in μFACS, using electrokinetic flow, and the cells are viable after sorting. Different ratios of wild-type to GFP-expressing *E. coli* cells were introduced into the input well (volume ranges from 10 to 30 μL of sample); the collection wells were filled with 10–30 μL of buffer with 10^{-5} M sodium dodecyl sulfate (SDS). After inserting the three platinum electrodes into the wells (with the ground electrode in the input well), the voltages were set for forward sorting. After sorting for 2 hr, cells were collected with a pipette and streaked onto LB plates (or

Table 5.1. Results of beads and *E. coli* cell sorting

	Input Well		Correction Well		Waste Well	
	Blue	Red	Blue	Red	Blue	Red
Forward bead sorting	0.925	0.074	0.160	0.840	0.998	0.002
Reverse bead sorting	0.988	0.012	0.043	0.957	0.999	0.001
Forward *E. coli* cell sorting	0.992	0.008	0.693	0.307	0.992	0.008

Preparation of *E. coli* cells for sorting is as follows: The *E. coli* cells (HB101) expressing GFP were grown at 30°C for 12 hr in LB liquid medium containing ampicillin (one colony was inoculated into 3 mL medium containing 50 μg/mL ampicillin). Wild-type *E. coli* HB101 cells were incubated for 12 h in LB-only medium. After incubation, HB101 and GFP-expressing HB101 *E. coli* cells were resuspended in PBS (ionic strength = 0.021) three times and stored at 4°C for sorting. Immediately before sorting, the cells were resuspended again in phosphate buffer (4.3 mM $Na_2HPO_4 \cdot 7H_2O$, 1.4 mM KH_2PO_4) containing 10^{-5} M SDS and diluted to a concentration of 10^9 cells/mL. The cells were filtered through a 5 μm syringe filter (Millipore Bioscience Inc., Bedford, MA) for elimination of elongated cells. A μFACS device with 10 × 4 μm channels was used. Fluorescence was excited by the 488 nm line of an Argon ion laser (6 mW into the objective). Coherent Innova 70 (Laser Innovations), and the emitted fluorescence was filtered with a 535DF20 filter.

other antibiotic-containing plates) and incubated overnight at 37°C for colony counting. We achieved enrichments of 30× with yields of 20%, where the yield is defined by the number of colonies on the plate divided by the number of positive fluorescence events detected in the device. The sorted cells show relatively constant viability in electric fields up to 100 V/cm, corresponding to velocities of 1–3 mm/s.

The electrokinetic μFACS system offers several advantages over traditional sheath flow methods. Because the channels in the device can be made with micron dimensions, the volume of the interaction region can be precisely controlled, and there is no need for hydrodynamic focusing. As fluid flows continuously through the system, there is no need for droplet formation, and a host of challenging technical issues can be sidestepped. Furthermore, no aerosol is formed because the system is entirely self-contained, allowing relatively safe sorting of biohazardous material. The disposability of the sorting devices obviates the need for cleaning and sterilizing the instrument and prevents cross contamination between samples.

An Integrated μFACS

Cell sorting has become an indispensable part in the studies of cellular metabolism at the single cell level (25, 33). These studies will demand immediate treatment of cells before or after sorting, with minimal time variation and sample loss. Thus, there is a need for μFACS to integrate more functionalities to perform a complete analysis "on-chip." Electrokinetic flow or direct pressure application alone will not be able to meet these demands.

Multilayer soft lithography is a new micromachining technique that exploits the elasticity and the surface chemistry of silicone elastomers to create monolithic microvalves within microfabricated devices (43). Using multilayer soft lithography, μFACS was integrated with microvalves and micropumps (37). Initial efforts to sort cells via electrokinetic flow demonstrated the sorting and recovery of bacterial cells in microfabricated devices in an automated fashion. However, the electrokinetic sorter suffers from the same drawbacks as all the electrokinetically actuated microfluidic devices such as buffer incompatibilities and frequent change of voltage settings resulting from ion depletion, pressure imbalance, and evaporation. The current cell sorter with microvalves and micropumps is a step closer to the realization of an integrated lab-on-chip. It has incorporated switching valves, dampers, and peristaltic pumps for sorting, sample dispensing, flushing, recovery, and absorbtion of any fluidic perturbation (see figure 5.8). Other microfluidic functions can be easily integrated for kinetic studies or treatment before or after sorting events. The active areas of these microvalves and micropumps are much smaller than those made by Unger et al. (43) and Chou et al. (9) to accommodate the size of a single bacterial cell. The active volume of one valve on this integrated sorter can be as small as 1 pL. The cell sorter also reduces concerns for buffer compatibility, automation, and viability of cells. Different algorithms of interrogation within this sorter, such as reversible sorting and cell trapping, were all exploited on these devices. The sorting accuracy and recovery efficiency using this integrated sorter were greatly improved relative to the electrokinetic μFACS. Finally, we have demonstrated the ability of μFACS to integrate various microfluidic functionalities into one chip to perform a complicated task in an automated fashion.

Design of an Integrated μFACS

Fabrication and design of the integrated, pneumatic-driven cell sorter adopted features from the electrokinetic sorter and novel engineering, using multilayer soft lithography. This integrated cell sorter has the shape of a "T" for sorting, as in the electrokinetic sorter, but with valves and pumps incorporated. As shown in the schematic layout of figure 5.8, the sorter has two layers. The top layer has the control lines for valves and pumps, and the bottom layer has the fluidic lines. The fluidic layer has channels of 30 μm in width, which narrow down to 20 μm and eventually tapers down to 6 μm at the T junction. Supports are lined up along both sides of the channels for visualization and alignment purposes. Fluidic holes are incorporated at the ends of the "T" for injection of cell samples and buffers. Collection wells 1 mm in diameter at the arms of the T are used for recovery. The control layer has distinct functionalities at different regions of the T for controlling fluid flow within the fluidic lines. Three valves, acting as a peristaltic pump, have a valve active area that is 80 μm long and 30 μm wide. These valves are 100 μm apart. Three dampers of similar dimensions are placed immediately following the peristaltic pumps to absorb any energy from fluidic perturbations introduced by pumping. Three pairs of switching valves are placed at the arms of the T, being 20, 30, and 50 μm wide. These valves have a valve active area of 20 $\times$ 20, 30 $\times$ 30, and 50 $\times$ 30 μm, respectively. The specific fabrication details of the integrated μFACS are as described (15).

Multilayer Soft Lithography

Multilayer soft lithography is a micromachining technique that is based on rapid prototyping and replica molding methods of soft lithography. A monolithic chip can be made of multiple layers of elastomeric channels, with each layer individually cast from a microfabricated mold. In a typical two-layer system, the bottom layer consists of the fluidic line, where the sample will be introduced and interrogated. The top layer has the control line, where the valves will be pneumatically actuated. When pressurized air or nitrogen is introduced into the control line, the thin membrane between the two lines

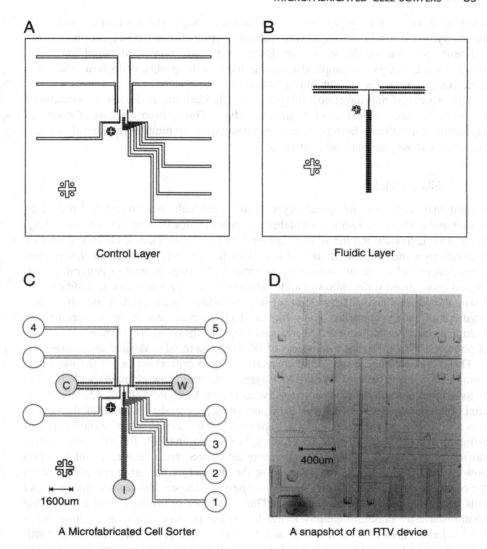

Figure 5.8. A schematic of the integrated cell sorter. This sorter is made of two different layers of elastomeric channels bonded together. (*A*) The control layer contains lines where pressurized nitrogen and vacuum are introduced to actuate the closing and opening of the valves, respectively. (*B*) The fluidic layer contains lines where the sample is injected. (*C*) In this integrated cell sorter, valves 1, 2, and 3 act as a peristaltic pump and valves 4 and 5 act as switch valves. Other two pairs of switch valves are not numbered. Holes, labeled as I, C, and W, are the input, collection, and waste wells, respectively. Patterns of 00001 and 00010 actuated by the AT-DIO-32HS card closes valves 1 and 2 respectively, where 0 indicates "valve open" and 1 indicates "valve close." (*D*) A snapshot of an integrated cell sorter made from GE RTV 615.

is deflected downward and seals off the fluidic line. A valve is created in this way. The simplicity and flexibility in multilayer soft lithography allows for integration of many different operations on the same chip. Unger et al. have fabricated switching valves and peristaltic pumps for sample dispensing and switching (43), and Chou et al. have made rotary pumps for mixing and incubation (9, 10).

The process of multilayer soft lithography is divided into two parts: fabrication of silicon molds and fabrication of elastomeric chips. This section discusses the crucial aspects of fabrication in both parts to successfully develop multilayer, microfluidic devices that can be pneumatically actuated.

Silicon Molds

In multilayer soft lithography, each layer is cast individually from a different microfabricated mold. There are two ways to fabricate molds from silicon wafers for soft lithography. One is to etch into the silicon, as mentioned in the previous section, and another is to pattern a thin layer of photoresist onto the silicon wafer. Although an etched mold is permanent and chemically resistant, the rapid prototyping process of patterning photoresist onto silicon wafers allows molds to be made within a few hours. In addition, different photoresists have different light sensitivity, surface chemistry, and viscosity. Combinations of these parameters allow the rapid prototyping process to be versatile for making a variety of different dimensions and shapes of the channels. Thus, a good choice of photoresist is critical for the success of fabricating a mold of desired dimensions.

The cross-sectional shape of the channels is also a critical factor for the ideal performance of the valves. According to Unger et al., only flow channels with a round cross section are able to close completely, as shown in figure 5.9 (11, 43). However, after obtaining the desired thickness from patterning photoresist on the mold, flow channels are usually of a rectangular or a trapezoidal shape as a result of ultraviolet light diffusion and the photolithography process. These flow channels fail to have complete closing of the valves. When under pressure introduced from above, a round-shaped flow channel is able to seal off a section of the channel by flattening completely from the center to the sides of the cross section, whereas a trapezoidal-shaped flow channel fails to seal completely from the sides. Thus, a further chemical modification of the photoresist after photolithography is needed. If the photoresist (an amorphous polymer) is heated above its glass-transition temperature, and given sufficient time, it will reflow to the edges and become rounded. This reflow process is used to tailor off any sidewall angles on the photoresist to completely round the flow channels.

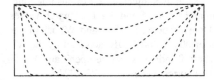 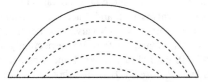

Figure 5.9. Cross sections of a trapezoidal-shaped and a round-shaped flow channels. Profiles of the channels when they are under actuation pressure. A flow channel with a round-shape cross section is able to completely seal off the channel under pressure. A channel with a trapezoidal-shape cross section will have leakage from the sides (43).

The aspect ratio of height over width of the flow channel is also an important issue for the fabrication of valves and pumps. Channels with too high of an aspect ratio will encounter problems in complete closing of valves in terms of geometry and pressure. Usually, the thin valve membrane ruptures first, before the valves can be closed completely. Oxygen plasma was used to isotropically etch the photoresist patterned on the silicon wafer. Hence, the height of the photoresist can be reduced with dry etch instead of using a higher spin rate. The etching also rounded the flow channel slightly because of the heat that was released. Further heating at low temperature may be needed to completely round the flow channel. The procedures of fabricating molds for different microfluidic functions, such as valves, peristaltic, and rotary pumps, can be found in references (9, 15, 28, 43).

Elastomers

Silicone elastomers retains much of the same chemistry as the natural organic polymers. Yet, their unique silicon–oxygen linkage provides much greater stability to high-temperature and chemical resistances. The commercial preparation of silicone elastomers is shown in table 5.2 (26). First, silica (sand) is reduced to elemental silicon metal. The elemental silicon metal is then grounded and reacts with methylchloride in the presence of a copper catalyst at 300°C. The products of this reaction are mono-, di-, and trimethylchlorosilanes (equation [10]). These silanes can be purified using fractional distillation. After distillation, the dimethylchlorosilanes1 are hydrolyzed to form silanols, which then condense to form cyclic silanes and other low molecular weight silanes (equation 11). These low–molecular weight silanes are reacted with base (KOH) to form dimethyl tetramers, a cyclic silane (equation 12). These tetramers are then linearized to become polydimethylsiloxane (PDMS) by an addition of a strong base and a monofunctional silane (equation 13). The monofunctional silane acts as a chain stopper and determines the viscosity of the linearized polymer.

Different organic side chains can be substituted in place of methyl on the silicon–oxygen backbone of PDMS. These substitutions serve to optimize a certain property of the elastomer, as required by a specific application. The inclusion of vinyl groups at various concentrations greatly increases the cross-linking efficiency of the polymer and yields elastomers with lower compression set or more rigidity. The substitution of phenyl groups allows the polymer to be more flexible at low temperatures, down to

Table 5.2. Polydimethylsiloxane manufacturing process [26] from silica (san) to PDMS.

$$Si + CH_3Cl \xrightarrow{Cu} \begin{array}{lll} Me_3SiCl & bp & 58°C \\ Me_2SiCl_2 & bp & 70°C \\ MeSiCl_3 & bp & 68°C \end{array} \tag{10}$$

$$2Me_2SiCl_2 + 4H_2O \longrightarrow 2Me_2Si(OH)_2 + 4HCl \tag{11}$$

$$HO[(CH_3)_2SiO]_mH \xrightarrow{KOH} Cyclic\ Dimethyl\ Tetramer \tag{12}$$
linear siloxane

$$Cyclic\ Dimethyl\ Tetramer + Chain\ Stopper \xrightarrow{KOH} -(Me_2SiO_2)_n - PDMS \tag{13}$$

−93°C. The inclusion of trifluoropropyl groups yields elastomers with higher resistance to many harsh chemical environments. An elastomer can be made by mixing different concentrations of these four functional groups—methyl, vinyl, phenyl and trifluoropropyl—to meet the demand of different applications (26).

Curing agents are added for the vulcanization of silicon rubber. Traditionally, curing agents are organic peroxides that can decompose to free radicals and react with the methyl and vinyl groups of silicone polymer. The vinyl groups have a much higher reaction rate, and thus have a much higher number of cross links. The number of cross links within the polymer determines the final physical profiling of the resulting silicone rubber. An alternative to vulcanization is called an addition cure. An addition cure uses a silicone hydride (SiH) as a cross-linking agent to the methylvinyl polymer. In the presence of a platinum catalyst, an addition reaction occurs in which the hydride is cross linked to the vinyl group of the silicone polymer. This addition reaction occurs without any by-products, such as water, and proceeds quite actively at room temperature (see figure 5.10). Thus, this type of silicone rubber is known as "room temperature vulcanization," or RTV.

Beginning in the 1990s, soft lithography has been slowly gaining recognition in fabricating silicone elastomer microfluidic devices for biological and chemical analysis. Many issues concerning the surface chemistry and the elastic properties of the silicon

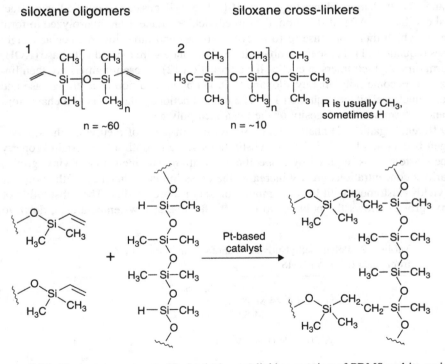

Figure 5.10. Room temperature vulcanization: cross-linking reaction of PDMS and its curing agent (5). In the presence of a platinum catalyst, an addition reaction occurs in which a silicone hydride is cross-linked to the vinyl group of the silicone polymer. This reaction occurs without any by-product and proceeds actively at room temperature.

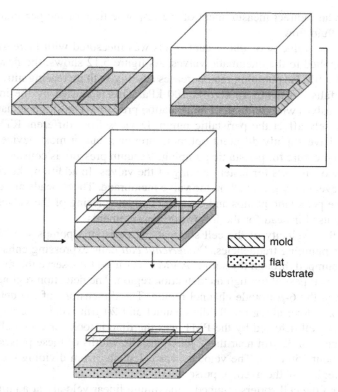

mold

flat
substrate

Figure 5.11. Multilayer soft lithography fabrication. Multilayer soft lithography fabrication creates a monolithic chip of multilayers of channels, having each layer individually cast from a microfabricated mold. When both layers are half-cured, the top layer is peeled off from the mold and bonded to the bottom layer. After the bonded layers are fully cured, a monolithic chip is created.

elastomers have been encountered. Especially in multilayer soft lithography, the composition of PDMS in each layer is different. The fluidic layer has an excess of one component, such as the curing agent, whereas the control layer has an excess of the other component, the silicone polymers. This is to enhance the surface bonding between the two layers to form a monolithic chip (figure 5.11) (43). One can design the fluidic layer to be either the top or the bottom layer, depending on the application of interest. The protocols of the elastomeric fabrication for different multilayer devices can be found in the literature (9, 15, 28, 43). The integrated μFACS was the first device made using multilayer soft lithography that demonstrated the automation and coordination of different valves and pumps on-chip to perform complex and decision-making functions.

Flow Velocity and Cell Trapping

Several parameters of the integrated sorter were characterized in preparation for sorting. These parameters are the optimum nitrogen pressure applied to the peristaltic pumps, the linear flow velocity of the cell sorter, and the mean reverse time. The mean reverse time is an indirect measurement of the response time of the peristaltic pumps and the actual fluid flow.

The linear flow velocity of fluorescent beads was measured with increasing nitrogen pressure applied to the pneumatic valves. As figure 5.12 shows, the flow velocity of the beads at a 100 Hz pumping rate increases steadily with increasing nitrogen pressure and then falls drastically to zero at 100 kPa. This is caused by the incomplete opening of the valves with too high of an actuation pressure. Thus high actuation pressure may adversely affect the peristaltic pump. Devices from different RTV batches were found to have slightly different optimum pressure, but in most devices, 60 kPa is the optimum pressure for peristaltic pumping. Vacuum pressure is constantly applied at the normally open ports for faster opening of the valves. In addition, the rigid polystyrene beads eventually get stuck on the valve membrane. These beads affect the performance of the peristaltic pumps and cause incomplete closing of the valves. GFP *E. coli* cells were used instead for the following measurements.

The linear flow velocity of the cell sorter using peristaltic pumps was also measured at various pumping frequencies. *Escherichia coli* cells expressing enhanced GFP (EGFP) were pumped through the sorter. A PMT was used to observe the fluorescence from each cell as it passed through the detection region. The detection region was near the T junction at the 6-μm-wide channel region. The dimensions of the detection region were 32 μm long along the fluidic channel and 20 μm wide. The width of the pulse from each cell detected by the PMT was the time it took for each cell to travel through a distance of 32 μm near the T junction. The widths of these pulses were averaged from about 150 cells. The velocity was calculated from dividing the length of the detection region by the average pulse width.

Cells through the cell sorters attained a maximum linear velocity at a pumping rate of 50 Hz (figure 5.13). Above this frequency, incomplete opening and closing of the

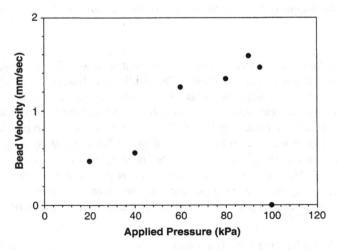

Figure 5.12. A plot of linear flow velocities versus applied pressure from the integrated cell sorter. The velocity of the beads increased as the pressure applied to the peristaltic pumping increased. Beads stopped flowing at 100 kPa because of the incomplete opening of the valves at such high pressure.

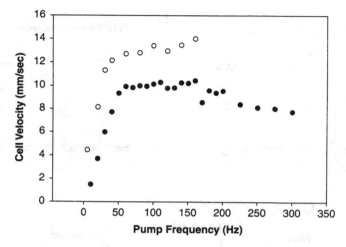

Figure 5.13. A linear flow velocity profile from the integrated cell sorter. The velocity of the cells increased as the frequency of peristaltic pumping increased, reaching to a maximum value at a certain frequency. (•) are the flow velocities recorded using an GE RTV 615-cell sorter with 30% SF-96, whereas (○) are the flow velocities from a Sylgard 184-cell sorter. Nitrogen pressure applied to the peristaltic pumps was 60 kPa at each frequency. Each value of the flow velocities is the mean velocity of measurements taken from 150 cells. Some data points were taken on separate days from the same devices.

valves occur with each pumping cycle. In two separate devices, one made of GE RTV 615 and another made of Dow Corning Sylgard 184, the values of maximum flow velocity are different. The RTV cell sorter has a maximum flow velocity of 10 mm/s, whereas the Sylgard cell sorter has a maximum of 14 mm/s. From several measurements done on different cell sorters (Fu, A.Y. and Quake, S.R., unpublished data), different maximum flow velocities were observed ranging from 6 mm/s to 17.5 mm/s. This indicates that the maximum flow velocities of the cell sorters may be tuned by altering the properties of the elastomer. Adding diluents or mixing different ratios of A and B component of the fluidic layer should allow us to fine-tune the stiffness of the valve membrane, which will affect the minimum closing pressure and the maximum pumping frequency. Changing the dimensions of the fluidic channel will also allow us to tune the flow velocity because different volumes will be moved with each actuation.

A novel method of interrogation was also demonstrated that consists of trapping a single cell within a region of detection. Analogous to reverse sorting for the electrokinetic μFACS, we devised an algorithm so that each time the sorter detects a fluorescent cell, it will reverse the direction of peristaltic pumping. Eventually, the cell falls out of the trapping region and flows into the output wells. Figure 5.14 shows the raw data recorded by the oscilloscope on the pattern of cell trapping. At 10 Hz pumping frequency, a single cell was redirected into the detection region more than 10 times before it fell out of the trap. At a higher pumping frequency of 75 Hz, multiple cell trapping instances were recorded. Measurements of the mean reverse time were taken for each pumping frequency. In figure 5.15, the mean reverse time was taken to be the

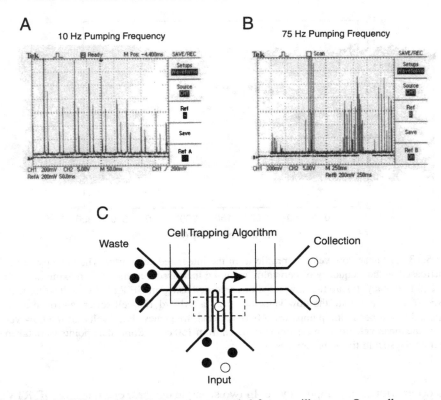

Figure 5.14. Cell trapping. (*A*) Raw data recorded from oscilloscope. One cell was trapped within the region of detection by reversing its direction each time it was detected. The difference in the periodicity between each detection may be a result of the variation in the distances the cell traveled away from the detection region before it reverted back. The frequency of the peristaltic pumps was at 10 Hz, and the nitrogen pressure applied to the valves was at 60 kPa. (*B*) Raw data recorded at 75-Hz pumping frequency. Multiple instances of cells trapping were recorded. (*C*) A schematic of the cell trapping algorithm. A cell can be trapped within the detection region by reversing the flow at each detection. Cell trapping scheme is as follows: A series of 0 and 1 patterns were used to digitally control individual valves on the chip, where 0 and 1 indicate "valve open" and "valve close," respectively. Forward peristalsis was actuated by the pattern 001, 011, 010, 110, 100, 101, whereas reverse peristalsis was actuated by the pattern 101, 100, 110, 010, 011, 001. These two peristalsis patterns alternate each time a cell's fluorescence reaches above a present threshold. Detection region is indicated by the dashed box.

time the cell travelled away from the detection region between the first detection and the second detection when the flow direction is reversed. This was measured as the time between the first and the second pulses read by the PMT. The mean reverse time gradually decreases as the pumping frequency increases. This is consistent with incomplete opening and closing of the valves, which will occur with an increasing pumping frequency.

This novel method of trapping cells and other bioparticles within a given region inside the sorter opens up new avenues to perform enzymatic kinetic studies on cells and

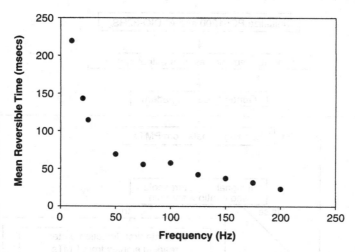

Figure 5.15. Mean reverse time as a function of the pumping frequency. This is the average time for the cells to flow back into the detection region after reversing the direction of the peristaltic pumping. The reverse time gradually decreases as the pumping frequency increases. These data were used as references to calculate the timing for the reverse sorting. The reverse peristaltic pumping pattern was the peristaltic pumping pattern in reverse exactly. Nitrogen pressure applied to the peristaltic pumps was at 60 kPa at each frequency.

beads. Multiple time-course measurements of the same cell can be taken to follow the kinetics of an enzymatic reaction. Sample dispensing can be done before or after the first interrogation and with each successive detection. Cell sorting can still occur after a certain number of detections. Sorted cells can also be redirected back into the sorting region to be sorted again, which allows new avenues for kinetic studies on a single-cell level that cannot be accomplished by commercially available conventional flow cytometers.

Sorting and Recovery

Sorting in this integrated cell sorter can be done in a variety of ways. Because each valve can be individually controlled in a coordinated and timely fashion, the pumping rate and the valve switching rate can be changed at any time in the course of a sorting event. However, to overcome the limitation of the switching speed, which is delayed by the intrinsic valve response time, a reverse sorting scheme was used to sort cells in this integrated pneumatic-driven sorter.

The reverse-sorting algorithm for the valves and pumps is as follows (see figure 5.16): Three patterns are generated for a sorting event: default pattern, reverse pattern, and recovery pattern. The default pattern pumps the cells into the waste channel at 100 Hz with 60 kPa valve pressure. Once a desired cell is detected, the reverse pumping pattern is generated at 10 Hz pumping frequency to bring the cells slowly back into the detection region to be detected once more. If there is no detection of any desired cell, the flow will reverse until the end of the reverse pattern and then generate the default pattern again. However, if the desired cell is detected again, the recovery pattern

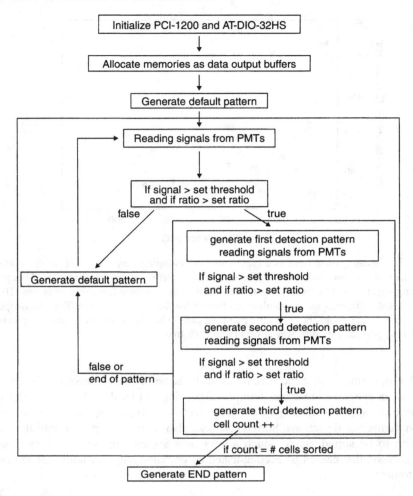

Figure 5.16. Sorting logic for reverse sorting. Reverse cell sorting scheme is as follows: With valve 4 as the collection valve, the default flow from the input to the waste was actuated by the pattern 01001, 01011, 01010, 01110, 01100, 01101 at 100 Hz. Once a target cell was detected, the reverse flow was actuated by the pattern 01101, 01100, 01110, 01101, 01011, 01001 at 10 Hz to slowly bring the cell back to the input channel. Once the cell passes the detection region again, all the valves were closed to stop the flow, 11111, and then the recovery of the cell was actuated by the pattern 10001, 10011, 10010, 10110, 10100, 10101 at 10 Hz to close the waste valve and direct the cell into the collection channel. The default flow was actuated again once the cell was in the collection channel. A pattern of 11111 was actuated at the end of the run to close all the valves for cell retrieval.

will be generated. This pattern will first close all the valves to stop any fluid flow. Then, with the waste valve remaining closed, the collection valve is opened and the pump sends the detected cell slowly into the opened collection channel, pumping at 10 Hz for a selected number of cycles. Following this pattern, the default pattern starts again until the next sorting event.

In a typical run, two populations of *E. coli* cells were separated, one expressing EGFP and the other expressing para-nitrobenzyl (pNB) esterase. The EGFP cells generate a fluorescence signal on the PMT, which triggers collection. The populations were mixed in a ratio of 1:2000 and introduced into the sorter first by nitrogen back pressure and then by peristaltic pumping when sorting began. After 3 hr, the cells at the collection and waste wells were retrieved using a pipetman into two microcentrifuge tubes. The contents of each microcentrifuge tube were then divided and spread out on two different antibiotic plates, ampicillin (amp) and tetracycline (tet). Because EGFP-expressing cells grow only on amp plates and the pNB esterase–expressing cells grow only on tet plates, the fraction of these two cells in each well can be easily counted on the different antibiotic plates. The plates were placed in a 37°C incubator overnight, and the colonies were counted. In the longest run, about 480,000 cells were sorted in 3 hr at a rate of 44 cells/s. The recovery yield is 40% in this run, and the enrichment ratio is about 83-fold. The enrichment ratio is calculated as the ratio of the fraction of EGFP in the collection well to the fraction of EGFP in the original mixture at the input well. Overall up to 50% recovery has been obtained, with as high as 90-fold enrichment. A throughput rate of 100,000 cells per hour has also been achieved.

The integrated sorter can run without interruption for 5 hr or more and can be used numerous times after proper cleaning. One device was used for 6 months continuously, with tens of millions of actuations on each valve and pump. Compared with our previous electrokinetic sorter (16), this integrated sorter alleviates many issues regarding buffer compatibility, surface chemistry, and cell viability. Different strains of *E. coli* cells and different types of bacteria, including magnetotactic bacteria, could be pumped in their own suspending media through the integrated cell sorter. By incorporating valves and pumps to control sorting by pneumatic actuation, the integrated sorter has a better capability of fine-tuning the flow and is less harmful to the cells than electrokinetic flow. Thus, although the throughput rate was increased only twofold, the recovery yield and the accuracy of sorting improved tremendously. Under the observation of the microscope, μFACS was able to capture most or all of the desired cells into the collection channel, even when they occurred in small numbers. Because we are only recovering most of the cells at the wells, we believe that we will be able to reach close to 100% recovery in future devices by incorporating a flushing mechanism to retrieve the remaining cells that are inside the collection channel.

The variability in the results of different runs can be attributed to factors such as initial cell concentration, fraction of EGFP-expressing bacteria at the input, time duration of each run, retrieval using pipetmen and device fabrication. Although conventional FACS machines still achieve higher throughput, recovery, and accuracy of sorting, the integrated μFACS serves as an alternative, inexpensive, robust, and easy way for sorting or manipulating single cells. Multiplexing the cell sorting channels can increase the overall throughput and allow for simultaneous measurements of cells in different compartments. Innovative sorting schemes can be implemented on the device to perform time-course measurements on a single cell for kinetic studies, which cannot be done by any conventional FACS. Sample dispensing and other chemical or enzymatic reactions, such as cell lysis or PCR, can also be carried out downstream immediately after the cell has been sorted (24,44). Moreover, a device of similar design has already been fabricated for sorting mammalian cells and other cell types by Fluidigm, Inc. (South San Franciso, CA).

Applications of μFACS to Digital Genetic Circuits

Genetic networks contain thousands of molecules interacting in various metabolic pathways for a cell to maintain proper metabolism. These networks consist of highly branched and interwoven "genetic circuits" that are analogous to complex, interconnected electrical circuits (31). Recent advances have led to the design of de novo genetic circuits inside *E. coli* cells to function as an oscillator (12) and a toggle switch (18). Moreover, when an autoregulatory, negative-feedback loop was added to a de novo genetic circuit, the noise or the variability in gene expression was dramatically reduced (2). These findings grant us not only a deeper insight into the regulatory mechanism of natural genetic networks but also a glimpse of how we can design de novo genetic circuits to program a cell or a group of cells to perform computational functions. Weiss et al. are laying down a general principle for designing genetic circuits that can implement the digital logic abstraction and thus is capable of programming cell behaviors that are complex, predictable, and reliable (45).

Digital logic, that is, physical chemical signals being translated to logical true (HIGH) or false (LOW) signals, can be engineered inside living cells using simple genetic elements, such as promotors, repressors, operator regions, and other DNA-binding proteins. Digital genetic circuits would allow us to reduce the inherent stochastic noise with high predictability and reliability. This also requires the presence of adequate noise margins; that is, the ability to produce a valid logical output signal from a physical input representation that is marginally valid or imperfect (23)—a signal restoration. These noise margins are critical for tolerating any noise or loss of signals within the circuit. By measuring the relation between the input and output signals in a steady state, that is, a transfer function, the gain and noise margins of a logic gate or a circuit can be calculated. The "forbidden zone," which corresponds to valid inputs but invalid outputs, can also be mapped. In any circuit design, the gain must be greater than one and be highly nonlinear.

Moreover, actual biological behaviors of genetic circuits display variations as a result of stochastic effects and other systematic fluctuations (2, 12, 45). These circuits are best characterized using flow cytometry to capture such inherent noise on the single cell level. The rest of this chapter presents the characterization of a simple genetic inverter, the $cI/\lambda_{P(R-O12)}$ genetic inverter (45). Using μFACS, we were able to measure its transfer function and fluctuations. A transfer band of the inverter, which encompasses actual biological fluctuations, was also plotted.

The $cI/\lambda_{P(R-O12)}$ Genetic Inverter

A simple inverter gives LOW output for HIGH input, and HIGH output for LOW input. The design of the $cI/\lambda_{P(R-O12)}$ inverter is described in the work of Dr. Ron Weiss (45). The $\lambda_{P(R-O12)}$ is a synthetic promoter that lacks the O_R3 operator of the wild-type λ promoter. Repressor dimers of cI, the λ repressor, bind to O_R1 and O_R2 almost simultaneously; this cooperative binding could lead to much higher gains from the drastic change of repression activity over a small range of repressor concentrations. This also would achieve a highly nonlinear behavior, as required in the circuit for a good digital performance. Two plasmids were eventually designed and modified to function as a genetic inverter in *E. coli* cells, pINV-112-R3 and pINV-107mut4. In this circuit, the repressor *cI* controls the output signal, the enhanced yellow fluorescent protein (EYFP), while it itself is controlled by another repressor, *lacI*. Under a constitutive promotor, *lacI* is always expressed, repressing the expression of *cI*. However, in the presence of an inducer, isopropyl-β-D-thiogalactoside (IPTG, which inhibits the binding of *lacI* to its operator region), *cI* is expressed with enhanced cyan fluorescent protein (ECFP) under the same promoter. This genetic inverter switches from "HIGH" (high EYFP fluorescence) to "LOW" (low EYFP fluorescence) output ranges, depending on the concentration of the inducer.

The transfer function of this logic gate, that is, the relation between the input and output signals, was then estimated by measuring several points on the curve. The mRNA level of the input protein, *cI*, represents a input signal in this circuit (45). By measuring the fluorescence intensities of ECFP in *E. coli* cells, the actual mRNA signal can be approximated (45). Similarly, the mRNA level of the output protein, EYFP, represents the output signal. This output signal level can also be approximated by measuring the fluorescence intensities of EYFP in cells. Because EYFP and ECFP are nearly identical in their decay rates and can have equivalent translation rates, the relative levels of these proteins can be normalized in terms of their fluorescence and protein numbers. The relationship between the ECFP and EYFP fluorescence intensities was approximated by comparing the fluorescence values of pINV-1022 and pINV-112-R3 cell populations induced at 1000 μM IPTG as they flowed through μFACS. The "absolute" values of their fluorescence intensities were obtained from their mean fluorescence val-

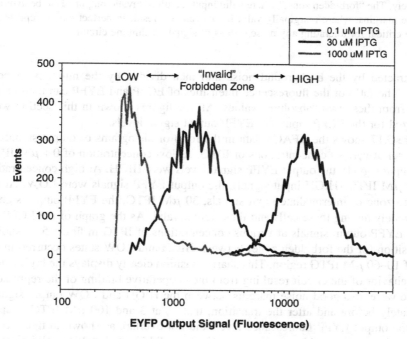

Figure 5.17. Histograms of EYFP output signals. At 1000 μM IPTG, the output of the inverter exhibited LOW EYFP fluorescence. At 0.1 μM IPTG, the output of the inverter exhibited HIGH EYFP fluorescence. At the forbidden zone of 30 μM IPTG, the EYFP output fluorescence signals were invalid.

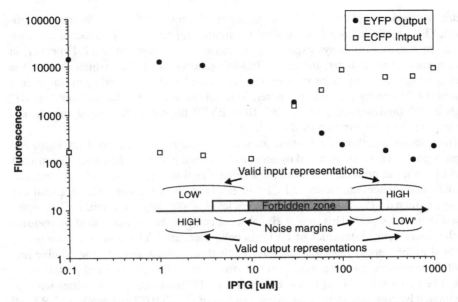

Figure 5.18. ECFP input (squares) and EYFP output (dots) signals at different IPTG concentrations. The forbidden zone, or the sharp transition, is mapped out in the range of 10–100 μM IPTG. The noise margins are immediately before or after the transition, at 3 and 100 μM IPTG, respectively. The "forbidden zone," where valid inputs result in invalid outputs, is to be avoided. The noise margins, where marginally valid inputs can still result in perfect output representations, are critical for tolerating any noise or loss of signals within the circuit.

ues subtracted by the background noise and then divided by the numbers of cells scanned. The ratio of the fluorescence intensities of ECFP and EYFP can then be estimated from these two "absolute" values. All the figures shown in this section were normalized for the ECFP input and EYFP output signal levels.

Figure 5.17 shows the μFACS data in the form of histograms of output signals of the inverter at various concentrations of IPTG. At low concentration of 0.1 μM IPTG (LOW input signal), the output EYFP signal levels were HIGH. At high concentration of 1000 μM IPTG (HIGH input signal), the output EYFP signals were LOW. At the forbidden zone of intermediate input signals, 30 μM IPTG, the EYFP output signals varied widely among these cells and thus were invalid. As the graph of the ECFP input and EYFP output signals at various concentrations of IPTG in figure 5.18 shows, the transition (or the forbidden zone) between HIGH and LOW states occurred in the range of 10–60 μM IPTG region. This sharp transition clearly displays the highly nonlinear behavior of the circuit resulting from the cooperative binding of the repressors.

There were also good noise margins between the HIGH and LOW output signals. Immediately before and after the transition, that is, at 3 and 100 μM IPTG, respectively, the output EYFP fluorescence levels did not overlap, as shown in figure 5.19. Thus, this inverter was able to restore marginally valid input signals to valid physical logic representations of HIGH or LOW output signals. This is critical for tolerating noise or loss of signals within the circuit. If some of the input signals are lost because of diffusion or interference with the host mechanism, these reduced input signals can

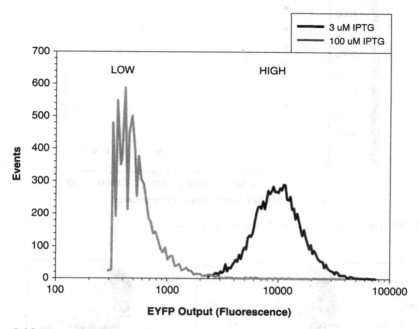

Figure 5.19. The noise margins of the $cI/\lambda_{P(R-O12)}$ inverter. Immediately before and after the transition, at 3 and 100 μM IPTG, respectively, the inverter can still output valid HIGH or LOW output signals from the marginally valid input signals.

still be valid and be restored through digital abstraction. By normalizing their fluorescence intensities, a transfer curve of the ECFP input signals and EYFP output signals is plotted in figure 5.20. These results demonstrate that this genetic inverter does exhibit fairly high gain and good noise margins for digital logic computation.

A Transfer Band

The transfer curve in Figure 5.20 is plotted using the means of the fluorescence intensities of the cell populations induced under different IPTG concentrations. This curve does not, however, describe the fluctuations that occur within biological systems. These fluctuations can be inherent because of stochastic effects, dead or damaged cells, and size distribution in addition to the systematic variations from μFACS. Scatter plots of cell populations at different IPTG concentrations of the $cI/\lambda_{P(R-O12)}$ inverter are shown in figure 5.21. These plots reflect the distributions of the cells at different input levels. An alternative way introduced by Weiss (45) to describe the transfer function of a genetic logic gate, the *transfer band*, is intended to capture such noise in signal levels within biological systems. This band maps out a region, which includes the maximum and minimum values of input and output signal levels, that encompasses these fluctuations. The transfer band of the $cI/\lambda_{P(R-O12)}$ inverter, shown in figure 5.22, is enclosed by a pair of transfer functions: one maps to the values of the mean plus one standard deviation of the input and output signals at each IPTG concentration, and the other maps to the values of the mean minus one standard deviation of the input and output

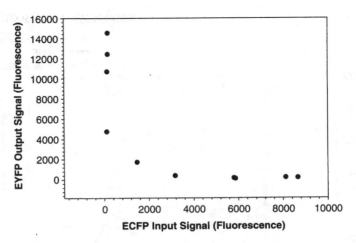

Figure 5.20. The transfer function of the $cI/\lambda_{P(R-O12)}$ inverter.

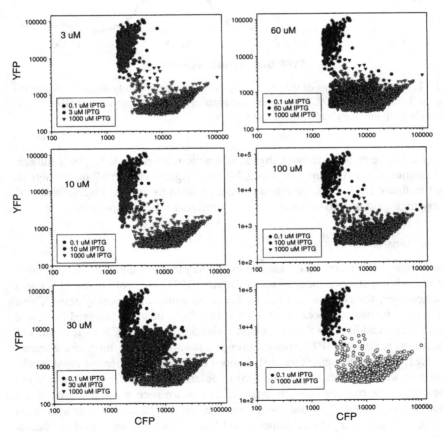

Figure 5.21. Scatter plots of cell populations at different IPTG concentrations of the $cI/\lambda_{P(R-O12)}$ inverter. Approximately 10,000 cells were measured in each concentration. The populations at 0.1 and 1000 μM concentrations are shown on every plot for comparison of HIGH and LOW signal ranges.

signals. This band thus represents about 70% of all the cell populations, as it includes all the cells that are within one standard deviation away from the mean value.

Artificial genetic networks will one day enable us to program living cells to perform computational functions and thus behave predictably and reliably. An important aspect of designing complex genetic circuits is having a resource of gates with different DNA-binding proteins and their respective binding regions. Although a number of natural DNA-binding proteins and their reaction kinetics are known, modifications of these proteins and their binding regions are pivotal to correctly program cell behaviors (12, 45). Novel genetic or protein engineering techniques could be used to fine-tune the existing genetic logic gates and to create novel genetic regulatory elements to construct faster and more complex circuits (23). Moreover, the transfer curve only describes the static behaviors of the circuit, not the dynamic behaviors. Hence, in the future, μFACS could be designed to take time-course measurements such that the dynamic behavior of the circuit can also be determined.

Future Prospects and Conclusion

Unlike the conventional aerosol flow cytometers, μFACS allows for flexibility in designing different ways of sample dispensing and methods of interrogation. This will enable single-cell studies to be achieved in a more precise and automated fashion. The

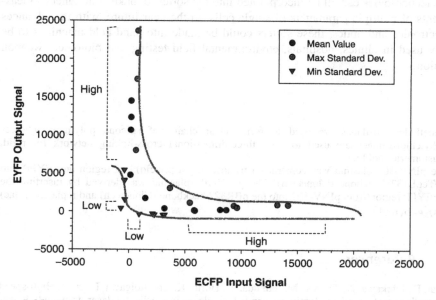

Figure 5.22. The transfer band of the $cI/\lambda_{P(R-O12)}$ inverter is plotted using two transfer functions. One transfer function (shaded dots) is plotted using the values of the mean plus one standard deviation of the input and output signals at each IPTG concentrations. The other transfer function (triangles) is plotted using the values of the mean minus one standard deviation at different IPTG concentrations. Some of the minimum values are in the negative range. The mean values of the transfer function are shown (•). The ranges of the HIGH and LOW inputs and outputs are also shown. This transfer band represents about 70% of the cells.

sorter is presently being used as a stand-alone device for molecular evolution and other biological applications. More fluidic functionalities, such as rotary pumps and incubation, can be integrated in the future. The electrokinetic μFACS allows for an easy fabrication process and disposal. The control of electrokinetic flow within these chips can be used for sensitive DNA and peptide studies in the fields of genomics and proteomics (8). In contrast, the integrated μFACS, which has a more complicated fabrication process, offers many versatile ways for sample dispensing, cell manipulation, and accommodation for different cell types.

A working μFACS system can be assembled for approximately $15,000. Most of this amount represents the cost of the external optics and detectors used to read out the chip, as the cost of the chip itself is negligible. Considerable cost savings can potentially be realized by fabricating the detectors and optical filters directly on the chip. We believe that this will be an important component of future integrated biomedical chip-based systems.

Future developments of the sorter may evolve in various directions. In terms of optics, the sorter can include other optical signals such as light scatter and absorbance measurements. There are also ways for the sorter to measure magnetic and electrical signals. We have demonstrated the feasibility of the sorter to detect and sort magnetotactic bacterial cells (17). Others also have demonstrated cytometric capacitance measurements on cells in microfluidic devices (40). In an exploratory experiment, reverse micelles containing E. coli cells were sorted on a similar μFACS setup (42). These technologies can all be incorporated into the sorter to make simultaneous measurements of multiple parameters of single cells. In the near future, with the advances of electronics and optics, these sorters could be made into hand-held machines to be readily used in clinics, hospitals, environmental field testing and biological weapon detection.

Notes

1. Dimethylchlorosilanes are used to form linear chains of silicone polymers, whereas trimethylchlorosilanes are used to form three-dimensional cross-linking network for rigid, nonelastomeric molding.
2. The pINV-102 plasmid was constructed by inserting a p(lacIq):lacI region and EYFP into pPROTet.E132 backbone (Clonetech). The pINV-107 plasmid was derived by inserting the p(lac):EYFP region from pINV-102 into the pBR322 backbone (Clonetech) and replacing p(lac) with $\lambda_{P(R-O12)}$ (45).

References

1. Arai, F., Ichikawa, A., Ogawa, M., Fukuda, T., Horio, K., and Itoigawa, K. 2001. High-speed separation system of randomly suspended single living cells by laser trap and dielectrophoresis. *Electrophoresis*, 22(2):283–288.
2. Becskei, A. and Serrano, L. 2000. Engineering stability in gene networks by autoregulation. *Nature*, 405(6786):590–593.
3. Bockris, J. O'M. and Reddy, A. K. N. *Modern Electrochemistry: An Introduction to an Interdisciplinary Area*. Plenum, 1977.
4. Brody, J. P. and Yager, P. Low reynolds number micro-fluidic devices. In *Proceedings of Solid-State Sensor and Actuator Workshop*, pages 105–108. Hilton Head, SC, June 1996.

5. Campbell, D. J., Beckman, K. J., Calderson, C. E., Doolan, P. W., Moore, R. H., Ellis, A. B., and Lisensky, G. C. 1999. Replication and compression of bulk surface structures with polydimethylsiloxane elastomer. *Journal of Chemical Education*, 76:537.

6. Carlson, R. H., Gabel, C. V., Chan, S. S., Austin, R. H., Brody, J. P., and Winkelman, J. W. 1997. Self-sorting of white blood cells in a lattice. *Physical Review Letters*, 79(11): 2149–2152.

7. Chee, M., Yang, R., Hubbell, E., Berno, A., Huang, X. C., Stern, D., Winkler, J., Lockhart, D. J., Morris, M. S., and Fodor, S. P. A. 1996. Accessing genetic information with high-density DNA arrays. *Science*, 274(5287):610–614.

8. Chou, H. P., Spence, C., Scherer, A., and Quake, S. January 1999. A microfabricated device for sizing and sorting DNA molecules. *PNAS*, 96(1):11–13.

9. Chou, H. P., Unger, M. A., and Quake, S.R. Nov 2001. A microfabricated rotary pump. *Biomedical Microdevices*, 3(4):323–330.

10. Chou, H. P., Unger, M. A., Scherer, A., and Quake, S.R. Integrated elastomer fluidic lab-on-a-chip—surface patterning and DNA diagnostics. In *Proceedings of Solid-State Sensor and Actuator Workshop*. Hilton Head, SC, June 2000.

11. Chou, Hou-Pu. *Microfabricated Devices for Rapid DNA Diagnostics*. Ph.D. Thesis, California Institute of Technology, 2000.

12. Elowitz, M. B. and Leibler, S. 2000. A synthetic oscillatory network of transcriptional regulators. *Nature*, 403(6767):335–338.

13. Feng, H. P. 2000. A protein microarray. *Nature Structural Biology*, 7(10):829–829.

14. Fiedler, S., Shirley, S. G., Schnelle, T., and Fuhr, G. 1998. Dielectrophoretic sorting of particles and cells in a microsystem. *Analytical Chemistry*, 70(9):1909–1915.

15. Fu, A. Y., Chou, H. P., Spence, C., Arnold, F. H., and Quake, S. R. 2002. An integrated microfabricated cell sorter. *Analytical Chemistry*, 74(11):2451–2457.

16. Fu, A. Y., Spence, C., Scherer, A., Arnold, F. H., and Quake, S. R. November 1999. A microfabricated fluorescence-activated cell sorter. *Nature Biotechnol.*, 17(11):1109–1111.

17. Fu, Anne Y. *Microfabricated Fluorescence-Activated Cell Sorters for Screening Bacterial Cells*. Ph.D. Thesis, California Institute of Technology, 2002.

18. Gardner, T. S., Cantor, C. R., and Collins, J. J. 2000. Construction of a genetic toggle switch in *Escherichia coli*. *Nature*, 403(6767):339–342.

19. Harrison, D. J., Fluri, K., Seiler, K., Fan, Z. H., Effenhauser, C. S., and Manz, A. 1993. Micromachining a miniaturized capillary electrophoresis-based chemical-analysis system on a chip. *Science*, 261(5123):895–897.

20. Huang, R. P. 2001. Detection of multiple proteins in an antibody-based protein microarray system. *Journal of Immunological Methods*, 255(1-2):1–13.

21. Jacobson, S. C., Hergenroder, R., Koutny, L. B., and Ramsey, J. M. 1994. High-speed separations on a microchip. *Analytical Chemistry*, 66(7):1114–1118.

22. Katayama, H., Ishihama, Y., and Asakawa, N. 1998. Stable capillary coating with successive multiple ionic polymer layers. *Analytical Chemistry*, 70(11):2254–2260.

23. Knight, T. F. and Sussman, G. J. 1997. Cellular gate technology. Pages 257–272 in *Unconventional Models of Computation*, eds. C. Calude, J.L. Casti, and M.J. Dinneen. Springer-Verlag Telos, Berlin.

24. Kopp, M. U., de Mello, A. J., and Manz, A. 1998. Chemical amplification: continuous-flow PCR on a chip. *Science*, 280(5366):1046–1048. 52.

25. Lee, C. L., Linton, J., Soughayer, J. S., Sims, C. E., and Allbritton, N. L. 1999. Localized measurement of kinase activation in oocytes of *Xenopus laevis*. *Nature Biotechnology*, 17(8):759–762.

26. Library, General Electric Technical. Product technology for silicone heat cured elastomers. Available at: http://www.gesilicones.com/silicones/americas/business/portfolio/h%ce/workshops/paperproducttech.shtml, Copyright 2001.

27. Lin, Y. C. and Huang, M. Y. 2001. Electroporation microchips for *in vitro* gene transfection. *Journal of Micromechanics and Microengineering*, 11(5):542–547.

28. Liu, J., Enzelberger, M., and Quake, S. 2002. A nanoliter rotary device for polymerase chain reaction. *Electrophoresis*, 23(10):1531–1536.

29. Lockhart, D. J., Dong, H. L., Byrne, M. C., Follettie, M. T., Gallo, M. V., Chee, M. S.,

Mittmann, M., Wang, C. W., Kobayashi, M., Horton, H., and Brown, E. L. 1996. Expression monitoring by hybridization to high-density oligonucleotide arrays. *Nature Biotechnology*, 14(13):1675–1680.

30. Martin, B. D., Gaber, B. P., Patterson, C. H., and Turner, D. C. 1998. Direct protein microarray fabrication using a hydrogel "stamper." *Langmuir*, 14(15):3971–3975.

31. McAdams, H. H. and Shapiro, L. 1995. Circuit simulation of genetic networks. *Science*, 269(5224):650–656.

32. McDonald, J. C., Duffy, D. C., Anderson, J. R., Chiu, D. T., Wu, H., Schueller, O. J. A., and Whitesides, G. M. 2000. Fabrication of microfluidic systems in poly(dimethylsiloxane). *Electrophoresis*, 21(1):27–40.

33. Meredith, G.D., Sims, C.E., Soughayer, J.S., and Allbritton, N.L. 2000. Measurement of kinase activation in single mammalian cells. *Nature Biotechnology*, 18(3):309–312.

34. Morgan, H., Green, N. G., Hughes, M. P., Monaghan, W., and Tan, T. C. 1997. Large-area travelling-wave dielectrophoresis particle separator. *Journal of Micromechanics and Microengineering*, 7(2):65–70.

35. Ostuni, E., Chen, C. S., Ingber, D. E., and Whitesides, G. M. 2001. Selective deposition of proteins and cells in arrays of microwells. *Langmuir*, 17(9):2828–2834.

36. Pethig, R. 1996. Delectrophoresis: using inhormogeneous AC electrical fields to separate and manipulate cells. *Critical Reviews in Biochemistry and Molecular Biology*, 16(4):331–348.

37. Quake, S. R. and Scherer, A. 2000. From micro- to nanofabrication with soft materials. *Science*, 290(5496):1536–1540.

38. Schasfoort, R. B. M., Schlautmann, S., Hendrikse, J., and van den Berg, A. 1999. Field-effect flow control for microfabricated fluidic networks. *Science*, 286(5441):942–945.

39. Schena, M., Shalon, D., Davis, R. W., and Brown, P. O. 1995. Quantitative monitoring of gene-expression patterns with a complementary–DNA microarray. *Science*, 270(5235):467–470.

40. Sohn, L. L., Saleh, O. A., Facer, G. R., Beavis, A. J., Allan, R. S., and Notterman, D. A. 2000. Capacitance cytometry: measuring biological cells one by one. *Proceedings of the National Academy of Sciences of the United States of America*, 97(20):10687–10690.

41. Soughayer, J. S., Krasieva, T., Jacobson, S. C., Ramsey, J. M., Tromberg, B. J., and Allbritton, N. L. 2000. Characterization of 54 cellular optoporation with distance. *Analytical Chemistry*, 72(6):1342–1347.

42. Thorsen, T., Roberts, R. W., Arnold, F. H., and Quake, S. R. 2001. Dynamic pattern formation in a vesicle-generating microfluidic device. *Physical Review Letters*, 86(18):4163–4166.

43. Unger, M. A., Chou, H. P., Thorsen, T., Scherer, A., and Quake, S. R. 2000. Microfabricated valves and pumps using multilayer soft lithography. *Science*, 288(5463):113–116.

44. Waters, L. C., Jacobson, S. C., Kroutchinina, N., Khandurina, J., Foote, R. S., and Ramsey, J. M. 1998. Microchip devices for cell lysis, multiplex PCR amplification, and electrophoretic sizing. *Analytical Chemistry*, 70(1):158–162.

45. Weiss, Ron. *Cellular Computation and Communications using Engineered Genetic Regulatory Networks*. Ph.D. Thesis, Massachusetts Institute of Technology, 2001.

46. Whitesides, G. M., Ostuni, E., Takayama, S., Jiang, X. Y., and Ingber, D. E. 2001. Soft lithography in biology and biochemistry. *Annual Review of Biomedical Engineering*, 3:335–373.

47. Xia, Y. N. and Whitesides, G. M. 1998. Soft lithography. *Angewandte Chemie–International Edition*, 37(5):551–575.

48. Ziaudinn, J. and Sabatini, D. M. 2001. Microarrays of cells expression defined cDNAs. *Nature*, 411:107–110.

6

Flow Cytometry, Beads, and Microchannels

TIONE BURANDA, LARRY A. SKLAR, AND GABRIEL P. LOPEZ

Introduction

Microfluidic devices generally consume microliter to submicroliter volumes of sample and are thus well suited for use when the required reagents are scarce or expensive. Because microfluidic devices operate in a regime in which small Reynolds numbers (50) govern the delivery of fluid samples, reagent mixing and subsequent reactivity has been a severe limiting factor in their applicability. The inclusion of packed beads in the microfluidic device repertoire has several advantages: molecular assemblies for the assay are created outside the channel on beads and characterized with flow cytometry, uniform populations of beads may be assured through rapid cytometric sorting, beads present a larger surface area for the display of receptors than flat surfaces, rapid mixing in the microcolumn is achieved because the distance that must be covered by diffusion is limited to the (≤ 1-μm) interstitial space between the closely packed receptor-bearing beads, and analytes are captured in a flow-through format and, as such, each bead can act as a local concentrator of analytes (29, 46).

Progress in the combined use of beads and microfluidic devices has been limited by the ability to pack beads at specific sections of microfluidic devices. A subsequent challenge associated with the packed microcolumns of beads is the substantial pressure drop that affects the fluid flow velocity. However, some of these challenges have been overcome in the design of simple model systems that have potential applications in DNA analysis (21, 36), chromatography (46), and immunoassays (2, 3, 8). It is the intent of this chapter to examine the recent emergence of small-volume heterogeneous immunoassays, using beads trapped in microchannels, while excluding other closely related applications such as capillary electrophoresis (20, 22, 25, 64, 67–69, 71) and flow injection–based approaches (54, 55). Of necessity, the authors' interests and availability of information pertinent to the specific discussions presented below impose additional restrictions.

Overview

To date, there are only a handful of applications that combine packed beads and microfluidic devices, and even fewer that make the overt connection between flow cytometry–based assays and beads. Harrison and coworkers (46) have provided one of the earliest conceptual demonstrations of the capability to incorporate packed beads in microfluidic devices for analytical purposes. In this effort, these authors fabricated a chip-based chromatographic device that functions on the principles of solid-phase extraction and capillary electrochromatography (17, 23, 52, 64, 70). The device consists of a chromatographic bed fabricated on a glass substrate, into which beads are packed with the aid of electro-osmotic pumping through connecting channels. The authors give an account of the packing and unpacking of beads into the chamber, in appreciation of problems often encountered in the fabrication and bead packing of similar devices by other authors. The device application is demonstrated by the solid-phase extraction procedure of a polar (fluorescein) and nonpolar (BODIPY) dye.

Much recent effort has been focused on the design and fabrication of micromachined chambers equipped with devices to hold the beads in place. Several options have been presented, where the common designs have comprised dams (46, 57), filter pillars (2, 3, 8, 37), and magnets (for use with magnetic beads; 21, 36). Recently, a technique involving microcontact printing and self-assembly on silicon or quartz substrates has been demonstrated (1). In this method, the internal walls of the channel are functionalized by microcontact printing, using an appropriate chemistry for the functional group of choice (e.g., biotin). When a chip thus treated is incubated in a suspension of streptavidin-coated beads, the resulting self-assembly of beads is confined to the areas covered with biotin. Because of the strength of the biotin–streptavidin interaction, the beads are held in position during the subsequent passage of pumped fluid through the channel. Related approaches have been described for cellular systems (16).

Implementation

Sample Delivery in Microchannels

A limiting factor in the utility and operation of microfluidic systems is the work that must be done against the loss in pressure associated with mass transport. In packed columns, the friction between the fluid and the high–surface area porous bed of beads is manifested as a pressure drop (ΔP) across the microfluidic channel (47, 59, 60). A simple approach to estimate the pressure drop in a packed bed of spherical particles is the Ergun equation, in which G is the superficial mass velocity, L is the length of the bead bed, μ is viscosity, L is the void fraction, and D_p is the particle diameter (24).

$$\frac{\Delta P}{L} = \frac{G}{\rho D_p} \frac{(1 - \varepsilon)}{\varepsilon^3} \left[\frac{150(1 - \varepsilon)\mu}{D_p} + 1.75G \right] \qquad (1)$$

Table 6.1. Estimated pressure drops through an empty and bead-packed microchannel

Volumetric Flow Rate (μL/min)	Velocity (cm/s)	ΔP in Empty Channel (Torr)[a]	ΔP in Packed Bead Bed (Torr)[b]
10.0	1.3	13.9	483
50.0	6.7	69.3	2414
100.0	13.3	138.5	4828
150.0	20.0	207.8	7241
200.0	26.7	277.0	9655
250.0	33.3	346.3	12069

Notes. Estimates of pressure drops (ΔP) are based on affinity microcolumn described in Buranda et al. (8). The typical channel dimensions used are length = 3.0 cm, width = 250 μm, height = 50 μm. The channel equivalent diameter is 83.3 μm; diameter of beads (D_p) = 6.2 μm, void fraction of bead section $\approx$0.50. The Reynolds number (Re) for these channels is on the order of 0.1–0.01 for the given fluid velocities: Re = $d_p U \rho / \mu$ where d_p is the diameter of the bead (6.2 μm); U is the velocity of the fluid through the microchannel $\approx$1 cms^{-1}; ρ is the density ($\approx$1 g/cm^3), and μ is the kinematic viscosity of the buffer ($\approx$10^{-2} cm^{2}s^{-1}).
[a]The ΔP is estimated using the following formalism: $\Delta P = \rho 2fLv_b^2/(g_cD_{eq})$ where P = pressure, ρ = fluid density, f = friction factor (= 16/Re), L = length of bed, v_s = superficial velocity, g_c = gravitational constant, and D_{eq} = equivalent diameter of channel.
[b]$\Delta P = [150 \mu Lv_s (1 - \varepsilon)^2]/[g_c D_p^2 \varepsilon^3]$ where: P = pressure, μ = viscosity, L = length of bed, v_s = superficial velocity, ε = void fraction, g_c = gravitational constant, and D_p = particle diameter.

From equation 1, it can be clearly surmised that the size of the beads and length of the column play an essential role in regulating the magnitude of ΔP. In table 6.1, we have modeled (at various flow rates) the level of pressure losses sustained by the microcolumns that we have used in the past (8). As indicated in the fourth column, the level of pressure loss can be substantial and can place mass transport (velocity and quantity) limitations on the way such a device is operated. The derived data do not include other factors that may or may not significantly affect the overall pressure drop across a channel. These factors can include rough surfaces in the channel, variations in channel height and width throughout the length, and sudden changes in diameter, such as the change from tubing to channel. Pressure loss measurements can be performed using an elementary setup such as that shown in figure 6.1. Such measurements are useful in defining the operational limits of such a device (3). We have performed some basic ΔP determinations in the microchannels fabricated from poly(dimethylsiloxane; PDMS; 8, 10). In that study, we found a reasonable correlation between experimental and empirically determined ΔP values in empty channels. However, in bead packed channels we observed increasingly large deviations between the smaller, experimental and larger, theoretical values with increasing volumetric flow rates. The smaller, experimental values of ΔP were likely a result of the forced expansion of the microchannel as the soft polymer yielded to increasing pump pressure.

Design and Fabrication of Microchannels

Microfluidic channels that incorporate beads are typically fabricated, with internal features created to act as filters to hold the beads in place. The usual fabrication methods have included photolithography (19, 37, 51, 65) deep-reactive-ion etching, and multi-wafer bonding (1–3, 39, 40, 46). Consideration of Equation 6.1 and table 6.1 indicates that a short, larger-diameter microchannel and the use of larger-diameter beads could

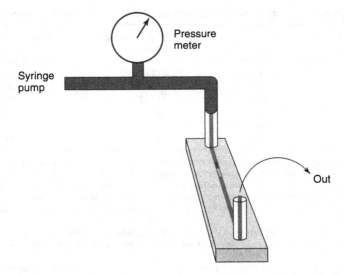

Figure 6.1. Basic measurement of pressure loss across a microfluidic channel. A syringe pump is used to push water through a microchannel. In our experiments, the "pressure meter" consisted of a vertical, open-ended tube connected through a "T" fitting between pump and poly(dimethylsiloxane) microchannel. At steady state, the height of the water was read (as cm of H_2O) in the vertical tube. The exit tube from the microchannel was open to atmospheric pressure. The height of the water is recorded versus the volumetric flow rate.

minimize the pressure drop. Such considerations are compatible with those used to design microfabricated packed bed reactors (40) and micromachined filter-chamber arrays (2, 3). Within the microchannel, weirs, or filter pillars, are patterned to hold beads. Because of limitations in resolution of photolithographic or micromachining tools, most common applications have features that are resolved down to 10-μm distances. This limits the size of beads that can be held by such filters to dimensions that are greater than 10 μm in diameter.

Packing of Microchannels with Beads

The packing of particles in channels is normally done with the aid of electrosmotic pumping (15, 46), syringe, or peristaltic pumps (8). Most surface-functionalized beads from commercial sources that are suitable for use in flow cytometry applications are typically 5–10 μm in diameter. As a consequence, such beads are smaller than the "mesh" size of the intrachannel filters. One way to get around this problem is to pack an inert layer of larger beads (e.g., 10–30 μm, depending on the size of filter mesh) to hold the smaller functionalized beads (8). Technical challenges often associated with the packing of beads into microchannels are the resistance to fluid flow during packing and uneven packing, which often produces observable voids within the packed column (46). Electrokinetic packing appears to provide the most facile way to pack naked beads because of exposed negative charges on silanol groups on beads (7, 46). Furthermore, this application can be enhanced by the use of organic solvents such as acetonitrile. This approach is of limited applicability in immunoassays in which the use of protein-coated beads and protein-coated channel walls results in the shielding of electrostatic charge in addition to the incompatibility of organic solvents and proteins.

Microcontact Printing Methods

A technique that enables the selective trapping of beads in microfluidic devices without the use of physical barriers has been recently presented (1). This approach involves microcontact printing and self-assembly and can be applied to silicon, quartz, or plastic substrates. A chip printed with functional groups of interest is submerged in a bead slurry in which the beads self-assemble and immobilize in the microchannel based on the surface chemistry on the internal walls of the channels. In one application, silicon channels were covered with monolayers of streptavidin-, amino-, and hydroxy-functionalized microspheres and resulted in good surface coverage of beads on the channel walls. High-resolution patterns of lines of self-assembled streptavidin beads, as narrow as 5 μm, were generated on the bottom of a 500-μm-wide and 50-μm-deep channel. The patterned bead assemblies are reported to have withstood the forces generated by flow through fluid in the channels. This application is suitable for applications in which monolayers of beads are desired over a three-dimensional bed. This approach provides for the precise definition and control of bead monolayers, mitigates the problem of pressure loss, facilitates the control of the space between beads.

Microimmunoassays

Immunoassays derived from the molecular recognition interactions of biological molecules have been the mainstay of many clinical, biochemical, and environmental research efforts (15, 27, 31, 32, 48, 56). Much of the work has been carried out using conventional titer plate assays or flow cytometry. Here, we will describe the recent emergence of flow-based microimmunoassays performed on beads in microchannels (8, 31, 56, 57). The design and optimal use of these microanalytical devices is governed by several universal considerations (31): first, the presence of beads in microchannels provides a high ratio of surface area to volume. As a result, the analyte diffusion distances are reduced in magnitude to those equivalent to the interstitial space between the beads. Second, the site coverage of the beads can be optimized for an assay outside the channel, with potential for prior quantitation by flow cytometry or spectrofluorometry (8, 11, 12). Third, the total number of binding sites within the detection scheme should be optimized to match the needs of detection limits and sensitivity (45). Fourth, the volumetric flow rates must be limited by the kinetic restrictions of the immunoreaction. Fifth, for closely packed beds, the flow rate may typically be limiting because of pressure loss, depending on the size of beads and length of bed as described here.

In some designs, the problem of limiting pressure losses has been circumvented by the trapping (in a weir) of a few relatively large beads (45 μm; 56, 57) or by using an orthogonal magnetic field to trap magnetic beads on one side of a channel (without total occlusion) and allowing the fluid to flow through at rates modulated to account for the diffusive interactions of beads and soluble proteins (31). Some schematic examples of recently developed flow-through microimmunoassay formats are shown in figure 6.2. The design

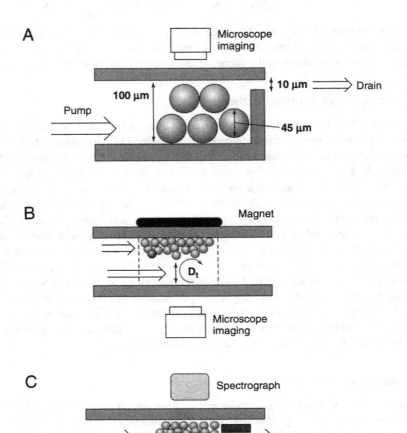

Figure 6.2. Schematic representations of flow-based microimmunoassay systems. (*A*) Antigen-bearing beads are sequestered in a dam inside a microchannel [adapted from (57) with permission from the American Chemical Society]. Subsequently, colloidal gold-tagged antibodies are captured on the beads in flow-through format. After a wash step, the antigen-antibody complex is detected by thermal lens microscopy. (*B*) Paramagnetic particles held in place within a channel by a magnet [adapted from (31) with permission from the American Chemical Society]. Reagents are introduced into the packed bed to perform standard immunoassays (either heterogeneous or sandwich assay), and the bed is imaged with an epifluorescence microscope with laser-induced excitation. The term D_t gives the time for radial diffusion of analyte to the surface of the bed. (*C*) Schematic of a microfluidic apparatus showing the configuration in which sample was delivered and fluorescence measurements taken with a spectrofluorimeter (8). Elastomeric silicone microchannel, mounted on glass slide with two openings for sample delivery and egress. Patterned features 20 μm apart act as filters for holding 30-μm borosilicate beads. Thirty thousand 6.2-μm streptavidin coated beads form a ≈600-μm-long affinity microcolumn.

and functional aspects of the schemes shown in figure 6.2A and 6.2B have more features in common than the approach shown in figure 2C. The first two approaches are minimally affected by pressure loss because of the use of beads as few as 11 in number (in the case of figure 6.2A) or a packed bed that does not occlude the channel (figure 6.2B). This has several implications for fluid dynamics, reagent interaction, and detection with the beads.

First, in figure 6.2B, most of the volume flow passes through the open section of . the channel, where the reagents interact with the bed by diffusion. Second, it is likely that most of the bead–analyte interactions involve those beads openly exposed to the flowing analyte fluid, with rapidly diminishing gradients of captured analyte on the inner layer beads. Third, the bed dwell time of the fluid should be much greater than the radial diffusion time (D_t) of the analytes. Fourth, because the detection is based on microscopy, only a section of the packed bed can be probed at a time, and thus the quantitation of analytes is limited in accuracy, as gradients are not well accounted for. Fifth, the ease with which beads can be dynamically packed in columns makes these approaches very amenable to rapid reuse of channels.

The scheme shown in figure 6.2C embodies several distinguishing characteristics in reagent interaction and detection that will be further developed here. This approach reproduces the essential elements of affinity chromatography with the advantage of direct and real-time analysis and miniaturization. Accurate quantitation is achieved by the use of known quantities of beads that have been previously characterized by flow cytometry and spectrofluorimetry to provide a known surface coverage of detection proteins. The detection relies on fluorescence resonance energy transfer. The transfer arises from the interaction between soluble analytes labeled with fluorescent acceptors and surface-bound fluorescent donors. Because the fluorescence of the beads corresponds to a known concentration of surface donors, the subsequent fluorescence changes define the amount of captured analytes without signal interference from unbound analytes (8, 11, 12). These assays could potentially be carried out in a few minutes, supplanting the need for time-consuming steady-state end-point assays. Simultaneous detection of a diverse group of analytes can be achieved by packing discrete segments of detector bearing beads in a single-affinity microsystem (49, 58). The biggest liability in the current design of the approach represented in figure 6.2C is the substantial pressure loss that occurs during fluid transport. As a result, the maximum length of a column is thus limited.

Analytical Characteristics of the Flow-Through Microimmunoassays

The flow-through microimmunoaffinity systems are governed by the same basic principles. However, important differences in design emphasis are presented where such variations are manifested in the performance characteristics of the devices. Table 6.2 summarizes the respective figures of merit of the approaches described in figure 6.2. Because of the nascent nature of these approaches, the necessary scaling laws (i.e., functional relationships that express the dependence of the design parameters on the operating conditions) have not yet been fully developed to optimize their universal working characteristics. The crucial factor from which the microsystems derive their significant operational advantage over the conventional titer plate assay is derived from the "specific interface," a parameter recently coined by Sato et al. (57) Specific interface (S/R) is defined as the ratio between substrate surface area and the fluid volume in the microchannel space occupied by the beads. The density of available binding sites

Table 6.2. Characteristics of recently developed flow-through microimmunoassays

Figures of Merit	Conventional[a]	Tokyo[b]	Arizona State[c]	New Mexico[d]
Dynamic range (orders of magnitude)	2	2	2	4
Reaction time	24 hrs	10 min	3 min	dynamic: sec-min.
Specific interface[1]	13 cm^{-1i}	487 cm^{-1ii}	1.3×10^4 cm^{-1iii}	1.3×10^6 cm^{-1iv}
Detection limits	1 ng/mL	≤ng/mL	≤ng/mL	≤ng/mL

Notes: Comparison is made between (a) conventional ELISA and approaches recently developed at (b) the Universiy of Tokyo (Sato et al.) (57), (c) Arizona State University (Hayes et al. (31), and (d) University of New Mexico (Buranda et al.) (8). Specific interface is defined as the surface area of the solid surface to solution volume ratio (S/V) (i) as defined by Sato et al. (ii) determined for 11; 45-μm-diameter beads in 100 × 100 × 200 μm channel space; (iii) as estimated by Hayes et al. (iv) determined for 30,000; 6.2-μm-diameter beads in 250 × 50 × 600 μm channel space; 4 nL fluid volume.

in the detection volume is referred to as the "reaction field" (57). It is useful to note that for the Tokyo (figure 6.2A) and Arizona State (figure 6.2B) methodologies, the detection area is smaller than the total bead-bed volume. Thus the definition of "reaction field" is different from the specific interface, and a relatively larger reaction field is inferred to mediate shorter reaction times. This argument has been used to rationalize the significantly longer reaction time (~24 hr) associated with conventional titer plate assays relative to the flow-through microimmunoassays (minutes). For a 6.5-mm-diameter titer plate containing analyte fluid (1.5 mm deep), the average time for an antibody to diffuse across the longest distance d (i.e., the depth of the analyte fluid) is $t = d^2/2D \approx 32$ hours, where D is the diffusion coefficient of the antibody molecule ($D \approx 10^{-7}$ cm^{2}s^{-1}). Conversely, in the packed columns, the magnitude of d is on the order of microns or less depending on the closeness in packing of the beads. There is a clear inverse correlation between the magnitude of the reaction time and the specific interface. It is important to note that in the cases discussed here, the flow rate, rather than D, limits the reaction times because of the micron-to-submicron magnitudes of the analyte diffusion distances present in the closely packed bead column.

Design, Fabrication, and Operation of a Flow-Through Microimmunoassay

In previous sections, we have given a broad overview of the current state of the field and various challenges and obstacles that must be overcome before the full potential of the microimmunoassays can be realized. For the reader wishing a more detailed approach on the practical considerations of design, microfabrication, and operation of a rudimentary flow-through microimmunoassay, we will develop a "how to" protocol, based on our research efforts.

Photolithography is the most common approach used to fabricate microchannels containing functional microstructures (weirs, filters, etc.) In this type of fabrication, the microchannel and its internal features are generated by a series of planar pattern-transfer steps. Most of our work in this area is based on the photolithographic methods developed by Whitesides and coworkers (16, 18, 19, 37, 38) for use with the elastomeric polymer, PDMS.

A prototypical microfluidic channel (as shown in figure 6.3) is 3 cm long, with typical dimensions of 250 by 50 μm in breadth and depth, patterned into a PDMS elas-

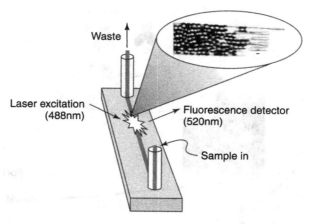

Figure 6.3. Schematic of a microfluidic apparatus showing the configuration for sample delivery and sample analysis by spectrofluorometry. Elastomeric silicone microchannel, mounted on glass slide with two openings for sample delivery and egress. The microchannel is 250 μm wide, 50 μm deep, and 3 cm long. Patterned features (inset) 20 μm apart act as filters for holding 30-μm borosilicate beads. Thirty thousand 6.2-μm streptavidin coated beads form a $\approx$600-μm-long affinity microcolumn [adapted from (8) with permission from the American Chemical Society).

tomer adhered to a glass slide support. Within the microchannel, obstructive features, 20 μm apart, are patterned as filters to hold 30-μm beads. A plug of cotton can be used as an alternative filter to hold the beads.

Packing Microchannels with Beads

We used streptavidin-coated beads displaying known quantities (8, 12) of either fluorescein biotin or a fluorescent peptide bearing the FLAG epitope (DYKDDDDK). In a typical experiment, a 30-μL aliquot suspension of 30-μm beads ($\approx$10^6 beads/mL) was injected into the channel with a Hamilton syringe. A peristaltic pump connected to the ("waste" port in figure 6.2) channel was used to remove the supernatant. A foundation layer of 30-μm beads was used to hold subsequent layers of smaller beads. The contents were allowed to settle during the peristaltic pump-assisted elution of several microliters of buffer. To minimize the dispersion of the 6.2-μm streptavidin-coated beads into the foundation layer of 30-μm beads, a layer of 10-μm glass beads was added to the microchannel as described above. Subsequently, 6.2-μm ligand-bearing polystyrene beads ($\sim$1.0 $\times$ 10^6 beads/mL, or $\sim$30,000 beads in the microchannel) were added to the microchannel in Tris buffer (pH = 7.5) containing 0.1% bovine serum albumin. Several microliters of the buffer were eluted through the column coating of the PDMS microchannel with bovine serum albumin. The coating minimized the potential for nonspecific adsorption of protein and peptides to the walls of the microchannels in subsequent assays.

Once packed, the column was ready to use. Two-microliter sample aliquots were used in a typical run analyzing for either a small molecule represented by biotin or a large molecule represented by Texas Red–tagged anti-FLAG antibodies. The time course of the interaction between the analyte solutions and beads was monitored on a Model Fluorolog-3 SPEX fluorimeter (Instruments S.A., Edison NJ) using 488-nm laser

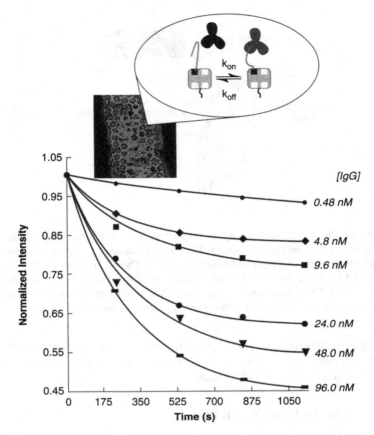

Figure 6.4. Binding curves of 2-μL plugs of Texas Red–labeled monoclonal anti-FLAG anti-bodies (IgG) through affinity microcolumns of fluorescein-labeled FLAG peptide-bearing beads. The points refer to the normalized intensity readings taken during each run. The lines represent simulations using parameters derived from fits of surface coverage data (cf. figure 6.5) [adapted from (8) with permission from the American Chemical Society].

excitation focused onto the 0.6-mm-long portion of the column containing the fluo-rescein-bearing beads. The typical intensity readouts are shown in figure 6.4, repre-senting the analysis of the binding of Texas Red–labeled anti-FLAG antibodies to FLAG peptide-bearing beads.

Real-Time Detection and Analysis of Analyte Capture in Affinity Microcolumns

General Considerations

In the existing format, the beads bear fluorescent ligands/receptors of known surface oc-cupancy. Thus, the subsequent changes from the initial intensity reading bear definite and known relationships to the amount of captured analytes without contribution from un-bound species and need for wash steps. From a mechanistic approach, the temporal and spatial destiny of target analytes, traversing through the affinity microcolumn, might be described in terms of convective and diffusive transport and reactive (binding and dissociation) processes (28, 33, 66). An additional point of detail in these analyses includes the generation of analyte concentration gradients (14, 35) in the transport fluid as well as those bound on beads. The potential complexity of an analysis that must account for such gradients can be circumvented by the irradiation of the beads with a laser whose spot size is larger than the reaction field. Thus, the passage of the antibodies through the column is analyzed as the change in intensity wholly integrated over the reaction field. In this approach, the analysis can be reduced to a model based on Langmuirian kinetics (4, 6). This model reproduces the basic elements of mass transport–dependent heterogeneous kinetics and provides sufficient account of the passage through different-affinity microcolumns of various concentrations of antibodies. The results are given in terms of kinetic binding and dissociation rate constants, where the derived equilibrium dissociation constant (K_d) has been found to be in good agreement with steady-state flow cytometry determinations (8).

Analysis at the Steady State

The intent of this section is to provide a detailed sketch of the analysis of the time-resolved changes in the fluorescence intensities of the reaction field with the passage of Texas Red–labeled antibodies. To perform a Langmuirian kinetic analysis, it is necessary to express the intensity data in terms of surface coverage (θ; 4, 6). A cursory examination of the normalized intensities (I_i) of the reaction field after the passage of antibody aliquots of various concentrations indicates that saturation coverage (i.e., $\theta <$ 1) is not achieved even with 96.0-nM concentration of the antibody because the data are still not close to an asymptotic limit. The values of θ can be determined from equation 2 under the condition that I_f (the intensity of the reaction field at saturation) is known either from experiment [using much higher concentrations of antibody (IgG)] or as a derived parameter from the law of mass action (equation 3).

$$\theta_i = \frac{(1 - I_i)}{(1 - I_f)} \tag{2}$$

$$(1 - I_i) = \frac{(1 - I_f)[IgG]_i}{K_d + [IgG]_i} \tag{3}$$

Figure 6.5A shows the least-squares fit to equation 3 of the normalized intensity of the reaction field versus [IgG] (table 6.3), with the result that $I_f = 0.38$ and $K_d \approx 14.0$ nM. It is important to note that the values of I_f and K_d derived from this fit are in good agreement with those obtained when surface coverage-saturating concentrations of IgG were used in flow cytometry (12) as well as affinity microcolumn experiments (8).

Kinetic Analysis

In figure 6.5B, the fluorescence resonance energy transfer binding data as shown in figure 6.4 are then expressed in terms surface coverage (θ), using equation 2 and subsequently fit to a kinetic model shown in equation 4 (6).

$$\frac{d\Gamma_{AB}}{dt} = k_f C_0 \Gamma_A - k_b \Gamma_{AB} = \frac{D}{\delta}(C^b - C_0) \tag{4}$$

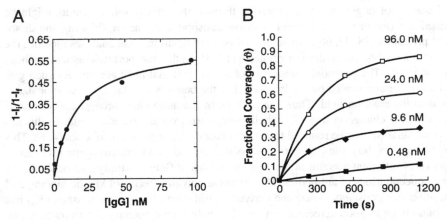

Figure 6.5. (A) Equilibrium binding isotherm for the interaction between Texas Red–labeled IgG and beads in the microcolumn ($K_d = 14.0$ nM). I_i refers to the intensity of the reaction field after flow through of the ith sample of IgG in microcolumn i, and I_f represents the intensity of the beads after the passage of a surface saturating concentration of the IgG. I_i and I_f are normalized with respect to the intensity of unquenched beads. (B) Kinetic analysis of surface coverage derived from fluorescence resonance energy transfer data (figure 6.4) using Equations 6.2–6.4. The lines represent least-squares fits to the data resulting, in the determination of the average association and dissociation rate constants: $k_f = (9.0 \pm 6.0) \times 10^4 \mathrm{M}^{-1}\mathrm{s}^{-1}$, $k_b = (1.2 \pm 0.8) \times 10^{-3}\ \mathrm{s}^{-1}$).

The variables C^b and C_0 represent the concentrations of antibody in the bulk and at the liquid–solid interface, respectively; Γ_{AB} is the surface concentration of FLAG peptides bound to antibodies; Γ_A is the surface concentration of unbound peptides; and k_f and k_b are the forward and reverse kinetic rate constants. The term D is the diffusion coefficient of the antibody, and δ is the thickness of the steady-state diffusion-convection boundary layer established by fluid transport in which we assumed a linear gradient in concentrations (between C^b and C_0). The parameter D/δ represents the effects of diffusive transport of analytes to the surface receptors. The integral form of this equation is shown in equation 5 in the dimensionless formalism given in the original reference (6):

$$\lambda\theta - \left(1 + \frac{\lambda}{1+\kappa}\right)\ln\left(1 - \frac{1+\kappa}{\kappa}\theta\right) = \frac{1+\kappa}{\kappa}\lambda\tau \qquad (5)$$

In this equation, $\theta = \Gamma_{AB}/\Gamma^s$ (where $\Gamma^S = \Gamma_{AB} + \Gamma_A$) represents the total surface concentration of FLAG peptides. The adsorption constant is $k = (k_f/k_b)C^b$. The diffusion

Table 6.3. Steady-state surface coverage data (θ) after passage of aliquots of antiFLAG monoclonal antibodies through affinity microcolumns

[IgG]$_i$ nM	0.48	4.8	9.6	24.0	48.0	96.0
I_i	0.927629	0.831504	0.766903	0.616634	0.545731	0.452819
θ_i	0.115966	0.269997	0.373514	0.614305	0.727919	0.876801

dependent rate of adsorption is $\lambda = k_f \delta \Gamma^s / D$. The dimensionless time normalized to diffusion time is $\tau = t/t_d$ (where the time that characterizes the diffusion process is $t_d = \delta \Gamma^s / D C^b$). For a 6.2-$\mu$m-diameter bead with 10^6 receptors per bead, $\Gamma_s = 1.38 \times 10^{-10}$ mol/dm^2. It is worth emphasizing that the surface coverage units must be expressed in moles per square decimeter for dimensional analysis considerations in equation 5.

The fits to the experimental data yield the following parameter values: $K_d = 13.3 \pm 2.0$ nM, $k_f = (9.0 \pm 6.0) \times 10^4$ M^{-1} s^{-1}, $k_b = (1.2 \pm 0.8) \times 10^{-3}$ s^{-1}, $D/\delta = (1.0 \pm 0.9) \times 10^{-9}$ dm s^{-1}, where the errors are the standard deviations for the constants determined from the fits of each experimental run. The derived binding and dissociation rate constants are in agreement with data reported in the literature on similar antibody–antigen interactions (13, 66). This approach is further validated by the conservation of microscopic reversibility by the close correlation of the affinity constant derived from kinetic data ($K_d = k_b/k_f = 13.3$ nM) and that derived from steady state analysis and flow cytometry (12). Although the dissociation constants derived from the microchannel data favorably compare to the flow cytometry determination (≈ 4.0 nM), it is worth mentioning that affinity microcolumn binding occurs under nonequilibrium conditions (28, 34). Thus, the apparently higher K_d value might be an artifact of mass transport limitations.

Concerning Transport Limitations

The rate (r) of bimolecular binding between the antibodies and the beads can be expressed as

$$r = k_f[\text{IgG}] + k_b \tag{6}$$

Thus, the time to half saturation is

$$t_{1/2} = \frac{\ln 2}{r} \tag{7}$$

An analysis of the data shown in figure 6.4, using our derived values of k_f and k_b and [IgG] = 96 nM obtains a $t_{1/2}$ value of 70 s. This value is in contrast to the experimental value, which is on the order of 200 s. This discrepancy is the result of mass transport limitations. The Damköhler number (Da) is a dimensionless quantity that is often used to represent the ratio between the rate of a reaction and the rate of delivery of reactants to the reaction interface (24):

$$Da = \frac{k_f[\text{IgG}]d}{(D/\delta)} \tag{8}$$

where the terms as used in our formalism are as follows: d is the interstitial space between the packed beads, and D/δ is the rate of convective transport of antibodies, [IgG], to the reaction interface, as has been previously defined. The Damköhler number is useful in that it allows a quick estimate of the degree of (reactant to product) conversion that can be achieved in flow-through systems. If $Da > 1$, then the process is mass-transport limited, and when $Da < 1$, the processes is reaction-rate limited. This approach can be applied to our data as follows: For the concentration range used (0.4–96.0 nM), the values of Da range from 0.48 to 96, using our experimentally derived values for k_f, [IgG], D/δ and an upper-limit estimate of 1×10^{-5} *dm* for *d*. This analysis indicates that increasing the flow rate would potentially reduce the reaction time of all but the lowest concentration of IgG used in this analysis. In the case of the 0.48-nM assay (*Da* < 1), it is clear that the binding rate (equation 6) is dominated by the dissociation rate constant k_b ($t_{1/2} \approx 552$ s). Thus, in the kinetic limited regime, increasing the flow rate rapidly approaches diminishing returns with decreasing concentration of the soluble analyte.

Multianalyte Approach

The concept described above can be expanded to a multianalyte model system comprising discrete segments of beads that bear distinct receptors for the simultaneous detection of diverse analytes. Pressure loss is mitigated by the use of larger beads (20 μm). Segments of inert spacer beads or filter pillars are used to separate adjacent sections of analyte beads. When used in the same detection format as the single-analyte scheme (figure 6.3), the microchannel can be mounted on a vertical translation stage, so that each analyte bead segment can be sequentially interrogated by the same excitation source (49). A more versatile approach could use separate excitation (e.g., using laser diodes) and detection (e.g., using optical fibers) dedicated to each analyte segment (figure 6.6). In another variation, a branched system of parallel channels can be used in a flow-through device. Sato et al. (58) have recently presented a multiple-sample bead-bed immunoassay system that is capable of simultaneously assaying up to four samples (in 50 min) on branching channels while driven by a single pump. The detection is based on thermal lensing microscopy analyzing for gold-coated antibodies.

A good example of the potential utility of the multianalyte approach is beautifully demonstrated in a microcapillary-based system. Recently, Phillips (48) has devised a

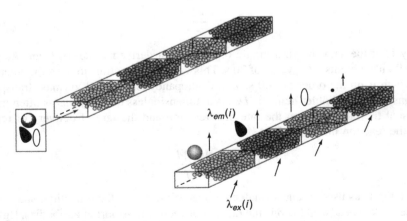

Figure 6.6. Model multianalyte detection array from a segmented affinity microcolumn comprising beads bearing receptors for different analytes. Each array may be associated with differently tagged receptors to be interrogated at given excitation wavelengths; $\lambda_{ex}(i)$ and $\lambda_{em}(i)$, respectively. This approach can be expanded to include parallel microfluidic networks, with individual sample delivery ports or a single one with several downstream branches [cf. Sato et al. (58)].

regenerable system for isolating and measuring up to 30 analytes in a single biological sample. The system is based on tandem-serpentine arrays of microcapillary columns packed with 10-μm glass beads, coated with different antibodies. The device was used to capture and analyze fluorescently labeled cytokines and chemokines in serum samples from patients with inflammatory allergic disease and autoimmunity and from patients on chemotherapy. A single analysis involves the serpentine-like passage of a 10-μL sample of biological fluid through the columns where immobilized antibodies (residing within each different column) can capture specific analytes. The column is then flushed with buffer to remove nonspecifically bound analytes before laser irradiation of the reaction fields for the analysis. The data collection and analysis has the potential to be fully automated. Data collection and analysis takes about 50 min with the detection limits being in the picogram/milliliter range, depending on the analyte.

Future Directions

Microfluidic devices fabricated out of PDMS are limited to pressures in the nominal range of less than an atmosphere. This failing limits the applicability of PDMS as a practical substrate in the fabrication of devices that are required to withstand pressures of several atmospheres, such as those that are associated with packed columns. Hasselbrink et al. (30) have recently developed a nonstick polymer formulation for creating moving parts inside microfluidic channels. This device can operate at pressures in excess of 297 atmospheres. Literature on other fabrication approaches is also available (5, 7, 26, 37, 40, 53). Once fabricated, a flow-through immunoassay device can be configured to analyze assays of choice. Several chapters in this book (chapters 9, 14, and 17) describe bead-based approaches that can be coupled to this format. Other literature on bead-based assays, which are amenable for use in microfluidic flow-through format, is also available (9, 41–44, 61–63). An optimally efficient device is likely to emerge from design criteria inherent in the examples given here (8, 31, 49, 56, 58) or described in the literature (5, 53).

Acknowledgments   This work was supported by grants from the National Science Foundation (National Science Foundation (MCB-9907611), National Institutes of Health Bioengineering Consortium (GM60799/EB00264), Office of Naval Research Grant N00014-95-1-1315, National Institutes of Health RR-01315 to the National Flow Cytometry Resource, New Mexico State Cigarette tax to the University of New Mexico Cancer Center, Air Force Office of Scientific Research (F49620-01-1-0168), and the Department of Energy through the US/Mexico Materials Corridor Initiative.

References

1. Andersson, H., Jonsson, C., Moberg, C., Stemme, G. 2001. Patterned self-assembled beads in silicon channels. *Electrophoresis* 22:3876–3882.
2. Andersson, H., van der Wijngaart, W., Stemme, G. 2001. Micromachined filter-chamber array with passive valves for biochemical assays on beads. *Electrophoresis* 22:249–257.
3. Andersson, H., van der Wijngaart W., Enoksson, P., Stemme, G. 2000. Micromachined flow-through filter-chamber for chemical reactions on beads. *Sensors Actuators B-Chem.* 67:203–208.

4. Andrieux, C.P., Saveant, J.M. 1986. In: *In Investigation of Rates and Mechanisms of Reactions Part II*, ed. C.F. Bernasconi, pp. 305–390. Wiley, New York.
5. Auroux, P.A., Iossifidis, D., Reyes, D.R., Manz, A. 2002. Micro total analysis systems. 2. Analytical standard operations and applications. *Anal. Chem.* 74:2637–2652.
6. Bourdillon, C., Demaille, C., Moiroux, J., Saveant, J.M. 1999. Activation and diffusion in the kinetics of adsorption and molecular recognition on surfaces. Enzyme-amplified electrochemical approach to biorecognition dynamics illustrated by the binding of antibodies to immobilized antigens. *J. Am. Chem. Soc.* 121:2401–2408.
7. Bruin, G.J.M. 2000. Recent developments in electrokinetically driven analysis on microfabricated devices. *Electrophoresis* 21:3931–3951.
8. Buranda, T., Huang, J., Perez-Luna, V.H., Schreyer, B., Sklar, L.A., Lopez, G.P. 2002. Biomolecular recognition on well-characterized beads packed in microfluidic channels. *Anal. Chem.* 74:1149–1156.
9. Buranda, T., Huang, J., Ramarao, G.v., Ista, L.K., Larson, R.S., Ward, T.L., Sklar, L.A., Lopez, G.P. 2003. Biomimetic molecular assemblies on glass and mesoporous silica microbeads for biotechnology. *Langmuir* 19:1654–1663.submitted.
10. Buranda, T., Huang, J., Schreyer, B., Sklar, LA.., Lopez, G.P. 2002, unpublished data.
11. Buranda, T., Jones, G., Nolan, J., Keij, J., Lopez, G.P., Sklar, L.A. 1999. Ligand receptor dynamics at streptavidin coated particle surfaces: a flow cytometric and spectrofluorimetric study. *J. Phys. Chem. B.* 103:3399–3410.
12. Buranda, T., Lopez, G.P., Simons, P., Pastuszyn, A., Sklar, L.A. 2001. Detection of epitope-tagged proteins in flow cytometry: fluorescence resonance energy transfer-based assays on beads with femtomole resolution. *Anal. Biochem.* 298:151–162.
13. Butler, J.E. 1992. In: *Structure of Antigens*, ed. M.H.V. Van Regenmortel. Boca Raton, FL: CRC.
14. Cabrera, C.R., Finlayson, B., Yager, P. 2001. Formation of natural pH gradients in a microfluidic device under flow conditions: model and experimental validation. *Anal. Chem.* 73:658–666.
15. Cheng, S.B., Skinner, C.D., Taylor, J., Attiya, S., Lee, W.E., et al. 2001. Development of a multichannel microfluidic analysis system employing affinity capillary electrophoresis for immunoassay. *Anal. Chem.* 73:1472–1479.
16. Chiu, D.T., Jeon, N.L., Huang, S., Kane, R.S., Wargo, C.J., et al. 2000. Patterned deposition of cells and proteins onto surfaces by using three-dimensional microfluidic systems. *Proc. Natl. Acad. Sci. USA* 97:2408–2413.
17. Colon, L.A., Burgos, G., Maloney, T.D., Cintron, J.M., Rodriguez, R.L. 2000. Recent progress in capillary electrochromatography. *Electrophoresis* 21:3965–3993.
18. Duffy, D.C., McDonald, J.C., Schueller, O.J.A., Whitesides, G.M. 1998. Rapid prototyping of microfluidic systems in poly(dimethyl siloxane). *Anal. Chem.* 70:4974–4984.
19. Duffy, D.C., Schueller, O.J.A., Brittain, S.T., Whitesides, G.M. 1999. Rapid prototyping of microfluidic switches in poly(dimethyl siloxane) and their actuation by electro-osmotic flow. *J. Micromech. Microeng.* 9:211–217.
20. Ekins, R.P. 1998. Ligand assays: from electrophoresis to miniaturized microarrays. *Clin. Chem.* 44:20–59.
21. Fan, Z.H., Mangru, S., Granzow, R., Heaney, P., Ho, W., et al. 1999. Dynamic DNA hybridization on a chip using paramagnetic beads. *Anal. Chem.* 71:4851–4859.
22. Figeys, D., Pinto, D. 2001. Proteomics on a chip: promising developments. *Electrophoresis* 22:208–216.
23. Fintschenko, Y., Choi, W.Y., Ngola, S.M., Shepodd, T.J. 2001. Chip electrochromatography of polycyclic aromatic hydrocarbons on an acrylate-based UV-initiated porous polymer monolith. *Fresenius J. Anal. Chem.* 371:174–181.
24. Fogler, H.S. 1999. *Elements of Chemical Reaction Engineering*. Upper Saddle River, NJ: Prentice Hall, pp.706–725.
25. Gao, Q.F., Yeung, E.S. 2000. High-throughput detection of unknown mutations by using multiplexed capillary electrophoresis with poly(vinylpyrrolidone) solution. *Anal. Chem.* 72:2499–2506.
26. Guzman, N.A., Stubbs, R.J. 2001. The use of selective adsorbents in capillary elec-

trophoresis-mass spectrometry for analyte preconcentration and microreactions: a powerful three-dimensional tool for multiple chemical and biological applications. *Electrophoresis* 22:3602–3638.

27. Hage, D.S. 1997. Chromatographic approaches to immunoassays. *Journal of Clinical Ligand Assay* 20: 293.

28. Hage, D.S., Thomas, D.H., Chowdhuri, A.R., Clarke, W. 1999. Development of a theoretical model for chromatographic-based competitive binding immunoassays with simultaneous injection of sample and label. *Anal. Chem.* 71:2965–12975.

29. Hanninen, P., Soini, A., Meltola, N., Soini, J., Soukka, J., Soini, E. 2000. A new microvolume technique for bioaffinity assays using two-photo excitation. *Nat. Biotechnol.* 18:548–550.

30. Hasselbrink, E.F., Shepodd, T.J., Rehm, J.E. 2002. High-pressure microfluidic control in lab-on-a-chip devices using mobile polymer monoliths. *Anal. Chem.* 74:4913–4918.

31. Hayes, M.A., Polson, N.A., Phayre, A.N., Garcia, A.A. 2001. Flow-based microimmunoassay. *Anal. Chem.* 73:5896–5902.

32. Holt, D., Rabbany, S.Y., Kusterbeck, A.W., Ligler, F.S. 1999. Advances in flow displacement immunoassay design. *Rev Anal. Chem.* 18:107–132.

33. Holt, D.B., Kusterbeck, A.W., Ligler, F.S. 2000. Continuous flow displacement immunosensors: a computational study. *Anal. Biochem.* 287:234–242.

34. Howard, M.E., Holcombe, J.A. 2000. Model for nonequilibrium binding and affinity chromatography: characterization of 8-hydroxyquinoline immobilized on controlled pore glass using a flow injection system with a packed microcolumn. *Anal. Chem.* 72:3927.

35. Jeon, N.L., Dertinger, S.K.W., Chiu, D.T., Choi, I.S., Stroock, A.D., Whitesides, G.M. 2000. Generation of solution and surface gradients using microfluidic systems. *Langmuir* 16:8311–8316.

36. Jiang, G.F., Harrison, D.J. 2000. mRNA isolation in a microfluidic device for eventual integration of cDNA library construction. *Analyst* 125:2176–2179.

37. Kenis, P.J.A., Ismagilov, R.F., Takayama, S., Whitesides, G.M., Li, S.L., White, H.S. 2000. Fabrication inside microchannels using fluid flow. *Acc. Chem. Res.* 33:841–847.

38. Kim, E., Xia, Y., Whitesides, G. 1995. Polymer microstructures formed by molding in capillaries. *Nature* 376:581–584.

39. Li, J.J., Tremblay, T.L., Wang, C., Attiya, S., Harrison, D.J., Thibault, P. 2001. Integrated system for high-throughput protein identification using a microfabricated device coupled to capillary electrophoresis/nanoelectrospray mass spectrometry. *Proteomics* 1:975–986.

40. Losey, M.W., Schmidt, M.A., Jensen, K.F. 2001. Microfabricated multiphase packed-bed reactors: characterization of mass transfer and reactions. *Ind. Eng. Chem. Res.* 40:2555–2562.

41. Nolan, J.P., Buranda, T., Cai, H., Kommander, K., Lehnert, B., et al. 1998. Real-time analysis of molecular assembly by kinetic flow cytometry. *Proc. SPIE* 3256:114.

42. Nolan, J.P., Lauer, S., Prossnitz, E.R., Sklar, L.A. 1998. Flow cytometry: a useful tool for all phases of drug discovery. *Drug Discov. Today* 4:178

43. Nolan, J.P., Lauer, S., Prossnitz, E.R., Sklar, LA. 1999. Flow cytometry: a versatile tool for all phases of drug discovery. *Drug Discov. Today* 4:173–180.

44. Nolan, J.P., Sklar, L.A. 1998. The emergence of flow cytometry for sensitive, real-time measurements of molecular interactions. *Nat. Biotechnol.* 16:633–638.

45. Ohmura, N., Lackie, S.J., Saiki, H. 2001. An immunoassay for small analytes with theoretical detection limits. *Anal. Chem.* 73:3392–3399.

46. Oleschuk, R.D., Shultz Lockyear, L.L., Ning, Y.B., Harrison, D.J. 2000. Trapping of bead-based reagents within microfluidic systems: on-chip solid-phase extraction and electrochromatography. *Anal. Chem.* 72:585–590.

47. Pfund, D., Rector, D., Shekarriz, A., Popescu, A., Welty, J. 2000. Pressure drop measurements in a microchannel. *Aiche J.* 46:1496–1507.

48. Phillips, T.M. 2001. Multi-analyte analysis of biological fluids with a recycling immunoaffinity column array. *J. Biochem. Biophys. Methods* 49:253–262.

49. Piyasena, M., Buranda, T., Huang, J., Wu, Y., Sklar, L.A., Lopez, G.P. 2004. Near simultaneous and real time analysis of multiple analytes in affinity microcolumns. *Anal. Chem.* 76:6266–6273.

50. Purcell, E.M. 1977. Life at low Reynolds Number. *Am. J. Phys.* 45:3
51. Quake, S.R., Scherer, A. 2000. From micro- to nanofabrication with soft materials. *Science* 290:1536–1540.
52. Rathore, A.S., Wen, E., Horvath, C. 1999. Electrosmotic mobility and conductivity in columns for capillary electrochromatography. *Anal. Chem.* 71:2633–2641.
53. Reyes, D.R., Iossifidis, D., Auroux, P.A., Manz, A. 2002. Micro total analysis systems. 1. Introduction; theory; and technology. *Anal. Chem.* 74:2623–2636.
54. Ruzicka, J. 2000. Lab-on-valve: universal microflow analyzer based on sequential and bead injection. *Analyst* 125:1053–1060.
55. Ruzicka, J., Scampavia, L. 1999. From flow injection to bead injection. *Anal. Chem.* 71:A257–A263.
56. Sato, K., Tokeshi, M., Kimura, H., Kitamori, T. 2001. Determination of carcinoembryonic antigen in human sera by integrated bead bed immunoassay in a microchip for cancer diagnosis. *Anal. Chem.* 73:1213–1218.
57. Sato, K., Tokeshi, M., Odake, T., Kimura, H., Ooi, T., et al. 2000. Integration of an immunosorbent assay system: analysis of secretory human immunoglobulin A on polystyrene beads in a microchip. *Anal. Chem.* 72:1144–1147.
58. Sato, K., Yamanaka, M., Takahashi, H., Tokeshi, M., Kimura, H., Kitamori, T. 2002. Microchip-based immunoassay system with branching multichannels for simultaneous determination of interferon-gamma. *Electrophoresis* 23:734–739.
59. Seguin, D., Montillet, A., Comiti, J. 1998. Experimental characterisation of flow regimes in various porous media: I: Limit of laminar flow regime. *Chem. Eng. Sci.* 53:3751–37561.
60. Seguin, D., Montillet, A., Comiti, J., Huet, F. 1998. Experimental characterization of flow regimes in various porous media: II: Transition to turbulent regime. *Chem. Eng. Sci.* 53:3897–3909.
61. Sklar, L.A., Edwards, B.S., Graves, S.W., Nolan, J.P., Prossnitz, E.R. 2002. Flow cytometric analysis of ligand-receptor interactions and molecular assemblies. *Annu. Rev. Biophys. Biomol. Struct.* 31:97–119.
62. Sklar, L.A., Seamer, L.C., Kuckuck, F., Posner, R.G., Prossnitz, E., et al. 1998. Sample handling for kinetics and molecular assembly in flow cytometry. *Adv. Optical Biophys.* 3256:144.
63. Sklar, L.A., Vilven, J., Lynam, E., Neldon, D., Bennett, T.A., Prossnitz, E. 2000. Solubilization and display of G protein-coupled receptors on beads for real-time fluorescence and flow cytometric analysis. *Biotechniques* 28:976–80, 982–985.
64. Slentz, B.E., Penner, N.A., Lugowska, E., Regnier, F. 2001. Nanoliter capillary electrochromatography columns based on collocated monolithic support structures molded in poly(dimethyl siloxane). *Electrophoresis* 22:3736–3743.
65. Unger, M.A., Chou, H.P., Thorsen, T., Scherer, A., Quake, S.R. 2000. Monolithic microfabricated valves and pumps by multilayer soft lithography. *Science* 288:113–116.
66. Vijayendran, R.A., Ligler, F.S., Leckband, D.E. 1999. A computational reaction-diffusion model for the analysis of transport-limited kinetics. *Anal. Chem.* 71:5405–5412.
67. Wang, C., Oleschuk, R., Ouchen, F., Li, J.J., Thibault, P., Harrison, D.J. 2000. Integration of immobilized trypsin bead beds for protein digestion within a microfluidic chip incorporating capillary electrophoresis separations and an electrospray mass spectrometry interface. *Rapid Communications in Mass Spectrometry* 14:1377–1383.
68. Wang, J., Chatrathi, M.P., Mulchandani, A., Chen, W. 2001. Capillary electrophoresis microchips for separation and detection of organophosphate nerve agents. *Anal. Chem.* 73:1804–1808.
69. Wen, E., Asiaie, R., Horvath, C. 1999. Dynamics of capillary electrochromatography II. Comparison of column efficiency parameters in microscale high-performance liquid chromatography and capillary electrochromatography. *J. Chromatogr. A* 855:349–366.
70. Xiang, R., Horvath, C. 2002. Fundamentals of capillary electrochromatography: migration behavior of ionized sample components. *Anal. Chem.* 74:762–770.
71. Yang, J., Rose, S., Hage, D.S. 1996. Improved reproducibility in capillary electrophoresis through the use of mobility and migration time ratios. *J. Chromatogr. A* 735:209–220.

7

DNA Fragment Sizing by High-Sensitivity Flow Cytometry: Applications in Bacterial Identification

BABETTA L. MARRONE, ROBERT C. HABBERSETT,
JAMES H. JETT, RICHARD A. KELLER, XIAOMEI YAN,
AND THOMAS M. YOSHIDA

Introduction

High-sensitivity, single-molecule detection in flow is a paradigm that has been defined at Los Alamos over the last two decades. A recent focus has been on applications of single-molecule detection for DNA fragment sizing using a compact, low-power, high-sensitivity flow cytometer (HSFCM). There are three key aspects of our approach that distinguish it from conventional flow cytometry and yield the high level of sensitivity that we achieve: a detector with high photon-detection efficiency, a small probe volume to reduce background noise, and slow flow to provide extended analyte dwell time in the probe volume. An additional factor for applications in DNA fragment sizing is a DNA stain with significant fluorescence enhancement when bound to double-stranded DNA, and low background fluorescence in the unbound state. DNA fragment sizing by HSFCM has important applications in bacterial species and strain identification, where it can replace the cumbersome and time-consuming pulsed-field gel electrophoresis (PFGE) approach routinely used by public health labs for bacterial identification. The revolutionary capability to interrogate single DNA molecules, as well as potentially other submicron-sized biological particles, in a high-sensitivity flow cytometer will provide new scientific insights into cellular and molecular biology and introduce high-sensitivity flow cytometry to a wide variety of new applications in biotechnology.

Flow cytometry has enabled major advances in the biomedical sciences by providing rapid, quantitative, and sensitive multiparameter measurements of individual cells and subcellular particles such as chromosomes. This analysis of individual entities produces information on population heterogeneity that is not revealed in ensemble measurements and that allows more precise quantitation of distinct attributes than is possible when measurements are done in bulk. However, one limitation of conventional flow cytometry is the inability to measure submicron-sized particles or weakly fluorescent particles labeled with fewer than several hundred fluorophores, primarily as a result of insufficient detection sensitivity. A wide variety of important biological particles, molecules, and molecular assemblies fall into these categories.

There have been many reports of bacterial measurement and characterization by conventional flow cytometry, dating back to 1947 (8, 18). In 1979, Steen developed a microscope-based system specifically for applications in microbiology (20). Many bacteria are large enough to generate a light-scatter signal, which is useful for their detection (19). In addition, specific dyes have been developed to successfully measure bacterial function (13, 17, 18). Although current methods for bacterial detection and characterization by conventional flow cytometry are useful for applications in environmental microbiology as well as clinical diagnostics, strain-level discrimination is not possible. To identify a bacterial strain, one must analyze genetic elements, which typically requires lysis of the cell and restriction fragment analysis of isolated DNA.

In 1993, we reported the first measurements of individual DNA fragment size, using a photon-counting flow cytometer capable of single-molecule detection (6). Our approach is a direct outgrowth of the single-molecule detection flow cytometry developed at Los Alamos (14). This was followed by several demonstrations of bacterial species (8, 11) and strain (12) discrimination, using similar high-power systems equipped with an argon-ion laser. These successful demonstrations indicated that bacterial identification by high-sensitivity flow cytometry might have immediate practical uses in many areas in which biotechnology innovations are being applied, including public health, biodefense, and environmental microbiology. The potential applications motivated us to develop a compact, low-power system for high-sensitivity flow cytometry and to optimize bacterial sample preparation methods for rapid, sensitive, and specific bacterial strain identification.

Implementation

High-Sensitivity Flow Cytometer Instrument

Our current HSFCM uses a solid-state diode pumped Nd:YAG laser that operates at a wavelength of 532 nm with a total available power of 20 mW (8). As seen in figure 7.1, the optical system is arranged in a folded U shape, with a series of optical components placed in the beam path to the flow cell that attenuate, steer, and focus the laser beam.

The laser is mounted so that its primary polarization vector is orthogonal to the vertical flow axis of the apparatus, but the final polarization angle is set by a polarizing beam splitter placed after the half-wave plate, typically oriented perpendicular to the flow axis. An acromatic-doublet lens (5 cm f) focuses the laser beam onto the sample stream in the flow channel of a fused-silica cuvette. On the emission side, a single-element 0.68-NA aspheric lens collects fluorescence emission from the sample stream and focuses it onto the detector through a 565–605-nm bandpass filter. A second aspheric lens (0.55 NA) is placed directly in front of the detector to reduce the spot size of the fluorescence emission on the 150-μ-diameter active area of an avalanche photodiode detector (APD), which has a photon-detection efficiency greater than 65%. The sample stream diameter is 7–8 μm and the calculated probe volume is about 1 pL (6).

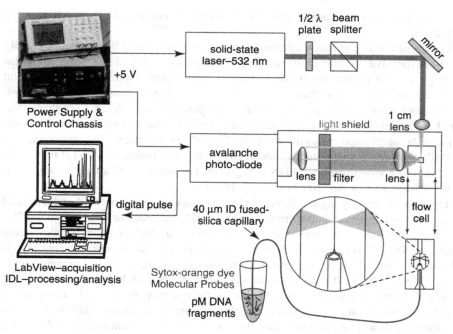

Figure 7.1. Diagram of the HSFCM apparatus. The optical system is arranged in a "U" shape, with the Uniphase laser at one end and the single-photon sensing avalanche photodiode detector at the other end. A half-wave plate and polarizing beam splitter serve to attenuate the laser power (16 mW maximum) down to 0.5–1.0 mW and to set the polarization vector orthogonal to the flow axis. One mirror and one lens, an acromatic-doublet, steer and focus the laser beam onto the sample stream inside the 250-μ square flow channel. An aspheric collection lens, optical filter, and second aspheric lens route the fluorescence emission to the active area of the avalanche photodiode detector (150 μ in diameter), which acts as a spatial filter. The system runs off a single 5-volt power supply, and the data acquisition system records the running count-rate history of the detector's activity in 100-μs time slices.

A gravity-fed fluidic system is used to provide very stable slow-flow conditions, with the sheath velocity at about 1 cm/s. The flow cell holder provides sample input, sheath, and flush connections to the flow cell. Sample is delivered directly into the flow channel through a 40-μ inner diameter fused-silica quartz capillary, inserted 3–4 mm into the 250-μ square channel of the flow cell. The inserted tip of the capillary (tapered at 14°) is positioned 300–400 μ below the laser beam. The sheath fluid is high-purity water fed from a 1-L bottle suspended 10–40 cm above the instrument. The sheath flows around the capillary in the corners of the flow channel from bottom to top. A drain line, connected to the upper end of the flow channel by a short piece of soft silastic tubing, creates a continuous column of water from the sheath bottle, through the flow cell, into a waste collection bottle (the height of the waste bottle is adjusted independent of the supply bottle). The height difference between the level of sheath fluid (in the 1-L bottle) and the liquid level in the waste container determines the volumetric flow rate through the flow cell. Consumption of the sheath fluid is about 1 mL/hr. A miniature, precision, electronic pressure regulator, set at 0.3–0.5

psi, is used to drive the sample through the capillary and into the flow cell at about 100 nL/hr.

The apparatus is compactly configured on a 2 by 1-ft. optical breadboard primarily using inexpensive, commercially available, components. All of the electronics, including the main power supply, the pneumatic components, and the data acquisition card are housed in a single chassis. The total electrical power consumption of the system is about 80 W. An inexpensive portable computer using LabView to control a National Instruments module (DAQPad-6070E, National Instruments, Austin, TX) performs data acquisition tasks through an IEEE-1394 interface. Measurements are typically performed under standardized conditions: laser power of 0.5–1.0 mW at an event rate of approximately 100 DNA fragments per second, and a transit time of individual fragments in the laser beam of 3–10 ms.

The data are entirely digital, detected photon pulses that are processed and analyzed using Los Alamos–developed software written in IDL (Interactive Data Language, Research Systems, Inc.). We collect the running count-rate history of the single-photon sensing APD in short time intervals (typically 100 μs) and then examine the entire digital record to identify individual events. When the count rate exceeds a predetermined level, an event begins. It ends when the count rate drops back below the threshold. If the event duration meets criteria established by the operator, the derived parameters of area, height, and duration are saved in correlated arrays for extensive analysis and presentation in a variety of graphical formats. For all DNA fragment-sizing applications, we have developed software to calibrate the photon burst data to read directly in kilobase pairs (kbp), using internal staining standards in a calibration process described below.

DNA Fragment Sizing by HSFCM

Our group and others have reported DNA fragment analysis by flow cytometry with increasing sensitivity and resolution (1–5, 15, 16). In the last few years, we have specifically targeted large DNA fragments of several thousand to several hundred thousand base pairs. For these large DNA fragments, our approach is more sensitive (sample size of picograms versus micrograms), faster (analysis time of 10 min vs. tens of hours) and more accurate (size uncertainty, 2%) than PFGE widely used for these measurements. We have applied this technique to the characterization of P1 artificial chromosome clones for DNA sequencing (10) as well as for bacterial identification (8, 11, 12).

In our approach, a sample of double-stranded DNA is stained with a fluorescent intercalating dye that binds stoichiometrically to the DNA such that the amount of dye incorporated is directly proportional to the fragment size in kilobase pairs (kbp). There is a large increase (500–1000 fold) in the fluorescence intensity of the intercalating dye on binding to the DNA, making it unnecessary to remove unbound dye from the solution before analysis. The stained fragments are diluted to 1–10 pM and introduced into the HSFCM. Fragments pass individually through the laser-illuminated detection region (probe volume ~1 pL) in the flow cell, with each fragment producing a fluorescence burst of photons as it transits the laser beam. Fluorescence bursts from individual fragments are detected by the APD and captured by the data acquisition system, and height, duration, and area parameters are extracted in software, with histograms generated on all three features. Although the integrated burst area is the measure of the fragment size, important information is contained in the height and duration parameters, which are used to gate the processed histograms to eliminate coincident events. A histogram of the burst sizes from a restriction digest of a bacterial genomic DNA is referred to as the bacterial DNA "fingerprint." Samples containing less than a picogram of DNA are analyzed in about 10 min, which represents an increase in sensitivity of about one million, and a reduction in analysis time of about 1000 over conventional gel electrophoresis. DNA fragments ranging from 0.125 to about 400 kbp have been sized by this technique at a rate of about 100 fragments per second. The resolution appears to be better than that of conventional gel electrophoresis for fragments larger than about 20 kbp (5).

Intercalating Dyes

The properties of a good intercalating dye for DNA fragment sizing in flow are binding independent of base-pair composition, high binding constant, high fluorescence intensity, significant fluorescence enhancement on binding to DNA, efficient excitation with readily available laser lines, and independence of fluorescence intensity to dye:bp ratio. Most of our original work used the bis-intercalating dyes YOYO-1®, TOTO-1®, and POPO-3®. However, a real breakthrough in our ability to size DNA fragments individually came about 4 years ago, with the availability of new mono-intercalating dyes from Molecular Probes: PicoGreen® stain (488/530 nm; excitation/emission), followed by SYTOX-Orange® stain (536/575 nm). We studied these two dyes to determine the stoichiometry of the DNA:dye binding interactions compared to their bis-intercalating predecessors (19, 20). Although PicoGreen stain has lower background fluorescence compared to SYTOX-Orange, both dyes have high binding affinities for double-stranded DNA in their primary binding mode of intercalation. Furthermore, both dyes have relatively low secondary binding affinities, which are presumably a result of electrostatic interactions of the dye with the outside of the DNA polymer. The binding of both dyes to DNA is independent of base-pair composition.

On the basis of our new understanding of the DNA:dye binding mechanism, SYTOX-Orange stain is currently our DNA dye of choice. This dye has high optical absorption at 536 nm, which matches perfectly with the Nd:YAG laser (532 nm) on the HSFCM. A simple DNA fragment-staining protocol, applicable to a wide range of DNA concentrations, has been developed, and the operation of the HSFCM instrument has been optimized for the analysis of SYTOX-Orange stained fragments.

The attributes of the current HSFCM instrument and its operating conditions for DNA fragment sizing of SYTOX-Orange stained fragments are summarized below.

Linearity

The signal from each DNA fragment is linearly proportional to the size of the fragment. Using several different dyes, excitation wavelengths, and instruments, this has been verified over the size range of from 0.125 to ~400 kbp. This characteristic is a direct result of the equilibrium binding of the dye to the fragments, with an estimated average of one dye molecule per four or five base pairs for mono-intercalating dye molecules. The linearity of the measurements allows careful calibration of the system response and results in an overall uncertainty of about 2% compared to gel electrophoresis methods, in which the uncertainty is about 10% (10).

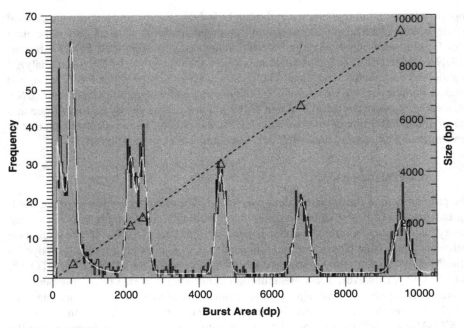

Figure 7.2. Hind III restriction digest of Lambda DNA. Six of the total of eight fragments produced in the Hind III restriction digest of Lambda phage DNA (48.5 kbp) are visible in this frequency distribution. The DNA sample, at 450 ng/mL, was stained with 0.2 uM SYTOX-Orange and analyzed on the HSFCM, with a total of 20,000 fragments represented in this data set. The figure shows the fitted sum of six Gaussians (white line) used to determine the centroid of each peak (Δ). The correlation coefficient of the signal (detected-photon burst areas) versus known DNA fragment lengths (right-hand y-axis) is greater than 0.999, with the linear regression shown by the dashed line. The smallest fragment detected was 564 bp (in the peak at about 500 detected photons); the 125-bp fragment was not observed, and the 23.1-kbp fragment is off the scale to the right.

Resolution

Figure 7.2 is the plot of a histogram of the DNA fragments produced by a Hind III restriction enzyme digest of lambda phage DNA stained with SYTOX-Orange. The peak at the low end of the distribution is caused by the 564-bp fragment, whereas a smaller fragment in the digest (125 bp) is masked by the background in this experiment and is not shown. Two fragments differing in size by only 295 bp (2027 and 2322 bp) are sufficiently resolved to simultaneously fit Gaussians to both peaks. Table 7.1 shows the corresponding signal sizes, coefficient of variation (CV), and frequencies of each of the measured fragments. Note that the CV improves dramatically as the fragment size increases.

Size Range

The smallest DNA fragment detected to date is the 125-bp fragment from the Hind III restriction digest of lambda DNA (data not shown). Although not seen in figure 7.2, the 125-bp fragment can be measured when the HSFCM is optimized for small fragments or single fluorescent molecules. The largest DNA fragment consistently detected

Table 7.1. Measurement data of DNA fragments from a Hind III restriction enzyme digestion of lambda phage DNA

DNA Fragment Size (bp)	Burst Area (detected photons)	Coefficient of Variation (%)	Observed Event Frequency (%)
564	559	15.9	22.9
2027	2164	4.0	13.6
2322	2514	3.3	12.9
4316	4642	2.9	11.9
6557	6829	1.9	12.5
9416	9526	1.8	9.8
23130	20564	1.0	6.9

Statistics on the data set represented in the histogram in Figure 7.2. Two fragments, which differ by only 295 base pairs, were resolved well enough to calculate respectable coefficients of variation for each peak. The smallest fragment produced in this digest (125 bp) was not detected in this data set.

using the current HSFCM is a 366-kbp fragment from a *Staphylococcus aureus* (*S. aureus*) genomic DNA digest.

Sensitivity

Detection of single phycoerythrin (B-PE) molecules has been achieved on the current HSFCM using crossed cylindrical lenses to focus the laser beam spot to about 11×50 μm, with 2 mW of laser power at 532 nm. The signal-to-noise ratio was 24.9, and the normalized number of photons detected averaged about 200 per B-PE molecule (data not shown). To put the sensitivity of the current HSFCM into perspective, one B-PE molecule is equivalent in brightness to 25 Rhodamine 6G molecules and is comparable to the signal from the 564-bp fragment stained with SYTOX-Orange (see figure 7.2).

Quantification

Because each fragment is analyzed and counted individually, the number of fragments represented in each peak is directly enumerated.

Sample Size

Very little DNA is needed for analysis. For example, 10 pg of a 50-kbp fragment contains approximately 200,000 molecules. The data in figure 7.2 are the result of the analysis of 20,000 fragments, all of which are smaller than 50 kbp. This is equivalent to about 2500 intact lambda DNA molecules, or about 0.1 pg. Thus, the amount of DNA analyzed is extremely small and not limited by the amount needed for detection but, rather, by sample handling considerations. With the current sample introduction arrangement, 5–10-μL samples can be analyzed, and we typically analyze samples at about 5 pM (fragments).

Stability

Sample analysis operating conditions have been optimized to achieve a stable flow of the sample stream even with large fragments. Recently, we developed an improved buffer consisting of 10 mM Tris-HCl (pH 8.0), 0.01% Tween-20, and 0.5% polyvinylpyrrolidone 1 Million MW (PVP) for all DNA solutions and found that it significantly enhances flow stability and improves the CV of individual fragment peaks. This is primarily because of the detergent, which reduces the surface tension in the capillary.

Bacterial Identification

Bacterial species and strain discrimination by DNA "fingerprinting" is typically based on restriction fragment–length polymorphism (RFLP) analysis by PFGE. Our primary focus over the last several years has been to develop the HSFCM as a replacement for the cumbersome and time-consuming gel electrophoresis measurements commonly used for bacterial identification while providing better resolution and linearity for fragments larger than 20 kbp. RFLP relies on the ability of restriction endonucleases to cut (digest) DNA at sequence specific sites, thereby producing a characteristic set of discrete fragments from each starting molecule.

Figure 7.3 shows a set of characteristic DNA fragments obtained from a Sma I restriction enzyme digest of *S. aureus* genomic DNA with the *x*-axis scale calibrated to read directly in base pairs. A total of 18 DNA fragments ranging in size from 3 to 300

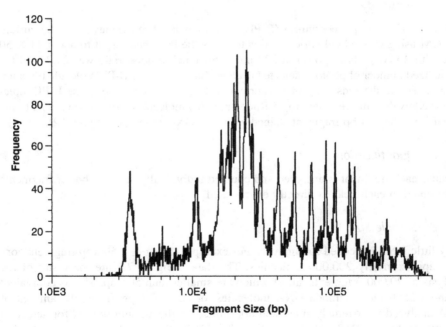

Figure 7.3. *Staphylococcus aureus* genomic DNA digest. This graph, a DNA "fingerprint" or DNA fragment pattern, represents the end-point of data collection and processing. In this case, the genomic DNA is from *S. aureus* (ATCC strain 25923), digested with Sma I, stained with SYTOX-Orange, and run on the HSFCM. The histogram of the detected photon burst areas on 15,400 events has been converted to read directly in base pairs fragment length, using the calibration procedure described in the text. The following sizes were measured on 18 resolved fragments: 3690, 10876, 16253, 18184, 19635, 21328, 25016, 27334, 31128, 41577, 54238, 71174, 89614, 105111, 133917, 144606, 243196, and 304402.

kbp are measured in this fingerprint. The analysis was completed in about 20 min, compared to 22+ hr for PFGE. Furthermore, fragments smaller than 50 kbp that are easily seen by the HSFCM are not seen in the gel electrophoresis method (data not shown). This graph represents the end point of the analysis and calibration procedure, in that the sample is run with and without added internal staining standards of known size, which provides the means to calibrate the histogram x-axis scale. Once calibrated, in a procedure described below the flow-based histogram provides direct quantitation of both fragment size and frequency.

The total analysis time for bacterial fingerprinting consists of both sample preparation time and fragment sizing time. One of the difficulties in bacterial identification by HSFCM has been the sample preparation. In initial studies we used protocols adapted from PFGE analyses of restriction fragment–length polymorphisms. RFLP analysis uses rare cutting enzymes (eight-base or six-base cutters) and yields a set of 10–20 fragments in the size range of 10–500 kbp. Effective sample handling requires that the complete procedure, including cell lysis and DNA digestion, be carried out in an agarose matrix to protect the large fragments from random shearing. The procedure took 7 days and was clearly the rate-limiting step, compared to the rapid DNA fragment sizing by HSFCM. Furthermore, for HSFCM analysis of DNA fragment size, the restriction fragments must be removed from the agarose matrix and stained before delivery to HSFCM for analysis. Therefore, we developed a sample preparation protocol that was dramatically shorter than the standard, 7-day protocol. Our short protocol for RFLP-based fingerprinting is generally applicable to bacterial typing by either HSFCM or PFGE. This shortened sample preparation procedure was accomplished by making several modifications to current published protocols, including the reduction of reaction times for cell lysis (1.75 hr), Proteinase K digestion (1 hr), and restriction enzyme digestion (2 hr on properly chosen restriction enzymes); a heat shock reaction (0.25 hr at 62°C) was added, and we optimized saturating enzyme concentrations. Compared with other rapid-preparation protocols, this 6-hr procedure was demonstrated to be reproducible and successful on over 30 different Gram-negative and Gram-positive bacteria including many difficult *Bacillus* species (Marrone, B.L., unpublished data). Therefore, it is suitable for use as a standardized alternative to the traditional 7-day protocol for bacterial fingerprinting. Moreover, the shortened procedure is now potentially adaptable to small volumes and faster sample handling protocols that can be customized for bacterial DNA fingerprinting using HSFCM.

Absolute Fragment Size Calibration

Through the use of DNA size standards, it is possible to calibrate the burst-area histogram scale to read directly in kbp. For the bacterial fingerprint analyses, three pre-stained size standards are used: a 17.3-kbp linear plasmid, lambda phage DNA (48.5 kbp), and T4 phage DNA (165.6 kbp). A fragment size distribution is recorded first on a DNA sample without standards added. Then a second distribution is recorded with the three size standards added to the initial DNA solution. We have shown that there is a rapid, subsecond equilibration of the dye among the DNA fragments in mixed samples. Therefore, analysis of the mixed sample containing bacterial DNA fragments and the size standards will yield a distribution with the standard peaks in the proper position relative to the fragments from the bacterial digest (data not shown).

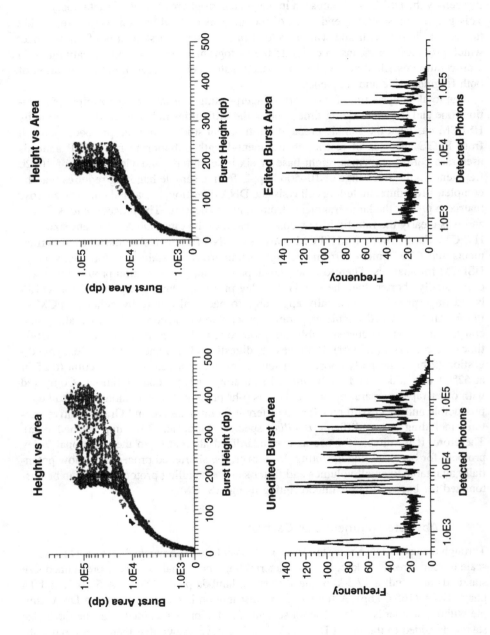

We use the following sequence of operations in our analytical software to calibrate each data set. This sequence will be illustrated with data from different *S. aureus* isolates. The first step is to edit each data set as illustrated in figure 7.4 to eliminate a specific artifact from the subsequent calibration steps. In the bivariate histogram of burst amplitude versus burst area, a plateau in amplitude is observed for fragments larger than about 50 kbp (~20 μm long, extended in flow by the hydrodynamic forces). This plateau is reached when a DNA fragment is long enough to fill the laser beam. The amplitude of an individual event increases as a fragment enters the laser beam, until the focused laser beam spot is filled by a sufficiently long fragment. At that point, the maximum number of intercalated dye molecules will be in the laser beam/probe volume–emitting fluorescence photons. As in conventional cytometry, where unwanted events such as doublets and debris are gated out of the data, we gate the DNA fragment sizing histograms to eliminate events in which the amplitude exceeds the plateau level that we know correlates with fully extended, double-stranded DNA fragments. Gating out this artifact reduces the apparent background of random-length fragments to more clearly reveal the peaks corresponding to discrete fragments.

After gating each data set, the second step is to determine the centroids of the peaks resulting from the internal standards by fitting a Gaussian profile to each standard peak. Linear regression analysis is then performed between the known fragment sizes and the corresponding peak centroids to produce a calibrated slope and intercept. Typically, the correlation coefficient is 0.9999 or greater. The calibrated slope and intercept are applied to a new copy of the processed data file. This calibrates the scale for the burst-area histogram to be read directly in fragment length (kbp), as seen in the lower left panel of figure 7.5. Finally, the uncalibrated data set without standards is overlaid with the calibrated reference set to ensure that the properly corresponding peaks in both data sets are aligned. This step is automated in the software, but the user can adjust the intercept and slope for the second data set to optimize the alignment. When the best alignment is achieved, the values are applied to a new copy of the second data set, thus completing its calibration.

To optimize calibration of a data set with a high background of random length fragments, which occurs in PFGE data sets as well, several additional processing steps can be applied, as shown in figure 7.6. When enabled, the software will fit a fifth-order polynomial to the entire burst-area distribution, which yields a reasonable estimate of

---

Figure 7.4. Effect of editing the list-mode data to eliminate coincidence artifacts. List mode parameters, as in conventional cytometry, are extracted from the raw count-rate record of the DNA fragment sizing system; amplitude, duration, and area measurements recorded for each event can be used to create two-dimensional displays. Furthermore, editing of the list-mode data can eliminate artifacts in molecular analysis, as a result of the coincidence of two fragments overlapping in the probe volume. A bivariate plot of burst amplitude versus burst area shows a region of greatly increased burst height (upper-left panel). These high-amplitude events make up 15% of the total events in this file. The corresponding burst-area histogram (lower left) has a relatively high background, especially for larger fragments (20,000–200,000 detected photons). When these high-amplitude events are gated out of the data (upper-right panel), the background in the edited histogram (lower right) is lower and more uniform, improving the resolution of discrete peaks from the background of random-length fragments.

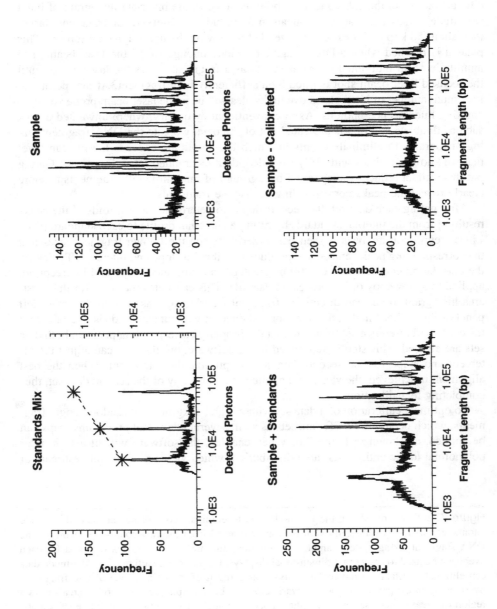

the background random-length fragments. Subtracting this background from the entire distribution helps to more clearly reveal the discrete fragment peaks. In some very noisy data sets, it can be difficult to clearly identify the corresponding peaks between the two data sets (with or without standards). Overlaying the histograms and shifting the uncalibrated histogram left or right, with respect to the calibrated histogram, to arrive at the best alignment helps to ensure that corresponding peaks have been properly identified. Once calibrated, the new data set can be compared directly to any other calibrated data set.

Matching Bacterial DNA Fingerprints

Methods for RFLP fingerprinting of a wide variety of bacteria have been developed and validated by parallel analysis using conventional PFGE methods. The Centers for Disease Control and Prevention uses PFGE for outbreak detection in a nationwide, networked Department of Health Laboratory system called PulseNet. Recently, we have developed software for performing statistical comparisons between individual data sets of RFLP fingerprints collected on the HSFCM. These comparisons are useful for matching and statistically relating histograms from unknown samples to known samples in a database of previously collected bacterial RFLP fingerprints in much the same way that the CDC PulseNet network matches PFGE patterns from local health departments to those residing in a central database. The greater speed and precision of the HSFCM measurement for DNA fragment sizing holds promise for eventually replacing PFGE for analysis of RFLP fingerprints. Planned improvements to the HSFCM to facilitate bacterial identification include automation of sample preparation and delivery to increase the analysis speed and throughput, as well as the ease of operation.

Future Directions

We have turned our experience with single fluorescence molecule detection by flow cytometry to the development of instrumentation for high-sensitivity flow cytometry having practical applications in bacterial species and strain identification. HSFCM has the potential to replace PFGE for bacterial strain identification, especially in the public health

---

Figure 7.5. Calibration of the burst-area histograms with internal standards. The burst area data, recorded initially as detected photons, are transformed to fragment length in base pairs, using a set of three internal size standards. A previously stained mix of the three standards is added to a portion of the unknown sample, also stained with SYTOX-Orange. The top-left panel shows the mix of three internal standards, without sample. The high correlation coefficient (0.999994) between the positions of the peaks and the known fragment sizes (left y-axis) is typical and shows the linearity of the measurement. The standards are: a 17.3-kbp linear plasmid; Lambda phage DNA, 48.5 kbp; and T4 phage DNA, 165.6 kbp. Each sample is run first without standards (top-right panel), and then with standards added (bottom-left panel; the tallest peaks are the standards). On the basis of identifying the peaks from the internal standards, a calibrated slope and intercept is assigned to the sample plus standards. Finally, by overlaying the calibrated histogram of the mix with the histogram of the sample alone to put the corresponding peaks into registration, a calibrated scale is applied to the sample histogram (bottom-right panel).

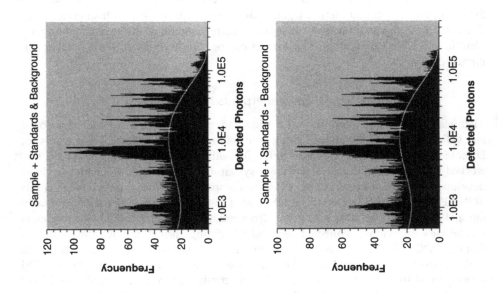

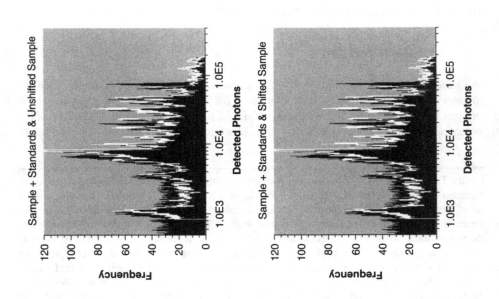

Figure 7.6. Background subtraction and histogram alignment to improve the calibration. Some bacterial genomic DNA digests have a large number of random length fragments, creating a high background continuum that can obscure the discrete peaks of the restriction fragments. Also, when many fragments are present in a digest, it can be difficult to unambiguously recognize corresponding peaks in the two data sets with and without standards. In those instances, the software can perform two additional steps: shifting one histogram with respect to the other to achieve proper registration, and background subtraction. The two left-hand panels show the slight effect of shifting the histogram of the sample file (white line) to better align with the corresponding peaks in the control histogram (with standards, black filled). As seen in the right-hand panels, it is possible to improve the resolution of the specific peaks by estimating the random fragment background and subtracting it from the initial histogram. The top-right panel shows the background estimate (white line) derived by fitting a fifth-order polynomial to the entire histogram. After subtracting the background from the sample file (upper right, black filled) and from the control file containing internal standards, the control histogram was properly aligned to the peaks in the sample histogram (white line, lower right).

area, where speed and accuracy are essential for pinpointing the source of an outbreak, tracking its course, and controlling its spread. Furthermore, the HSFCM instrumentation used routinely in our laboratory for DNA fragment size analysis will be improved and expanded to enable other new biological applications, such as analyses of lipid vesicles, individual mitochondria, and other subcellular organelles. Our ability to push the limits of detection in flow cytometry will continue to enable the next generation of important applications of flow cytometry methodology in biotechnology.

Acknowledgments   The writing of this chapter was supported by the Department of Energy, the National Institutes of Health (RR-01315), and the Federal Bureau of Investigation. The statements and conclusions herein are those of the authors and do not necessarily represent the views of the FBI.

References

1. Agronskaia, A., Schins, J.M., de Grooth, B.G., and Greve, J. 1999. Two-color fluorescence in flow cytometry DNA sizing: identification of single molecule fluorescent probes. *Anal. Chem.* 71:4684–4689.
2. Ambrose, W.P., Cai, H., Goodwin, P.M., Jett, J.H., Habbersett, R.C., Larson, E.J., Grace, W.K., Werner, J.H., and Keller, R.A. 2003. *Flow Cytometric Sizing of DNA Fragments in Topics in Fluorescence*, Vol. 7, *DNA Technology*, J.R. Lakowicz, ed. Kluwer Academic/Plenum Publishers, New York, pp. 239–270 (2003).
3. Castro, A., Fairfield, F.R., and Shera, E.B. 1993. Fluorescence detection and size measurement of single DNA-molecules. *Anal. Chem.* 65849–65852.
4. Chou, H.P., Spence, C., Scherer, A., and Quake, S. 1999.A microfabricated device for sizing and sorting DNA molecules. *Proc. Natl. Acad. Sci. USA* 96:11–13.
5. Ferris, M.M., Yan, X., Habbersett, R.C., Shou, Y., Lemanski, C. L., Jett, J.H., Yoshida, T.M., and Marrone, B.L. 2004. Performance assessment of DNA fragment sizing by high sensitivity flow cytometry and pulsed field gel electrophoresis. *J. Clin. Microbiol.* 42:1965–1976.
6. Goodwin, P.M., Johnson, M.E., Martin, J.C., Ambrose, W.P., Marrone, B.L., Jett, J.H., Keller, R.A. 1993. Rapid sizing of individual fluorescently stained DNA fragment by flow cytometry. *Nucleic Acids Res.* 21:803–806.

7. Gucker, F.T. Jr., O'Konski, C.T., Pickard, H.B., and Pitts, J.N. Jr. 1947. A photoelectronic counter for colloidal particles, *J. Am. Chem. Soc.* 69:2422–2431.
8. Habbersett, R.C., and Jett, J.H. 2004. An analytical system based on a compact flow cytometer for DNA fragment sizing and single-molecule detection. *Cytometry* (Part A 60A): 125–134.
9. Huang, Z., Jett, J.H., and Keller, R.A. 1999. Bacterial genome fingerprinting by flow cytometry. *Cytometry* 35:169–175.
10. Huang, Z., Petty, J.T., O'Quinn, B., Longmire, J.L., Brown, N.C., Jett, J.H., and Keller, R.A. 1996. Large DNA fragment sizing by flow cytometry: Application to the characterization of P1 artificial chromosome (PAC) clones. *Nucleic Acids Res.* 24:4202–4209.
11. Kim, Y., Jett, J.H., Larson, E.J., Pentilla, J.R., Marrone, B.L., and Keller, R.A. 1999. Bacterial fingerprinting by flow cytometry: bacterial species discrimination. *Cytometry* 36:324–332.
12. Larson, E.J., Hakovirta, J.R., Cai, H., Jett, J.H., Burde, S., Keller, R.A., and Marrone, B.L. 2000. Rapid DNA fingerprinting of pathogens by flow cytometry. *Cytometry* 41:203–208.
13. Nebe-Von-Caron, G., Stephens, P.J., Hewitt, C. J., Powell, J.R., and Bradley, R.A. 2000. Analysis of bacterial function by multi-color fluorescence flow cytometry and single cell sorting. *J. Microbiol. Methods* 42:96–114.
14. Nguyen, D., Keller, R., Jett, J., and Martin, J. 1987. Detection of single molecules of phycoerythrin in hydrodynamically focused flows by laser-induced fluorescence. *Anal. Chem.* 59:2158–2161.
15. Petty, J.T., Johnson, M.E., Goodwin, P.M., Martin, J.C., Jett, J.H., and Keller, R.A. 1995. Characterization of DNA size determination of small fragments by flow cytometry. *Anal. Chem.* 67:1755–1761.
16. Schins, J.M., Agronskaya, A., de Grooth, B.G., and Greve, J. 1998. New technique for high-resolution DNA sizing in epi-illumination. *Cytometry* 32:132–136.
17. Shapiro, H.M. 1995. *Practical Flow Cytometry*. Third edition. Wiley-Liss, New York.
18. Shapiro, H.M. 2000. Microbial Analysis at the single-cell level: tasks and techniques. *J. Microbiol. Methods* 42:3–16.
19. Steen, H.B. 1980. Further developments of a microscope-based flow cytometer: light scatter detection and excitation intensity compensation. *Cytometry* 1:26–31.
20. Steen, H.B, Lindmo, T. 1979. Flow cytometry: a high resolution instrument for everyone. *Science* 204:403–404.
21. Yan, X., Grace, W.K., Yoshida, T.M., Habbersett, R.C., Velappan, N., Jett, J.H., Keller, R.A., and Marrone, B.L. 1999. Characteristics of different nucleic acid staining dyes for DNA fragment sizing by flow cytometry. *Anal. Chem.* 71:5470–5480.
22. Yan, X., Habbersett, R.C., Cordek, J.M., Nolan, J.P., Yoshida, T.M., Jett, J.H., and Marrone, B.L. 2000. Development of a mechanism-based DNA staining protocol using SYTOX orange nucleic acid stain and DNA fragment sizing flow cytometry. *Anal. Biochem.* 286:138–148.

8

A Guide to Informatics in Flow Cytometry

ADAM TREISTER

Introduction

Flow cytometry is a result of the computer revolution. Biologists used fluorescent dyes in microscopy and medicine almost a hundred years before the first flow cytometer. Only after electronics became sophisticated enough to control individual cells and computers became fast enough to analyze the data coming out of the instrument, and to make a decision in time to deflect the stream, did cell sorting become viable.

Since the 1970s, the capabilities of computers have grown exponentially. According to the famed Moore's Law, the size of the computer, as tracked by the number of transistors on a chip, doubles every 18 months. This rule has held for three decades so far, and new technologies continue to appear to keep that growth on track. The clock speed of chips is now measured in gigahertz—billions of instructions per second—and hard drives are now available with capacities measured in terabytes.

Having computers so powerful, cheap, and ubiquitous changes the nature of scientific exploration. We are in the early steps of a long march of biotechnology breakthroughs spawned from this excess of compute power. From genomics to proteomics to high-throughput flow cytometry, the trend in biological research is toward mass-produced, high-volume experiments. Automation is the key to scaling their size and scope and to lowering their cost per test. Each step that was previously done by human hands is being delegated to a computer or a robot for the implementation to be more precise and to scale efficiently.

From making sort decisions in milliseconds to creating data archives that may last for centuries, computers control the information involved with cytometry, and software controls the computers. As the technology matures and the size and number of experiments increase, the emphasis of software development switches from instrument control to analysis and management. The challenge for computers is not in running the cytometer any more. The more modern challenge for informatics is to analyze, aggregate, maintain, access, and exchange the huge volume of flow cytometry data. Clinical and other regulated use of cytometry necessitates more rigorous data administration techniques. These techniques introduce issues of security, integrity, and privacy into the processing of data.

Thankfully, all significant manufacturers of flow cytometers produce files that conform to a standard file format. The basic structure and capabilities of the flow cytometry standard (FCS) format are explained below. With standardized files, the next logical step is to create a standard library to read them. In this chapter, we will outline a new programmer's library that has been created to simplify writing programs to handle data.

Although the existing format is widely accepted and satisfies most of the needs of the lab scientist, it does have limitations in combining multiple data samples into a single file. In many high-throughput experiments, the number of events per sample is relatively small, but the number of samples is in the hundreds or thousands. Depending on the homogeneity of the samples and preparation, it may make sense to collect data from all of the samples into a single file and to be able to differentiate them afterward. A strategy to accomplish this is discussed.

Not long ago, being able to read and write files was most of the battle. Now an informatics plan must anticipate how it will access data over the Web, or how it will interact with a database server. There are no current standards on how this is to be done in flow cytometry, but the architecture for structured documents on the Internet has been established, and it is clear what the next-generation standards will look like.

Central to the organization of a laboratory is its set of protocols. The same principles apply to informatics. High-level automated processing will come as a result of carefully organized input and well-structured data. There is relatively little information needed to link a data file generated in an experiment to the protocol that describes the overall structure of the experiment. Yet in current practice, protocol panels are not often used effectively. The majority of files collected today are labeled FL1, FL2, and so forth. Analysis relies on the scientist going back to the lab notebook to map the files to their role in the experiment. We propose a simple format for defining protocol information and for applying it to flow cytometry data, either before or after acquisition. This will provide a mechanism for users to design experiments before running them on the instrument. When the users log in to acquire data, the operator has a full diagram of the experiment and of a panel set-up in the cytometer's settings format, so the operator can annotate the data properly as it is collected.

Software plays critical roles in most aspects of the scientific process: acquisition, analysis, administration, management, and publication. Dozens of people at different times and locations contribute to the experiment, making the efficient and flexible management of an experiment extremely complex. It is crucial to the integrity of the scientific work that steps be accurately documented and presented. This document format will enable separate tools to cooperate in the various steps of data processing. As the data file format allows different instruments to interoperate on the sample level, we are looking for an open metadata standard to facilitate description of high-level experiments and studies.

FCS File Specification

Support of data formats and standardization in the flow cytometry community is quite good. All of the vendors write essentially compatible files, and there is a published specification of how the files must be structured. Users are able to move between multiple instruments, produced by different vendors, without losing the ability to compare data. Although there are definitely issues in trying to compare data collected on instruments made by different manufacturers, the ability to read the data files is rarely a concern.

The FCS was first published in 1984 (11) and then revised to FCS 2.0 in 1990 (3). The standard was updated to FCS 3.0 in 1998 (14). Most data files written by cytometers in use today are still in the FCS 2.0 format, though the software associated with some recent instruments has started writing FCS 3.0 files.

This standard file format is endorsed by the International Society of Analytic Cytometry (ISAC). The society maintains a Data Standards committee to deal with issues that may arise over the standard. Acceptance of the standard has been harmonious and unanimous among instrument manufacturers, and their transgressions of the letter of the standard are generally innocuous. Compatibility between instruments and vendors cannot be underestimated as a critical factor in the evolution of the technology. Having a well-adhered-to, open standard file format is a critical factor in the design of the data processing.

FCS files all start with a first line that tells the version of the file and the start and end locations of key sections of information located in the document. There are two sections that are commonly used: the TEXT segment and the DATA segment. The TEXT segment contains the textual keyword value pairs to describe the structure of the DATA segment. The DATA segment is the binary data containing the actual values collected by the instrument. If you open an FCS file in a word processor, you will see that it starts something like this:

```
FCS2.0          58      1761      2048      64628
```

This tells you that the file conforms to the FCS version 2 standard, its TEXT section goes from the 58th to the 1761st bytes, and the DATA section extends from the 2048th to the 64,628th bytes.

The TEXT segment of an FCS file contains a list of strings, keyword and value pairs. The first character of the segment is the delimiter that is used in the file to separate successive strings. If the delimiter character exists in any of the keywords or values, it must be doubled to distinguish it from a delimiter. Empty keywords or values are explicitly prohibited from being included in the TEXT segment. Keywords defined as part of the standard always start with a dollar sign ($). Other keywords are allowed and often used, but nonstandard keywords must not start with a dollar sign.

The keywords fall into two groups: those that describe the entire file, and those that describe a single parameter. (The term *parameter* in flow cytometry does not follow the normal mathematical definition of the term. Instead of referring to a single value that affects an equation or distribution, a parameter in flow cytometry is a vector of data values measured off a single collector. If you consider a FCS file to be a spreadsheet, with each row containing the measurements for one cell, a parameter is all of the data in one column.) Examples of standard keywords that apply to the entire file are included here:

$DATE   The date of acquisition.
$TOT   The total number of events collected.
$CYT   The cytometer used to collect the file.
$FI   The name of the file.
$PAR   The number of parameters in the file.

The other group of keywords contains those that describe the individual parameters. Because there will be several instances of these keywords, corresponding to the several parameters collected in the file, the syntax of these keys is $P followed by a number, followed by a final letter to specify the function. The name of the first parameter is found in the keyword $P1N, the range of the second parameter is found in $P2R, and so forth. When speaking of a generic keyword, the convention is to use "n" instead of the number, as in $PnS, the keyword to define the stain of an unspecified parameter. The important parameter keywords are

$PnN   The name (fluorochrome) of the nth parameter.
$PnS   The stain (antibody) of the nth parameter.
$PnB   The number of bits used to store the value of the nth parameter.
$PnR   The number of possible values (range) of the nth parameter.
$PnG   The gain (amplification) of the nth photomultiplier tube.

The DATA segment contains the actual data collected in the run. This is a table of binary data forming a matrix of values. The rows in the table are the data collected from any single event (cell). The columns in the table are the data collected from any photomultiplier tube or sensor. In theory, it is possible for each parameter to have a separate length ($PnB) and for the lengths to be any number of bits. In practice, because the length is a function of the PMTs built into the cytometer, the parameters are virtually always the same length. Most often they are either 8 or 16 bits, making it very easy to read most FCS files.

One aspect of this data standard relevant to high-throughput flow cytometry is the issue of how to deal with many small data sets at a time (1). Either each well in the plate becomes its own FCS file, or one file can contain all of the wells in one plate (see chapter 3). Although there will be experiments in which the wells are distinct enough to warrant separate files, most experiments in high-throughput cytometry are structured so that the cells in the wells are similar in nature or preparation. We will often want to analyze all of the wells as a collective population, in addition to viewing individual wells to find their important differences.

In addition, storing data by the plate instead of by the well will make it significantly easier to organize. Highly automated processes demand highly structured data definitions, and treating each well as a separate file, though clearly necessary in certain cases, means the generation of 100-fold (or up to almost 400-fold, as 384-well plates supplant the current 96-well variety) the clutter of files. Except where the disparity of data between wells requires otherwise, it seems worthwhile to store the data in fewer, larger files representing multiple wells within the sample.

Although the FCS standard supports the concept of multiple data sets through its $NEXTDATA keyword, the linked list approach to many segments within a file is problematic. The primary problem is that no software exists today that reads this aspect of the standard. Although the standard document has been approved for several years, the dynamics of the industry have not required this feature, so it has rarely been implemented. This means that a high-throughput system that generated legal FCS files containing multiple segments would probably not be compatible with any analysis software other than what was written for that system.

Fortunately, there are several alternative approaches that will store multiple wells of data within a single file and yet have them be readable by existing software. This simplest approach is to collect all of the samples sequentially in a single stream but provide enough information to be able to break the samples apart in the analysis. This could be accomplished by inserting a special marker event, such as all zeros, into the data stream. Software unaware of the intersample marker would interpret all of the samples to be a single data set (with a small number of extraneous events in a corner of the graph), yet a program capable of looking for this marker could easily use it to divide the data for individualized processing.

If this marker adds unwanted extra events, or if there is a possibility that collected events might coincidentally have all zero values (or whatever is being used as the marker), another approach would be to record the time stamp for each event and to divide samples on the basis of the largest differences in successive time stamps. Again, naïve analysis programs would treat the file as if they represented a single tube, but more sophisticated software could choose to break up the samples based on the flow rate or continue to treat the file as a single population.

A more powerful approach is to include an additional tube identification parameter to identify the well number within each event (13). This parameter could then be used in an analysis gate to look at just one well at a time as long as we have a convention to recognize the parameter that contains the tube or well identifier and keep the well information in that field. With this information, files can easily be merged or split. To support this, an analysis program that encounters a keyword $PnN of the tube identification parameter should understand that the parameter is an attribute of each event identifying the origin of the event and not a collected value. This approach does require that the identifier be an integer, so the values would have to range, for example, from 1 to 96, as opposed to from A1 to L8.

The major drawback of any of these methods is that the same annotation information is presumed for all of the samples. In reality, there are often going to be differences in the cell origin, preparation, or treatment, which traditionally would be kept in the text section of the FCS file. If many tubes are multiplexed into a single file, that information will have to be stored elsewhere and accessed through the tube number and keywords shared by all of the data sets in that file.

Application Programming Interface

Having an accepted file format and using it uniformly across the industry is important in lowering barriers to entry for new tools. Yet it does not negate the fact that it is still a significant amount of work needed to write the software to read these files, and there is little to be gained from reproducing that process each time. Not only is it an unnecessary wheel to reinvent, but it introduces a significant potential for subtle errors that could easily be avoided if there were a standard set of tools for reading and writing FCS files.

The creation of such a library is the current project of the ISAC Data Standards committee—the same group that has authored the FCS file format. The intent is to provide source code that can serve both as a working tool set and a didactic example to anyone working with the FCS files. There is a document defining the FCS Application Program Interface (API) (12), and initial implementations in Java and C++ exist (10). This API abstracts the actual file format to a set of requests, simplifying the process of writing software that reads FCS data and providing servers the latitude to optimize storage parameters without sacrificing accessibility.

The FCS API is object oriented. The objects, their public functions, and the relationships between them are shown in figure 8.1. The key objects defined in the interface are the System, DataSet, Data, Parameter, Keyword, and Error. Given this small set of objects, a programmer can create simple programs to access and manipulate data files.

Following is a section of code taken from the FCS API definition written by the Data Standards committee that shows how these objects can be used to manipulate the data files. The code is written in Java, which was the initial language to implement the API. Other object-oriented languages, such as C++ or Perl, can also be easily supported via proxy classes. This code illustrates that, using the public domain library, it is very straightforward to open, read, and manipulate flow cytometry data with a minimum of programming:

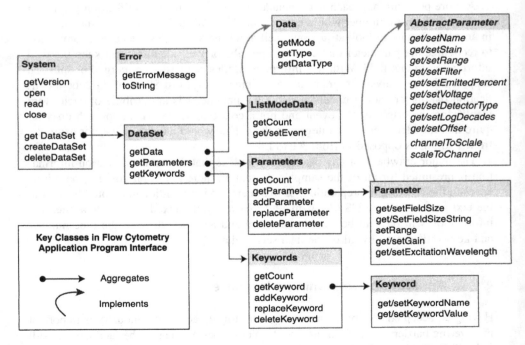

Figure 8.1. Class relationship diagram of the objects defined in the current application programming interface. All objects start with the prefix CFCS, which was omitted here for readability.

```
// Copy an FCS file, removing all values of P1 < 100 channels
for all // listmode datasets in the file (skipping any
non-listmode datasets)

program main(argv)
{
    CString inputFile = argv[0], outputFile = argv[1];
    CFCSSystem readSystem = new CFCSSystem();
    UInt32 nSets;

    error = readSystem.Open(inputFile);
    if (FailIfError(error) != 0) exit;
    error = writeSystem.Create(outputFile);
    if (FailIfError(error) != 0) exit;

    readSystem.GetCount(&nSets);
    for (UInt32 iset = 0; iset < nSets; iset++)
    {

        CFCSDataSet readSet;
        CFCSData readData;
        CFCSParameters paramList;
        UInt32 readDataType;

        readSystem.GetDataSet(iset, &readSet);
        readSet.GetParameters(&paramList);
        readSet.GetData(&readData);
        readData.GetType(&readDataType);
        if (readDataType == CFCSData.ListModeData)
        {

            CFCSDataSet writeSet;
            CFCSData writeData;
            CFCSListModeData sourceList, destinationList;
            UInt16 *event;
            UInt32 nCells, nParam, nGatedEvents = 0;

            writeSystem.CreateDataSet(&writeSet, readSet);
            writeSet.GetData(&writeData);
            readList = (CFCSListMode) readData;
            writeList = (CFCSListMode) writeData;

            // allocate an array to hold one event
            paramList.GetCount(&nParam);
            event = (UInt16 *) calloc(nParam, sizeof(UInt16));
            readData.GetCount(&nCells);
            for (UInt32 idx = 1; idx <= nCells; idx++)
            {

                sourceList.GetEvent(idx, event);
                if (event[0] < 100)
                {
                    destinationList.AddEvent(++nGatedEvents, event);
                }
            }
        }
    }
    readSystem.Close();
    writeSyste.Close();
}
```

Having such a library provides access to FCS files, opens up to a wider variety of tool builders, and simplifies the work for new companies or individuals building better ways to handle the files. In the process, it encourages everyone to use standardized tools that will interoperate. As long as the community is able to avoid restrictions imposed by legacy data or proprietary formats, the software will evolve quickly.

Flow Cytometry Markup Language

The major shortcoming of the ongoing API approach is that it does not go far enough. Programmers can use this library to access the raw data, but there is no way to ask for the percentage of CD4+ cells within the T-cell population. For that, there needs to be a richer language to access more of the breadth of functionality that is central to flow cytometry.

In the few years since the FCS file format was last revised and the API was proposed, much has changed. Over the last 5 years, the World Wide Web has emerged to become of such importance that it directs current programming techniques. Undoubtedly, the Internet is the dominant influence in all computing, networking, scientific, and telecommunications fields. Therefore, looking at it from today's perspective, the obvious technology to structure cytometric data is the one used in the World Wide Web.

This technology is XML, the eXtensible Markup Language (19, 20). It is the next generation of HTML, the HyperText Markup Language, designed to fix design flaws in the original HTML definition while maintaining compatibility with existing documents (8, 15). XML is the officially sanctioned language of the Internet, and as such, it is the irrefutable choice as the document architecture in any informatics plans these days. The language is just completing an exhaustive standards committee evaluation and is not expected to be undergoing changes in the near term. The W3C consortium is the ruling body of the World Wide Web, and hence as authoritative an organization as one could hope to find in the high-tech industry. The current versions of Web browsers support XML already, and there are hosts of tools for creating, editing, and validating documents (see xml.com/resourceguide/).

XML brings an assortment of client service tools, but more important, it brings the ability to use commercial server products. Every transaction service, information request, or other service provided over the Internet is using a markup language, so it seems obvious that industrial strength database tools, validation tools, and other utilities are available and supported. Very little customization is required to extend tools to a particular set of elements, so teaching browsers to display contour plots of flow cytometry data is a relatively simple operation, especially when compared to writing a stand-alone application to perform the same function (2).

Importantly, IT departments (the information technology specialists within an organization) are comfortable with XML. Any project that is large enough to involve many people or resources will ultimately have to work with the group of people responsible for the computer infrastructure. If you request support running scientific data servers, you are likely to undergo long meetings explaining all sorts of irrelevant minutia and to receive assessments that your project is (depending on whether you are in academia or industry) either impossible or prohibitively expensive. If the same support staff members are asked to support a database of XML structured data, they are more likely to nod enthusiastically and set you up an account on a system that they already use for other purposes.

Within the medical community there is also strong support for XML (9). The HL7 is developing standards for the electronic exchange of medical data, for clinical and administrative purposes, that is entirely based on XML (6, 18). Using this structured document technology for the description of flow cytometry data is essential to being able to manage it in a clinical environment (5).

XML is a small set of syntax rules that allows users of a particular domain or interest group to embed custom tags (7). If a browser reads an XML document, either from a file or off of the Internet, it will display the document as it does a Web page. If the browser encounters a tag that it does not understand, it has a mechanism to search for a plug-in that might be able to render it. Therefore, using XML to express the objects defined in the FCS API will make it simple to view FCS files from within a Web browser. This will reduce the cost of making custom clients to a minimum expense and will leverage an infrastructure driven by a massive consumer market. Using XML, any object can be expressed as a text string. By definition, these strings are encoded using the Unicode standard (17), so all major languages and alphabets, including all of the Asian character sets, are supported in XML documents.

A canonical example of an XML structure is the bibliographic reference. Any published scientist knows the importance of managing a database of references. At present, you have to use a special tool to maintain a detailed description of the references to be able to format them for each different publication. Using XML, a single reference can contain information with tags describing it, and the publisher can reformat it to fit the style requirements of the journal.

In the future, a reference might be submitted as follows:

```
<reference title ="Effects of Dilithium Crystal Diffraction
in High Throughput Flow Cytometry Experiments">
   <author lastName="Scott"
           firstName="Montgomery"
           homePlanet="Earth"
           email=scottie@enterprise.starship.gov />
   <author lastName="Spock"
           homePlanet="Vulcan"
           email=spock@enterprise.starship.gov />
</reference>
```

This version clearly contains more structured information than a traditional reference and, hence, would be preferable to any program processing its content. The journal publisher could display it as it looks best in print, Web pages, tricorder, or whatever the medium in which it is being viewed.

Just as the bibliographic reference can be stored as a structured element in XML, markup language elements can be created for objects in the domain of flow cytometry. The first proposal for a Flow Cytometry Markup Language (FCML) was presented at the ISAC conference in 2000 (16), and all recent versions of our commercial software supprt XML-based workspaces. The definition is still in development and there are no cross-vendor standards in place, yet the advantages of this architecture are undeniable, and several XML implementations are currently under development. A consortium of academic and commercial researchers has been formed to referee the process to arrive at a single common language.

The FCS API provides concepts such as *DataSet* and *Parameter*, but it does not provide any means for them to be handled outside the context of a complete data file. XML will facilitate those objects being transmitted in e-mail, stored in databases, and embedded into Web pages. It is easy to extend the set of objects in the FCML vocabulary. FCML contains definitions for concepts such as *Gates*, *Graphs*, *Statistics*, and *Reports*, which are important building blocks of flow cytometry analysis that are not included in either the FCS file format or the application programming interface.

Elements defined in FCML can be combined with other dialects of XML to leverage standards that are already in use. Each XML dialect defines its own name space, so the terminology from one field can reference other dialects or redefine the same terms without danger of ambiguity or confusion.

A key element in FCML is the *Gate* element. The concept of gating, subsetting events on the basis of the values of one or two parameters at a time has long been used as the key operation in flow cytometry analysis. Yet the algebraic definition has never been formalized, and there is no standard terminology to describe the gate. There is no public test suite of sample files and gates so that a new tool can test against known results. As a result, clinical approval of flow cytometry software is based on exhaustive examination of the source code, instead of the much simpler automated execution of a test suite. A common understanding of the definition of the gate will make the creation of such a test suite very simple.

Most elements defined in FCML are involved in a request for a graph or statistic. The encompassing element for containing the component of that request is called a *Query*. A *Query* contains three elements: a reference to a data set, a definition of the subset of events that are of interest, and the details of what statistic or graph you want to display. An example *Query* is shown here:

```
<Query id="foo">
   <DataSet id="myfile" uri=file://C:\flow\files\exp1.001/>
   <Population id="lymphocytes" >
      <Gate>
         <RectangleGate x="ForSc" y="OrthSc" left="200"
             right="300" top="120" bottom="40">
      </Gate>
   </Population>
   <Statistic type="median" parameter="CD3">
</Query>
```

The *DataSet* tag points to the source of the data. Conventionally, this is an FCS file, but it could conceivably be a uniform resource identifier Web location, or a key by which a database references the data. This is an important extension, as it allows handling files that are not on the local computer.

The second element is a *Population*, which specifies the information needed to delineate the subset of events that is processed in the graph or statistic. A *Population* is closely related to a *Gate* but is slightly different. A *Gate* is an operation, and a *Population* is a result of that operation. A *Population* can contain zero or more *Gate* elements. Examples of *Gates* include *RectangleGate, RangeGate, PolygonGate*, and *EllipseGate*, to define population subsets based on their geometrical position in one or two dimensions. *AndGate, OrGate*, and *NotGate* are logical combinations of other gates, so that a *Gate* can be composed of any number of other *Gates*. The conjunction ("And-ing") of successive gates is the common means of expressing populations, because the scientist is generally interested in the events that express all of the characteristics in a branch of an analysis tree.

The third element names the test or visualization to perform on the data set. This may be a *Graph*, or it may be a *Statistic* element. Each of these have more specific definitions as subelements; *ScatterPlot, ContourPlot, Histogram*, and so forth, in the case of the *Graph*, or *Mean, Median, Count, Frequency*, and so forth, for the *Statistic*. Statistics and graphs have elements detailing the one or two parameters that they visualize, as well as often other incidental arguments such as the number of contour levels, the color of dots, and so on.

Every graph and statistic of any arbitrarily complex report in flow cytometry can be broken down into a trio of data elements: the pointer to the FCS file, the criteria for the relevant events within the data set, and the description of the operation to perform. With these pieces of information, an analysis engine can locate the data, gate it to the population of interest, and compute the statistic or draw an image.

All traditional reports, tables, and plots can be built from a collection of single requests for graphs or statistics. There may be optimization reasons to build more complex queries, but with requests for a single graph and a single statistic, an arbitrarily complex report can be built, so most of the elements defined in FCML are found inside of a *Query* element.

To gain the best use of the vast amount of data being gathered, there needs to be a mechanism to map data files to a laboratory protocol (4). A *Protocol* is an XML element containing a set of keywords associated with the file. This might be as simple as the name of the operator, the date, or the stain names. Each one has a pointer to that file's location, a set of parameter specifications, and a map directing the parameters to specific wells. This information can be created in advance of the experiment and can contain any information that you want to store with the data files.

An *Experiment* is an FCML element that serves as a container for the other elements discussed so far. The example below shows an *Experiment* containing *Protocol* elements and *DataSetsRef* elements to show how different *Protocols* can be applied to *DataSets* within a single element. This serves to describe the organization of the staining panels for a multiwell experiment. Note that the Experiment contains *DataSetRef* elements instead of *DataSets*. The *DataSetRef* is a reference to a *DataSet*, which may or may not exist at the time the Experiment is created. This provides the capability to talk about the structure of a laboratory experiment without having or knowing the location of the actual data. This is crucial functionality because it provides the ability to create reusable templates that can describe future or abstract experiments, and a template could be handed to the acquisition program to annotate the data as it is collected. As the acquisition program creates the files, it would replace the references with the new data files and write a new version of the Experiment containing the newly acquired information.

```
<Experiment date="15-Jan-2002" id="Exp.3239">
   <Protocol name=" system">
      <Keyword word="$OP" value="BOB">
      <Keyword word="$CYT" value="Sparky2000">
   </Protocol>
   <Protocol name="panel1">
   <Protocol name="system" />
      <Keyword word="$P3N" value="FITC">
      <Keyword word="$P3S" value="CD3">
      <Keyword word="$P4N" value="PE">
      <Keyword word="$P4S" value="CD4">
   </Protocol>
   <Protocol name="panel2">
   <Protocol name="system" />
      <Keyword word="$P3N" value="FITC">
      <Keyword word="$P3S" value="CD3">
      <Keyword word="$P4N" value="PE">
      <Keyword word="$P4S" value="CD8">
   </Protocol>
   <DataSetRef id="1" well="A1">
<ProtocolRef name="panel1" /> </DataSetRef>
   <DataSetRef id="2" well="A2">
<ProtocolRef name="panel1" /> </DataSetRef>
   <DataSetRef id="3" well="A3">
<ProtocolRef name="panel1" /> </DataSetRef>
   <DataSetRef id="4" well="B1">
<ProtocolRef name="panel2" /> </DataSetRef>
   <DataSetRef id="5" well="B2">
<ProtocolRef name="panel2" /> </DataSetRef>
   <DataSetRef id="6" well="B3">
<ProtocolRef name="panel2" /> </DataSetRef>
</Experiment>
```

Taking it a step further, it is also possible to embed *Populations* within *Experiments* so that the gating information can also be embedded in a template. This addition establishes templates of analysis; gates that predate the data they are filtering. This is familiar to anyone who has sorted cells, as the sort gate is defined before acquisition, but the predefined gates can also be important in repetitive analysis. Because the nature of most high-throughput experiments is to alter the biology and test for effect, it is common to have gates defined on control samples and then apply the identical gates to the test samples. FCML provides the abstraction layer to implement this templated batch analysis.

Additional FCML elements can be created as necessary to specify the structure of new analysis operations, such as cell cycle or proliferation analyses. The beauty of the XML syntax is that it allows documents to contain many different types of information that are each encapsulated inside of angle-bracketed tags. Any program trying to read a FCML document can easily ignore tags it does not understand without limiting its ability to read the data types it does comprehend.

Summary

Software may be complex to create, but once written, it is extremely malleable. In building instruments or designing biological experiments, everything has to be right from the beginning, or it is very difficult to fix after the fact. Software, though it can suffer to some extent from poor design, is much more amenable to extension over time. Software also has a relatively smooth evolutionary track. Minor changes can be made and released incrementally via the Internet. There is no physical medium necessary, so the cost of distributing improvements is negligible. Best of all, software is scalable. As high-throughput experiments grow, the cost of reagents and instruments will continue to climb. The cost of software (depending on license terms) will not. To go from three-color experiments to six-color experiments is a major leap in the laboratory and the setup of instrumentation but requires virtually no change in the programs used to collect, analyze, or manage the data.

The future will undoubtedly bring more impressive applications of using software to automatically analyze, gate, and categorize cell populations without human intervention. The continuous growth of computing power will undoubtedly deliver the cycles to make that intelligence possible. This chapter has focused on relatively mundane tasks of managing data because, for the present, that is the important challenge. We are going to generate more and more data. and unless we invest in methods to annotate, store, and control it, we will not be able to understand it.

References

1. Coder, D.M. Data management and storage, in Diamond, R.A. and DiMaggio, S., eds. *In Living Color Protocols in Flow Cytometry and Cell Sorting*, Springer-Verlag, Berlin and Heidelberg, 2000.
2. Daconta, M. and Saganich, A. *XML Development with Java 2*, SAMS, Indianapolis, IN, 2001.
3. Dean, P.N., Bagwell, C.B., Lindmo, T., Murphy, R.F., and Salzman, G.C. Data file standard for flow cytometry. *Cytometry* 1990;11:323–332, 1990.
4. Dean, P.N. Methods of data analysis in flow cytometry, in Van Dilla M.A., Dean, P.N, Laerum O.D., and Melamed M.R., eds. *Flow Cytometry: Instrumentation and Data Analysis*, Academic Press, Orlando, FL, 1985.
5. Dolin, R.H., Rishel, W., Biron, P.V., Spinosa, J., and Mattison, J.E. SGML and XML as interchange formats for HL7 messages. *JAMIA Fall Symp Suppl* 1998:720–724.
6. Dolin, R.H., Alschuler, L., Beebe, C., Biron, P.V., Boyer, S.L., Essin, D., Kimber, E., Lincoln, T., and Mattison, J.E. The HL7 clinical document architecture. *J Am Med Inform Assoc* 2001;8(6):552–569.
7. Fallside, D.C., editor. *XML Schema Part 0: Primer*, World-Wide Web Consortium Candidate Recommendation, 24 October 2000. Available at: www.w3.org/TR/xmlschema-0.
8. Goldfarb, C.F. and Prescod, P. *The XML Handbook*, Prentice Hall PTR, Upper Saddle River, NJ, 1998.
9. Kahn, C.E., Jr. A generalized language for platform-independent structured reporting. *Methods Information Med* 1997;36(3):163–171.
10. Lane, C.D., CFCS library. Available at: http://flowjo.com/cfcs/cfcsjava.pdf
11. Murphy, R.F. and Chused, T.M. A proposal for a flow cytometric data file standard. *Cytometry* 1984;5:553–555.
12. Murphy, R.F. FCS API Development Project Home Page. Available at: murphylab.web.cmu.edu/FCSAPI/.
13. Robinson, J.P., Durack, G., and Kelley, S. An innovation in flow cytometry data collection

and analysis producing a correlated multiple sample analysis in a single file. *Cytometry* 1991;12:82–90.

14. Seamer, L.C., Bagwell, C.B., Barden, L., Redelman, D., Salzman, G.C., Wood, J.C., and Murphy R.F. Proposed new data file standard for flow cytometry, version FCS 3.0. *Cytometry* 1997;28:118–122.

15. St. Laurent, S. *XML Elements of Style*, McGraw-Hill, Columbus, OH, 2000.

16. Treister, A. and Briggs, D. *Draft Proposal for a Flow Cytometry Markup Language*, presentation to ISAC XX, Montpelier, France, 2000.

17. The Unicode Consortium, *The Unicode Standard, Version 3.0*, Addison-Wesley Longman, Reading, MA, 2000.

18. Van Hentenryck, K. Health Level Seven: a standard for health information systems. *Medical Computing Today*, June 1998; also available at: www.medicalcomputingtoday.com/0ophl7.html.

19. Bray, T., Paoli, J., Sperberg-McQueen, C.M., editors. World Wide Web Consortium, *Extensible Markup Language (XML)*, 1.0. (W3C Recommendation 10 February 1998). (REC-xml-19980210). Available at: www.w3.org/TR/xml11

20. World Wide Web Consortium, *XML Pointer Language (XPointer)*, working draft, 3 March 1998a. Available at: www.w3.org/TR/xptr-framework/.

9

Molecular Assemblies, Probes, and Proteomics in Flow Cytometry

STEVEN W. GRAVES, JOHN P. NOLAN, AND
LARRY A. SKLAR

Introduction

The many proteins and nucleic acids encoded in the genome predominantly perform their functions as macromolecular assemblies. In fact, modern biomedical research often targets the interactions of individual molecules of these assemblies, usually by disrupting or enhancing specific contacts, to provide treatment for many different diseases (15). Therefore, efficient pharmaceutical design requires knowledge of how macromolecular assemblies are built and function. To achieve this goal, sensitive and quantitative tools are essential.

This chapter will discuss the use of flow cytometry as a general platform for sensitive measurement and quantification of molecular assemblies. First, this chapter will introduce general methods for analysis of molecular interactions along with a comparison of flow cytometry with these methods. Second, an overview of current flow cytometry instrumentation, assay technologies, and applications in molecular assembly analysis will be given. Third, the implementation of the above approaches in molecular assembly will be discussed. Finally, potential future directions of flow cytometry in molecular assembly analysis will be explored.

At present, the analysis of macromolecular assemblies is performed by a wide variety of techniques that are chosen for the target molecules under study (proteins, DNA, lipids, etc.), the type of measurement required (kinetic or equilibrium), and whether the assembly of interest needs to be studied in vivo or in vitro. This continuum of techniques can be divided into the heterogeneous assays, which require a separation step to resolve products from reactants, and homogeneous assays, which can measure interactions without a separation step.

Heterogeneous assays, in general, use radioisotopes, which are not perturbing; offer excellent sensitivity; and provide accurate quantification. The products are quantified after a separation step such as gel filtration, gel electrophoresis, or centrifugation. Rapid quench methods can provide subsecond kinetic resolution (49); however, the added separation steps are tedious and make collection of kinetic time courses difficult, as each time point must be separated and measured individually. Furthermore, in the time it takes the separation to occur, the interaction of interest can dissociate, which is a problem specific to low-affinity assemblies. Nonetheless, by using rapid chemical quench techniques, reaction times as short as a few milliseconds can be observed (77).

Homogenous assays can be separated into solution- or surface-based assays. Solution-based assays measure an optical signal generated by the assembly to quantify an interaction. High component concentrations (micromolar) allow changes in intrinsic molecular properties, such as protein fluorescence or circular dichroism, to be used to study molecular assemblies. For greater sensitivity (nanomole component concentrations), resonance energy transfer (86) or polarization assays (48) using exogenous fluorescent labels can be used. In combination with stopped-flow spectroscopy methodologies, solution-based assays allow reactions to monitored in a continuous fashion with submillisecond dead times (49).

Surface-based homogenous assays provide information on molecular assemblies by measuring the properties of the surface to which the assembly is attached. Typically, one component is attached to the surface, and another solution-based component changes the surface's optical properties on binding. In the case of surface plasmon resonance and resonant mirror systems, changes of mass in molecules attached to a surface as a result of binding events result in a measurable change of the surface's refractive index (25, 64, 84, 96). Kinetics measured on such systems are not straightforward; rather, they are based on complex modeling from which the kinetic constants are extrapolated, and the range of rates measured are limited by mass transport issues (84). The scintillation proximity assay uses β-particle-emitting radio-labeled molecules to generate luminescent signals from immobilized molecules on a bead containing a scintillant, which occurs when the labeled molecule is near the bead (16). This system provides micromolar sensitivity with kinetic resolution on the timescale of a minute, using standard techniques, or a few seconds, when combined with freeze-quench methodologies (76).

Flow cytometry has long been recognized for its ability to measure ligand–receptor interactions on cell surfaces (7). Flow cytometry's inherent ability to discriminate free from bound probe on a particle surface makes it a homogeneous assay format that is sensitive enough to measure a single fluorescent molecule and routinely measure femtomole concentrations of fluorophores (68). The addition of high-resolution time collection made flow cytometry a platform that could perform high-sensitivity cellular kinetic and equilibrium studies with microsecond temporal resolution (62, 69). Combined with improved mixing technologies, this has allowed subsecond dead times for kinetic measurements on a flow cytometer (70). In addition, the development of microspheres and the corresponding surface chemistries has opened flow cytometry to particle-based molecular assembly studies (72). In combination with high-throughput technologies, such as multiplexed microsphere sets (52) and high-throughput sample handling (55), flow cytometry is poised to become the premier platform for molecular assembly analysis, having the ability to perform homogenous assays with subsecond kinetic resolution and subfemtomole sensitivity on thousands of assemblies daily.

Overview

Instrumentation

Flow cytometry uses hydrodynamic focusing to generate a narrow sample stream (generally containing cells or microspheres) that is interrogated by a tightly focused laser beam. Multiple optical parameters, such as light scatter, along with several wavelengths of fluorescence, are collected simultaneously from this interrogation volume, using high-sensitivity detectors. Because the interrogation volume is small, background from the unbound fluorescent probe is low, which allows the measurement of particle-associated probes without separation steps. Fluorescence emission, light scatter, and time are recorded separately for each particle, enabling the observation of population distributions of particles in a mixture with continuous kinetic resolution. The principles behind this remarkable instrument are well reviewed in other works (87, 93). However, for the purposes of molecular assembly, particular emphasis needs to be put on the mechanism by which a flow cytometer discriminates free probe from bound probe, on mixing devices for kinetic flow cytometry, and on thermodynamic regulation of flow cytometry samples.

Resolution of Free Versus Bound

The discrimination of free versus bound probes in a flow cytometer depends on specifically measuring the particle-associated fluorescence as the laser beam interrogates it. Typically, light scattering generated by the particle triggers data acquisition. The fluorescence reported depends on the amount of fluorophore present in the interrogation volume, which is defined by the intersection of the laser and the sample stream (figure 9.1).

When low concentrations of free probe are used, discrimination of free versus bound probes is easily performed because the ratio of free probe relative to bound probe is small. However, when low-affinity interactions are studied, higher concentrations of free probe are required for binding. In this situation, the unbound probe present in the interrogation volume will cause a high direct current background that obscures specific probe binding and limits the range of probe concentrations at which flow cytometry can discriminate free versus bound (figure 9.1). The quantitative aspects of this effect have been considered, and it was shown that the theoretical sensitivity of a flow cytometer depends on the number of binding sites on the particle, the probe's affinity for those sites, and the interrogation volume as shown by the following equation (67):

$$\frac{S}{B} = \frac{N_b}{(V_i - V_p) \cdot C_{free} \cdot N_0}$$

where N_b is the number of fluorescent molecules on the surface of the particle, V_i is the interrogation volume, V_p is the volume displaced by the particle is the sample stream, C_{free} is the concentration of free fluorescent probe in the sample stream, and N_0 is Avogadro's number. In general, one or more of these factors limit the observation of low-affinity interactions by flow cytometry to those with dissociation constants less than 100–200 nM (68). Reductions in the interrogation volume, via smaller laser spots and stream diameters, will improve the resolution of free versus bound probe if the back-

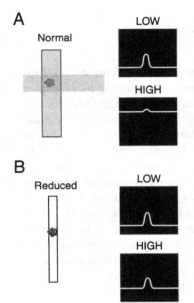

Figure 9.1. Resolution of free versus bound probe. (*A*) Normal laser and stream dimensions, where the sample stream is a vertical rectangle and the laser beam a horizontal rectangle. A particle labeled with several fluorophores is shown in the stream. The black boxes represent the direct current signal in either a low or high background of free fluorophore. (*B*) The same as (*A*), except with reduced laser and stream dimensions, resulting in a smaller interrogation volume.

ground signal is predominantly the result of free probe (figure 9.1). As the laser spot size and stream diameter are reduced, a particle with low fluorescence could result in a negative signal. The generation of negative signals by particles in a high-fluorescence background is the basis for a volumetric sizing technique (35), and this effect could potentially used to quantify particles with low fluorescence in a high background.

Kinetic Instrumentation

The addition of time as a parameter in the 1980s made flow cytometry an experimental platform capable of kinetic measurements (62). Current commercial instruments provide electronic time stamps for each event with microsecond resolution. However, current fluidics and optics limit instruments to analysis rates of about 5×10^5 particles per second. Because 100 particles are generally required to obtain meaningful particle statistics, the temporal resolution is limited to between 10 and 100 ms, which is sufficient to observe most molecular assemblies.

However, introduction of accurately mixed samples to the flow cytometer is more problematic. When manual kinetic experiments are performed, they require removal of the sample tube, addition of reagent, mixing of the sample, and replacement of the tube. This process results in a dead time of at least 10 s, which is too long to observe most molecular assemblies.

To shorten this dead time, both time window and continuous analysis approaches to kinetic mixing on a flow cytometer have been employed. Time-window methods mix the sample at a set distance from the analysis point, and this distance determines the analysis time (6, 21, 83, 99). Coaxial mixers placed within the sample stream are the best example of this method (figure 9.2). The small size and controlled mixing of these devices limits the disturbance to fluidics of the flow cytometer and allows close

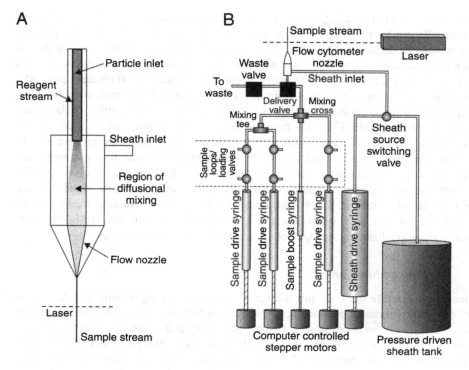

Figure 9.2. Stopped flow and coaxial mixing schematics.

placement of the mixer to the analysis point (83). Flow cytometers equipped with this technology have measured reactions, with dead times as short as 100 ms (6). By varying the distance of the mixer relative to the analysis point reaction, times up to 3 s have been observed (6). However, as with all time-window methods, this method only collects a discreet time point of an overall reaction. To observe the complete reaction, the mixer must be moved to many locations to collect multiple time points that permit the entire reaction to be extrapolated. Although this process provides little disturbance to the fluidics of the system, as with all time-window methods, collection of the complete time course can be complicated.

As an alternative, continuous analysis methods mix the sample as rapidly and closely as possible to the analysis point (34, 50, 51, 58, 70, 75, 85). After mixing, the sample is delivered to analysis, and data are continuously recorded using the time of mixing as time zero for the experiment. The continuous data acquisition allows the complete reaction to be monitored from a single mixing event, which increases convenience and decreases sample consumption. However, because the entire time course is monitored using this method, the volume of mixed sample is larger. This larger volume requires vigorous mixing, which can disturb the fluidics of the flow cytometer, which would ordinarily prevent the observation of fast reactions. However, the use of high-speed valves and computer-controlled stepper motor–driven syringes has reduced this problem, as illustrated by the rapid-mix flow cytometer (figure 9.2). In this configuration, mixing occurs with the waste valve open and the cytometer access valve closed, isolating fluidic disturbances from the analysis point during mixing. After mixing, the valves reverse positions followed by a rapid boost of the sample to the analysis point, whereupon the syringes slow the delivery rate to allow analysis of the sample. Initial iterations of this instrument could analyze reactions with a 300-ms dead time (70), and lower-cost versions have been constructed that provide nearly as good performance (500 ms dead time; 85). The fluidic disturbance originating from the boost of sample to the analysis point is the primary limitation to shorter dead times (34). The magnitude and duration of the fluidic disturbance is the result of pressure built up during the boost phase, which is directly related to the geometry of the flow nozzle (34). Using careful nozzle design and a high-resolution stepper motor syringe drives, the most recent rapid-mix flow cytometer has achieved dead times as low as 100 ms (34). Furthermore, the multiple syringes allow for multiple mixing steps for complex experimental protocols (pulse-chase, rapid dilution, etc.) and direct control of sheath flow rates, which has the potential to further reduce dead times (34, 85). Such systems offer a great deal of versatility and are likely to become the sample handling methods of choice for kinetic flow cytometry.

Temperature Control

Kinetic and equilibrium constants are greatly affected by temperature; therefore, control of this experimental parameter is critical. Nonetheless, temperature control on flow cytometers is not well developed. Historically, sample thermoregulation in flow cytometry has been accomplished with simple water-bath devices, which are relatively static, expensive, and bulky (51, 75). More recently, Peltier modules have been used to regulate sample temperature (32). These devices are advantageous because they are small, inexpensive, and dynamic (figure 9.3). These devices can be used in a static manner to provide simple regulation or in a dynamic manner to obtain thermodynamic values for reactions, such as nucleic acid denaturation, or to control the specificity of nucleic acid hybridizations (32). Peltier modules can control the temperature of the sample line between 0° and 100°C and heat the sample from 20° to 95°C in less than a minute. These capabilities, along with their other advantages, are likely to make Peltier modules the preferred thermoregulation element for flow cytometers.

Assay Technologies

Flow cytometry's analytical features have made it a powerful tool for molecular assembly analysis using simple binding assays, as well as more complex cellular and assays. However, the scatter properties of a flow cytometer (forward and side) are generally insensitive to molecular assembly formation (with some exceptions discussed below), which necessitates the use of fluorescently labeled molecules to monitor molecular assembly formation. Furthermore, as flow cytometry measures particle-associated fluorescence, molecules of interest must be either attached to a microsphere or displayed on a cellular surface for the assembly of interest to be observed.

Fluorescent Labeling Strategies for Biomolecules

Nucleotides and proteins are the most common biomolecules involved in molecular assemblies, and as such, much work has been done on the fluorescent labeling of these

Figure 9.3. (*A*) Schematic of the dynamic inline thermoregulation unit. A, sample inlet; B, titanium sample line; C, copper heat transfer block; D, titanium sample line; E, sample delivery line to the nozzle; F, flow cytometry nozzle; G, nozzle sample line; H, holding bracket; I, sheath inlet; J, focusing nozzle; K, sample stream; L, analysis point; M, fan; N, heat sink; O, Peltier module. (*B*) Melting curve of a DNA oligonucleotide from a microsphere generated using the dynamic inline thermoregulation unit.

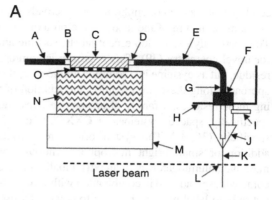

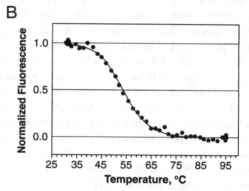

molecules. In the case of nucleotides, labeling is not generally performed by the average biochemist because of the large number of fluorescent nucleotides available commercially. As most of these nucleotides are available as phosphoramidites, it is possible to simply order the fluorophore of choice in the desired position within a polynucleotide sequence. However, this is not the case for proteins, and historically, peptides and proteins have been labeled using chemical conjugation techniques involving derivatized fluorophores directed at amino groups (46). Amine reactive fluorescein conjugates have been the most common choice for protein labeling, as this fluorophore has good water solubility, extinction coefficients, and quantum efficiency (39). However, this fluorophore suffers from pH-sensitive fluorescence, a broad emission spectrum (problematic for multicolor experiments), photobleaching, and self-quenching when high levels of labeling are used (39). Several other common dyes improve on these features; most notably the BODIPY and Alexa series of dyes from molecular probes, which have derivatives that cover the entire visible spectrum (39). Amine-reactive cyanine and rhodamine dyes are also available if longer wavelengths of fluorescence are needed (39). Although this approach allows for many different fluorophores to be used, it is nonspecific with regards to fluorophore location on the protein, which can interfere with protein interactions.

Recombinant DNA technology has provided several avenues to improve on the above scenarios, and site-specific mutation can be used to direct the location of a fluorophore. For example, biotinylation sites or epitopes can be added to direct the binding of fluorescently labeled avidin or antibodies, respectively (91). Although this is convenient, the size or chemical nature of the label on the protein could interfere with potential interactions. To specifically locate fluorescent labels, specific amino acids can be engineered to allow for novel chemistry (39). This approach is commonly attempted using a single cysteine residue, but as cysteine is a common amino acid, this can be problematic. A powerful new approach along these lines requires the addition of two cysteines at specific spacing. Using these residues, a family of arsenoxide dyes known as FLaSH probes, which require specifically spaced cysteines (CCXXCC), can be attached with a high degree of specificity (36, 37). This system offers great promise, as it allows for sequence specific addition of small bright fluorophores (http://www.invitrogen.com/content/sfs/manuals/mammalianlumiogateway_man.pdf). Finally, the use of fluorescent fusion proteins has become widespread, and in combination with flow cytometry, they offer a convenient method not only to label proteins but also to potentially use the labeled proteins in living cells to study in vivo interactions. The most commonly used fluorescent proteins are green fluorescent protein (GFP with 520 nm emission) from *Aequorea victoria* and its variants: enhanced green (eGFP, 30-fold brighter than GFP), enhanced blue (eBFP, emits at 430 nm), enhanced cyan (eCFP, emits at 475 nm), and enhanced yellow (eYFP, emits at 540 nm; 100). In addition to these, red fluorescent proteins from *Discosoma* sp. and *Heteractis crispa* provide proteins that fluoresce in the red region (dsRED, 580 nm emission, and HcRED, 645 nm emission; 38). Although GFP-related proteins are about 27 kDa in size, their addition to either end of a protein seldom has an effect on protein function, which allows them to be used for both in vivo and in vitro molecular interaction studies (100). However, the red proteins oligomerize strongly, which limits their effectiveness in molecular assembly studies. Monomeric variants have been constructed that have greater promise in these applications (11).

Fluorescence Analysis Techniques

Once labeled, the fluorescent reporter molecules can be used in conjunction with the inherent resolution of free versus bound probe of a flow cytometer to simply monitor particle-based fluorescence as an indication of a molecular interaction. Alternatively, fluorescence resonance energy transfer offers sensitive detection of molecular interactions that bring two well-matched fluorophores into close spatial proximity with one another. This technique with its stringent distance dependence allows for discreet detection of specific molecular interactions. Fluorescence resonance energy transfer (FRET) has been successfully used in cellular microscopy and fluorescence spectroscopy, providing sensitive detection of a wide variety of molecular assemblies (63). FRET combined with flow cytometry has also determined cell surface interactions of receptors, using fluorescently labeled antibodies (3). Fusion proteins with GFP variants have been used in a FRET system in combination with flow cytometry to assess the interactions of receptor subunits (12). Finally, a two-tiered FRET system on microspheres has been used to detect multivalent interactions between cholera toxin and ganglioside GM1, using a flow cytometer (92).

Attachment Strategies

In vitro molecular assembly studies require the attachment of biomolecules to a microsphere surface. In the case of oligonucleotides, this is relatively simple, as they are readily synthesized with terminal amino groups or specifically located biotin tags. Using these groups, nucleotides either can be attached to surfaces in a covalent fashion with a known orientation via chemical conjugation techniques or bound to avidin- or streptavidin-coated microspheres (10).

However, proteins pose a more difficult problem because the chemical conjugation of proteins to surfaces by their amine groups results in a random conjugation location within proteins and orientation of proteins on the surface, which can greatly affect function and interaction studies. To overcome this problem, proteins are often expressed with affinity tags such as FLAG, biotinylation sequence, c-Myc, GST, His, or HAT and then attached to microspheres that have the corresponding binding molecule (antibodies, glutathione, avidin, etc.) on their surface (72). These tags bind with relatively high affinity and allow for known orientation of binding, but each suffers potential problems. For example, large protein tags, such as GST and c-Myc, can interfere with protein interactions and structure, biotinylation of artificially incorporated biotinylation sites is often inefficient, and small epitope tags (FLAG) require the use of large antibody molecules on the surface of the microsphere that can interfere with molecular interactions. Alternatively, small metal binding tags (His and HAT) can be used, but these tags only bind chelated metals with nanomole affinity, which allows significant dissociation and exchange of molecules from the surface of the microsphere (57). Finally, glycosylated proteins can be attached to microspheres coated with lectins, as demonstrated by the capture of crude membrane fractions onto microspheres for analysis by flow cytometry (5).

Microsphere Arrays

Multiplex microsphere sets are now available commercially from several sources. Microspheres are multiplexed by differential staining with a fluorescent dye that results in a set of microspheres with several distinct levels of fluorescence. Using differential staining with two fluorescent dyes, each at individual levels, multiplex microsphere sets with 100 elements have been created (Luminex Corporation, Austin, TX). Smaller sets of 10 are available from other manufacturers and are created by either using one dye (Duke Scientific, Palo Alto, CA) or staining two different-sized microsphere sets with five different levels of an individual dye (Bangs Laboratories, Fishers, IN). Multiplex microsphere arrays have been critical in the burgeoning single-nucleotide polymorphism field (73; see chapter 14), and they are anticipated to provide a path for molecular assembly analysis by flow cytometry to enter the wider field of proteomics, discussed below in the Applications section.

Display Strategies

The in vitro study of discreet molecular interactions has largely been the domain of microspheres. However, the use of protein libraries displayed on the surface of bacterial and yeast cells has recently been used in the high-throughput screening of antibody binding, T-cell receptor binding, and enzyme interactions (13, 26, 45, 53, 54, 74, 97). This approach has the potential to be immensely powerful, as it eliminates the need for high-level expression and purification of recombinant proteins that are required for attachment to microspheres. In the yeast system, the proteins are expressed as fusions with the N terminus of the yeast cell wall protein Aga2p (97). These fusions are then expressed on the yeast cell surface, where they have been shown to be available for molecular interactions in the form of binding studies. In the case of bacterial display, the proteins are expressed as N-terminal fusions of the *Escherichia coli* Lpp-OmpA′ system (26). This results in the display of the protein on the bacterial cell surface. This system has been used successfully in both binding and enzyme interactions (19, 26, 74). Although these systems have primarily been used to display protein libraries for relatively simple binding assays, it is not difficult to conceive of their use as the basis of display for more complex assays on single or small subsets of molecular assemblies.

Applications

At present, the two common aspects of all flow cytometer molecular assembly assays are that they occur on or in a particle and that they use optical reporting systems. The particles can be cells or microspheres, and the reporting system can be the simple presence of a fluorescent dye, the increased scatter caused by cell aggregation, or the interaction of multiple fluorescent dyes. Within these bounds, a wide variety of molecular assembly assays can be performed on the flow cytometer, ranging from straightforward binding assays to complex mechanistic assays using multiple probes and FRET. Furthermore, with the addition of multiplex technologies, it is possible to measure many interactions simultaneously.

Cell-Based Studies

Flow cytometry has been used to study a wide variety of molecular interactions that occur on or in the cell. The inherent resolution of free versus bound probe offered by flow cytometry has made it an ideal tool to study many different ligand receptor interactions including lipoproteins (41), peptides (14), and nucleotide aptamers (81). Ligand binding to G-protein coupled receptors is an excellent example of such investigations into cellular molecular assemblies (see chapter 17) and reviews elsewhere (68, 89). In addition to studying ligand–receptor interactions, flow cytometry has been used to elucidate more complex cellular topics such as multivalent ligand binding (42, 43) and cellular adhesion (22, 23), which subjects are reviewed in chapter 18 and elsewhere (68, 69, 89).

Beyond the analysis of specific molecular interactions on the surface of the cell, significant progress has also been made in the analysis of internal cellular processes. Signal transduction has been investigated by the analysis of multiple kinase states, using multiple monoclonal antibodies labeled with varying fluors (78). Gene regulation has been inspected by monitoring levels intracellularly expressed GFP fusions of transcription factors (2). Intracellular enzymatic studies measuring the conversion of a fluorogenic substrate to monitor the kinetics and presence of several hydrolytic enzymes have also been performed (59, 60, 82). Intracellular measurements of molecular interactions, as shown here, demonstrate the potential of flow cytometry to measure a host of in vivo molecular assemblies given the right reporter systems.

As discussed above, recent pioneering work combining flow cytometry sorting and cellular display technologies has resulted in a powerful new field capable of the rapid screening of single-chain antibodies, T-cell receptors, and functional enzymes. These topics are again reviewed here (see chapter 12) and elsewhere, but this technique has the potential to greatly facilitate in vitro molecular assembly analysis by eliminating the need to purify and attach proteins to microsphere surfaces.

In combination with a flow cytometry sorter, it should be possible to screen a library of displayed proteins for interaction with a fluorescently tagged target protein by mixing cells expressing the library with the protein and then sorting labeled cells for recovery. The interacting proteins could then be decoded from the DNA of the sorted cells. In this way, very large protein libraries could be rapidly screened for molecular assembly interactions.

Protein–Protein and Protein–Peptide Interactions on a Microsphere

Flow cytometry immunoassays are the most common application of microsphere-based molecular assembly analysis, as they offer a rapid and sensitive detection method for almost any molecule to which an antibody can be raised (65). Recently, traditional immunoassays have been combined with multiplex microsphere arrays to evaluate antibody response to specific epitopes of *Bacillus anthracis* Protective Antigen (80). In addition, immunoprecipitations analyzed by flow cytometry are becoming an alternative to Western blotting (61).

Microspheres have also been used as the platform for the analysis of protein–protein interactions, as highlighted by work with receptor proteins and receptor ligands. N-formyl peptide receptors bound to Ni^{++}-NTA silica labeled microspheres were used to measure the binding constants of fluorescent peptide ligand (90) and were used in conjunction with FRET system to measure and confirm the binding affinities of antibodies for fluorescent β-endorphin and flu antigen peptides (9). Similar studies on heterotrimeric G protein subunits, in which detergent-solubilized G protein subunits were immobilized on streptavidin microspheres to determine kinetic and equilibrium constants such as K_D, k_{on}, and k_{off}, demonstrate the utility of taking this approach with proteins (4). Enzyme interactions with their protein substrates have also been studied. Fluorescently labeled gelatin attached to microspheres was used to monitor Gelatinase-B degradation activity and evaluate potential inhibitors for this enzymatic function (94). Finally, the evaluation of the equilibrium binding constants for a number of interactions was performed simultaneously by covalently coupling 34 coactivator peptides for nuclear receptors to a muliplex microsphere set and mixing the microspheres with fluorescently labeled receptor proteins (47). These examples demonstrate that flow cytometry can measure kinetic rates and equilibrium constants as well as make these measurements in multiplex fashion. Such features of flow cytometry position it to become a general platform for high-throughput analysis of protein–protein interactions such as proteomics.

Microsphere-Based DNA–Protein Interactions

DNA–protein interactions on a microsphere in combination with the rapid-mix flow cytometer discussed earlier (figure 9.1B) provided the seminal examples of subsecond kinetics performed on a flow cytometer (71). In this work, FEN-1, the human flap en-

A

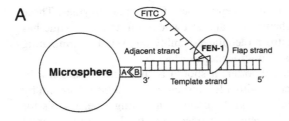

Figure 9.4. Cleavage of FLAP substrate by FEN-1. (*A*) Schematic of FEN-1 cleaving FLAP endonuclease. (*B*) Cleavage of FLAP substrate by FEN-1 versus time. Closed circle time points were measured with the rapid mix flow cytometer; open circles by hand mixing.

B

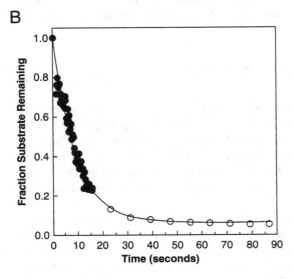

donuclease that is critical in recombination and repair, was mixed with fluorescent DNA substrates covalently attached to a microsphere. Cleavage of the fluorescent substrate resulted in a loss of microsphere-associated fluorescence (figure 9.4). This assay, in combination with a series of Mg^{++} jump experiments performed on the rapid-mix flow cytometer, determined the binding affinity and cleavage rates for FEN-1 on a DNA flap substrate (71). This work was further extended to evaluate the effect of point mutations on the structure and activity of FEN-1 (28, 88). Similar approaches have also been used with DNA polymerases and ligases (10), as well as with gelatinase, as described earlier (94).

Lipid Bilayers on Microspheres

A primary goal of in vitro biochemical experiments is to simulate the natural environment of a reaction in as realistic fashion as possible while maintaining control over all reaction constituents. This is difficult to do in the case of membrane-bound components such as receptors and lipids, as basic solution experiments do not adequately mimic the natural environment. However, microspheres allow the creation of lipid bilayers on microspheres, which provides for accurate imitation of the cellular environment of membrane molecules. Proteins attached to lipid bilayers have been constructed on microspheres to analyze multivalent binding by cholera toxin with its lipid receptor GM1 and to observe the resultant receptor clustering (56, 92). In this system, cholera toxin was unlabeled, but GM1 was present in the bilayers in both red and green fluorescent labeled forms. Cholera toxin binding resulted in receptor clustering, whereupon FRET occurred, providing a FRET signal for detection. One picomole of cholera toxin was achieved using a two-tiered FRET system (92). G-protein coupled receptors have also been reconstituted in lipid bilayers on microspheres (66). In addition, there have also been many studies of lipids on microspheres that do not require lipid mobility, which include studies on the binding interaction of factor VIII to determine the number and type of phospholipids involved in binding (30), measurement of the binding characteristics for the phospholipid binding protein Annexin V (95), and quantification of the affinity of different phospholipid antibodies, using microspheres using lipid bilayers (20). Finally, recent work demonstrated the value of lipid bilayers in the incorporation of ligands, fluorescent dyes, and transmembrane proteins (8). In this work, lipid bilayers were formed on solid glass beads to demonstrate the receptor density dependence of FRET between molecules within a supported lipid bilayer. Importantly, the use of mesoporous silica beads was explored both as a method of displaying molecules on the surface of a bilayers but also as a way for a microsphere to contain a "cytosolic" space in which molecules can be incorporated. Such technology could be extremely valuable not only in the display of transmembrane proteins but also in the creation of artificial storage compartments for biological molecules, which would greatly increase the versatility of microspheres in the molecular assembly and membrane transport applications (8).

Implementation

The study of molecular assemblies by flow cytometry has many applications, but the implementation of the methods follows common threads such as labeling, display/ attachment, quantification, binding artifacts, mixing, and experimental environmental control. Each of these threads has practical considerations for optimal implementation, which are discussed below.

Fluorescent Labeling

Fluorescent labeling of biomolecules is discussed in detail in the overview section above, and this section highlights aspects important to molecular assembly. Fluorescent labeling can be divided into methods for nucleotides and methods for proteins. In the case of nucleotides, small oligomers are best labeled chemically during synthesis with the chosen fluorophore, a biotin or a 5' primary amine for the attachment of a custom fluor (all of these modifications are readily available from a number of commercial sources). If increased sensitivity is required, very bright protein fluors (e.g., phycoerythrin [PE]) can be bound via the biotin group. However, PE and similar fluors are very large (300+ kDa) and potentially disturbing. Therefore, their use requires time-consuming secondary staining methods that also limit kinetic analysis. For DNA too large to synthesize, nucleotide triphosphates with a variety of groups and fluors already incorporated are available (39). RNA poses a more difficult synthesis challenge than DNA, but analogous methods exist (39).

When labeling proteins for molecular assembly, the main concern is how the labeling will affect the molecular assembly. Of the methods described above, genetic labeling techniques, along with covalent attachment of organic fluors, are the most common. The other methods have advantages and have recently become commercially available (http://www.invitrogen.com/content/sfs/manuals/mammalianlumiogateway_man.pdf). The appropriate method will largely depend on the molecular components. In vivo labeling of cellular protein will assuredly dictate genetic methods, whereas microsphere-based studies on an easily purified protein may be better suited to chemical labeling methods. The above examples are extreme cases, and a researcher is likely to face a less obvious choice, weighing the difficulties in creating a genetically labeled protein or the random labeling provided by chemical techniques, which, if appropriately performed, can be minimized (see below).

Once the molecule has been labeled, the degree of labeling (DOL) must be calculated. For genetically labeled proteins, the degree of labeling is 1, but the absorbance of a chemically labeled molecule and its attached fluor must be measured to calculate the degree of labeling. Precise protocols have been described (39), but the general formula for proteins is listed here:

$$DOL = \frac{A_{\lambda max}}{\varepsilon_{dye} \cdot \frac{[A_{280} - A_{\lambda max} \cdot \varepsilon_{dye@280}/\varepsilon_{dye})]}{\varepsilon_{protein}}}$$

The term $A_{\lambda max}$ is the absorbance of the labeled protein at the dye's optimal wavelength, A_{280} is the absorbance of the labeled protein at 280 nm, $\varepsilon_{protein}$ is the molar extinction coefficient of the protein at 280 nm, ε_{dye} is the molar extinction coefficient of the dye at its optimal wavelength, and $\varepsilon_{dye@280}$ is the molar extinction coefficient of the dye at 280 nm.

Analogous calculations can be made for nucleotides by substituting measurements at 260 nm for those made at 280 nm. The degree of labeling necessary will be a balance between sensitivity and potential interference. For example if a fluorophore/protein ratio of 1 is achieved, on average, each protein will have a single fluor attached, which will minimize interference potential. If sensitivity is an issue, higher F/P ratios may be desirable. However, extremely high values have the potential for self-quenching, which has been explored in detail using DNA oligomers as substrates (79). Furthermore, high degrees of labeling may interfere with potential assemblies, as the attached fluor may sterically hinder critical interactions.

Attachment/Display

Covalent coupling and biotin-avidin/streptavidin are the most commonly used attachment strategies for both proteins and nucleotides. For oligonucleotides, orientation specific attachment can be achieved relatively easily by synthesizing the DNA with an amino group and attaching it to carboxy-labeled microspheres or with a specific biotin and attaching it to avidin or steptavidin microspheres.

In the case of proteins, covalent coupling has the advantage in that no dissociation occurs but orientation-specific covalent attachment of protein is not currently possible. The protocol for this procedure is relatively simple (40) but without specific orientation, a percentage of the proteins on the surface of microspheres will be sterically hindered from forming assemblies.

Proteins can also be chemically biotinylated (40), as well as expressed with specific biotinylation tags (17). Biotin-avidin attachments are not permanent (though they are long lasting), but they do allow for orientation-specific attachment of proteins that are expressed in vectors containing biotinylation tags in *E. coli*. Therefore, it may be possible to attach recombinant proteins in an orientation-specific manner to avidin/ streptavidin microspheres (see chapter 17). However, the biotinylation tag is large (>100 aa; 17) and may interfere with molecular interactions. Chemical biotinylation of proteins, like chemical labeling, results in random attachment of biotin to the protein, thereby providing potential steric hindrances. HIS and HAT tags (discussed earlier) represent a promising technology, but as of yet, with the relatively rapid dissociation of HIS/HAT-tagged proteins from chelating resins, they are not always suitable for use in molecular assembly studies (57). An ideal method of fluorescent labeling would allow site-specific labeling with very bright, small, and nonperturbing fluors of many wavelengths. Clearly there are many choices for protein attachment each with potential pitfalls; it is a bit like picking your poison.

Quantification

Quantification is critical to any study on molecular assemblies. This issue is of particular importance to studies using the flow cytometer because the primary measurement is fluorescence, which is a relative value. Conversion of this relative value to an actual value representing the number of molecules requires the use of a standard. Fortunately for many of the widely used fluorophores such as fluorescein, PE, and various PE conjugates, there are commercially available standard microspheres with a precalibrated amount of a fluorophore present on their surface (Bangs Laboratories, Fishers, IN). Using such microsphere sets, it is relatively easy to develop standard curves that relate the fluorescence of a particle to the number of fluorescent molecules present (with a few qualifications, discussed later). In the absence of commercially available standards, it is possible to use a solution of fluorophore with a precisely known concentration to calibrate a set of uniformly fluorescent microspheres such as Spherotech Rainbow or BDIS Quantibrite microspheres (44). However, it is important to note that these microspheres will only be valid when used in the same conditions as when they were standardized. The experimental variables include wavelengths for excitation and emission, pH, ionic strength, and so forth. Using standards generated with the same fluor (e.g., fluorescein with fluorescein beads) ensures that environmental changes from experiment to experiment are taken into account. Nevertheless, both the commercial standards and the solution-calibrated standards can be used to provide microspheres that have fluorescence responses that represent the mean equivalent soluble fluorophores (MESF) per particle.

Once a set of calibrated microspheres is available, the relative fluorescence of the fluorophore conjugated to the reporter molecule (generally the ligand in the reaction) to the free fluorophore must be determined. If the labeled ligand was used to calibrate the uniformly fluorescent microspheres, this step is unnecessary. However, because the fluorescence of the fluorophore is tightly tied to its local microenvironment, conjugation to other molecules can change its relative fluorescence efficiency. For example, FITC conjugated to a small peptide has a relative yield of 0.82 relative to that of free fluorescein (24). The relative yield (Q_r) of the free fluorophore versus the conjugate is readily determined by measuring the intensity of the fluorescence from both species in a spectrofluorometer.

$$Q_r = \frac{F_{conjugate}}{F_{free}}$$

In addition, the binding of the conjugated fluorescent ligand to its receptor (or other binding molecule on the particle) may alter its fluorescence either up or down. Therefore, the fluorescence of the free ligand (I_f) relative to that of the bound ligand (I_b) must also be taken into account. When all of the above is considered, the actual number of fluorescent particles per cell (N_f) can be calculated from the mean fluorescence channel value of the particle (M) using the formula

$$N_f = \frac{M \times MESF \times \dfrac{1}{Q_r} \times \dfrac{I_f}{I_b}}{DOL}$$

This quantitative approach was first used for fluorescein-labeled ligands in studies of N-formyl peptide receptors (24) and has been extended to other fluorophores (98). This approach shows the minimal considerations one must give in determining the number of bound molecules, but proper experimental procedure must be followed to account for artifacts such as nonspecific binding and rebinding.

Binding Artifacts

Although experimental procedures that account for nonspecific binding (e.g., blocking, subtraction of signals from negative controls, etc.) are well understood, rebinding, which is another phenomena critical to measurements on a surface, is less appreciated. Most often, rebinding is observed as a reduction of the dissociation rate of ligand or protein from its receptor or binding partner. These effects are most likely to be seen with a high surface density of receptor or other binding partner. Theory predicts that high receptor densities on a surface will slow both the binding of a ligand to the receptors on the surface and the dissociation of the ligand from the surface (31). Dissociation will be slowed because as a ligand dissociates, it will encounter a high concentration of surface receptors, which will make it likely that it will rebind to another receptor rather than dissociate from the surface and into bulk solution. Binding is slowed because as a ligand binds, nearby receptors compete for the same ligand. The magnitude of these effects can be defined as (31)

$$k_f = \frac{k_a}{1 + RAk_a/k_+}$$

$$k_r = \frac{k_d}{1 + RAk_a/k_+}$$

Where R is the concentration of free receptors on the surface in terms of receptors per unit surface area, A is the surface area of the particle, k_+ is the diffusion-limited forward rate constant for the binding of a ligand to the particle, k_f is the apparent forward-binding rate, k_r is the apparent dissociation rate, and k_a and k_d are the solution phase association and dissociation rates, respectively. These effects must be taken into account when measuring interactions on a particle surface. Rebinding occurs because a ligand has an increased probability of binding to a receptor at high receptor densities, which can be described by the ratio of k_d/k_r or

$$\frac{1}{1 + RAk_a/k_+}$$

The diffusion-limited forward rate constant for binding of a ligand to the particle is defined as

$$k_+ = 4\pi Da$$

(where D is the diffusion coefficient of the ligand and a the particle radius). Therefore, it is predicted that that rebinding will occur under the following conditions (31):

$$\frac{RAk_a}{(4\pi Da)} \gg 1$$

Although this equation is difficult to extrapolate in all circumstances, it predicts that low receptor densities can prevent rebinding. As a rule of thumb, rebinding plays a significant role for small molecules for which there are roughly 10^5 or more receptors present on particles of radii approximately 10 μm.

Experimentally, the presence or absence of rebinding can be verified by measuring dissociation rates as a function of receptor density on the particle. If the rates are independent of the receptor density, then rebinding is not an issue. If the rates are dependent on receptor density, then investigators should do one of the following: work at lower receptor densities, include additional solution phase receptors, use unlabeled ligands for competition in dissociation experiments, or fit data with the appropriate model.

An example of the rebinding artifact is shown in figure 9.5. In this figure, a DNA melting experiment was performed in which an oligo was attached to a microsphere and a fluorescently labeled complimentary oligomer was hybridized to the microsphere. The microspheres were passed through the dynamic inline thermoregulation unit (figure 9.3A). With low numbers of oligos on the microsphere (open circles, ~19,000 oligos/microsphere), the DNA rapidly dissociates for the microsphere, resulting in a faster loss of fluorescence. With high numbers of oligos on the microsphere (closed circles, ~250,000 oligos/microsphere), the dissociation rate of the DNA is slowed by rebinding, and a slower loss of fluorescence is shown. All experimental details (flow rate, temperature ramp rate, etc.) were identical except for the number of oligos on the microsphere. This clearly demonstrates the potential dramatic effects of rebinding.

Mixing and Environmental Control

As described above, kinetic measurements on a flow cytometer with rapid mixing are a challenge. Hand mixing, at best taking nearly 10 s, is too slow to observe the fast reactions that are common for low-affinity complexes and enzymatic reactions. There are several custom approaches one can take for subsecond mixing, but the Time-Zero system from Cytek Development (Fremont, CA) is a potential solution for mixing re-

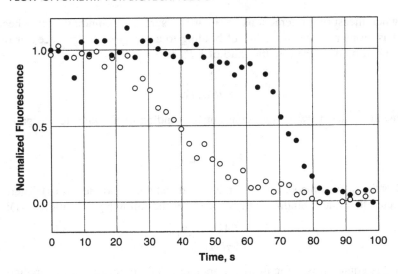

Figure 9.5. Rebinding effects on temperature-based dissociation of fluorescently labeled microspheres.

quirements on the scale of a few seconds. The route chosen will in large part be influenced by the financial and technical resources of the investigator, as the Rapid Mix Flow Cytometer described above is completely assembled from commercial parts with a few custom circuits (34). Such instruments require significant technical and financial resources, and as such, the National Flow Cytometry Resource (bdiv.lanl.gov/NFCR/) maintains such advanced instrumentation (including rapid mix and temperature instruments) for general use in collaborations.

In kinetic and equilibrium experiments, the effects of temperature are well appreciated. Temperature control is desirable, and some commercial instruments provide limited thermoregulation of the sample before they are introduced into the flow cytometer. Every effort should be made to control sample temperature throughout the system, as it will improve experimental precision and allow for additional experimental parameters (e.g., temperature jump experiments). Conventional water baths could be used, but Peltier modules are more desirable, and a detailed explanation of their use in a flow cytometer has been published (33).

Future Directions

Overview

Flow cytometry is rapidly becoming a highly valuable instrument in modern biochemical and biophysical laboratories. As commercial instrumentation becomes more user friendly and available, flow cytometry will make even greater inroads into the laboratory, with uses in the analysis of molecular assembly in genomics, proteomics, and combinatorial chemistry. In combination with the increasing availability of labeling technologies and the development of high-throughput sampling instrumentation (see chapter 3), flow cytometry will become further entrenched as an important tool in the study and screening of molecular assemblies. Furthermore, with the development of microsphere array and display technologies, flow cytometry will become a major investigative platform within the field of proteomics.

Instrumentation

Improved instrumentation will be critical to the success of flow cytometry as a platform for molecular assembly analysis. Efforts are ongoing to develop a flow cytometer with improved resolution for free versus bound probe by reducing the laser spot size, reducing the sample stream diameter, using confocal optics, and modifying the data collection system to collect both positive- and negative-going pulse shapes. Such improvements will enhance homogeneous analysis of low-affinity molecular assemblies.

Flow cytometers with improved sensitivity could eliminate the need to display proteins on a particle surface for analysis. Longer particle transit times in combination with photon-counting detectors make it possible to detect single molecules in a flow cytometer by increasing the number of photons collected from the sample and increasing the efficiency of photon collection (1). Multicolor, single-molecule flow cytometers are on the horizon, which could allow direct detection of individual molecular assemblies in solution. Furthermore, this technology has been constructed to be compatible with microfluidic technology that could be adapted for high-throughput applications (27; see chapter 5)

Finally, the use of array detectors such as intensified charge-coupled (CCD) devices could allow either complete spectral collection or imaging of the flow stream. Imaging of the flow stream has matured to the point at which commercial instruments capable of high-resolution particle imaging, at rates of thousands per second, will soon be available (Amnis Corporation, Seattle, WA; see chapter 4), offering the possibility of direct observation of the result of a molecular assembly, such as a change in cell morphology. In the case of spectral collection, similar arrangements could be made with appropriate dispersion elements to collect the entire spectrum simultaneously with nanometer resolution. Initial efforts at such spectral collection have been made previously (29), but CCD technology is now at the point that high-speed analysis on a particle-by-particle basis is possible.

Assay Technologies

Cellular display libraries will revolutionize how molecular assembly screens and studies are performed. By expressing proteins of interest on the surface of the cell, it will be possible to create a "smart particle" that eliminates the need for purification and attachment strategies. Such particles could be used in sorting schemes to screen for molecular assemblies or to simplify particle production by eliminating the need for surface attachment chemistry steps. Such technology, in combination with genetic labeling methods using GFP variants or other proteins, could also be the basis of completely in vivo molecular assembly studies. By expressing two protein libraries as either fusions with a FRET donor protein or those with a FRET acceptor protein, specific interactions would result in FRET, analogous to the yeast two-hybrid screen.

Applications

Although flow cytometry excels at multiparameter measurements, molecular assembly measurements using flow cytometry have predominantly used only a few parameters. Flow cytometry is currently capable of measuring up to 13 wavelengths of fluorescence simultaneously (see chapter 10). This capability has been underused in flow cytometry, and in the future it should be possible, in combination with the increasing labeling choices, to monitor an entire pathway from receptor binding to the signaling complex to the final cellular result (such as a change in cellular adhesion) simultaneously. Alternatively, the multiple parameters could be used to observe the assembly of a complex multiprotein assembly, such as a eukaryotic transcription complex, by labeling each participant with a unique fluor.

Microsphere arrays in combination with the analysis techniques described above will make flow cytometry a powerful analysis platform for large-scale screening and detection of in vitro molecular assemblies. Examples of this potential are found in the multiplex analysis of nuclear receptors (47) and cytokine expression levels (18), which are very analogous to the multiplex single nucleotide polymorphism analysis described in the chapter by Nolan. At present, multiplex array technology is being used commercially to analyze a variety of molecular assemblies (DNA hybridization, immunoassays, cytokine expression, etc.). At this time, these assays are only using a small subset of the possible 100-element array, but full use of the array could allow large-scale screening of in vitro molecular assemblies. For example, in the simplest form, an array of 100 unknown proteins could be reacted with an individual fluorescently tagged protein. By screening the array of proteins serially against a set of labeled proteins, detection of interacting assemblies could be accomplished quickly. Simply increasing the number of arrays would allow for the rapid screening of whole libraries of proteins. Moreover, because of the multiparameter nature of flow cytometry, screens that measured function along with binding could be devised. Such technology will be critically important in the field of proteomics.

Sorting is another aspect of flow cytometry that could greatly assist molecular assembly screens. In the above examples, multiplex analysis was accomplished with differentially stained microspheres. In the case of cellular display, multiplexing may not be necessary, as each positive assembly can sorted, and the interacting protein's identity can be determined by genetic methods. This has obvious implications for molecular assembly screens in that the size of the library screened will be limited by the sorting speed. Detailed consideration of such screens is given in chapter 11. In the future, high-speed sorting of protein libraries genetically labeled for FRET studies could be a powerful method for high-throughput molecular assembly screens. Again, these methods will be of great assistance to the field of proteomics.

References

1. Ambrose, W.P., Goodwin, P.M., Jett, J.H., van Orden, A., Werner, J.H., Keller, R.A. 1999. Single molecule fluorescence spectroscopy at ambient temperature. *Chem. Rev.* 99:2929–2956.
2. Anderson, M.T., Tjioe, I.M., Lorincz, M.C., Parks, D.R., Herzenberg, L.A., Nolan, G.P., Herzenberg, L.A. 1996. Simultaneous fluorescence-activated cell sorter analysis of two dis-

tinct transcriptional elements within a single cell using engineered green fluorescent proteins. *Proc. Natl. Acad. Sci. USA* 93:8508–8511.

3. Batard, P., Szollosi, J., Luescher, I., Cerottini, J.C., MacDonald, R., Romero, P. 2002. Use of phycoerythrin and allophycocyanin for fluorescence resonance energy transfer analyzed by flow cytometry: advantages and limitations. *Cytometry* 48:97–105.

4. Bennett, T.A., Key, T.A., Gurevich, V.V., Neubig, R., Prossnitz, E.R., Sklar, L.A. 2001. Real-time analysis of G protein-coupled receptor reconstitution in a solubilized system. *J. Biol. Chem.* 276:22453–22460.

5. Bieri, C., Ernst, O.P., Heyse, S., Hofmann, K.P., Vogel, H. 1999. Micropatterned immobilization of a G protein-coupled receptor and direct detection of G protein activation. *Nat. Biotechnol.* 17:1105–1108.

6. Blankenstein, G., Scampavia, L.D., Ruzicka, J., Christian, G.D. 1996. Coaxial flow mixer for real-time monitoring of cellular responses in flow injection cytometry. *Cytometry* 25:200–204.

7. Bohn, B. 1980. Flow cytometry: a novel approach for the quantitative-analysis of receptor-ligand interactions on surfaces of living cells. *Mol. Cell. Endocrinol.* 20:1–15.

8. Buranda, T., Huang, J., Ramarao, G.V., Ista, L.K., Larson, R.S., et al. 2003. Biomimetic molecular assemblies on glass and mesoporous silica microbeads for biotechnology. *Langmuir* 19:1654–1663.

9. Buranda, T., Lopez, G.P., Keij, J., Harris, R., Sklar, L.A. 1999. Peptides, antibodies, and FRET on beads in flow cytometry: a model system using fluoresceinated and biotinylated beta-endorphin. *Cytometry* 37:21–31.

10. Cai, H., White, P.S., Torney, D., Deshpande, A., Wang, Z.L., et al. 2000. Flow cytometry-based minisequencing: a new platform for high-throughput single-nucleotide polymorphism scoring. *Genomics* 66:135–143.

11. Campbell, R.E., Tour, O., Palmer, A.E., Steinbach, P.A., Baird, G.S., et al. 2002. A monomeric red fluorescent protein. *Proc. Natl. Acad. Sci. USA* 99:7877–7882.

12. Chan, F.K.M., Siegel, R.M., Zacharias, D., Swofford, R., Holmes, K.L., et al. 2001. Fluorescence resonance energy transfer analysis of cell surface receptor interactions and signaling using spectral variants of the green fluorescent protein. *Cytometry* 44:361–368.

13. Chen, G., Hayhurst, A., Thomas, J.G., Harvey, B.R., Iverson, B.L., Georgiou, G. 2001. Isolation of high-affinity ligand-binding proteins by periplasmic expression with cytometric screening (PECS). *Nat. Biotechnol.* 19:537–542.

14. Chigaev, A., Blenc, A.M., Braaten, J.V., Kumaraswamy, N., Kepley, C.L., et al. 2001. Real time analysis of the affinity regulation of alpha(4)-integrin: the physiologically activated receptor is intermediate in affinity between resting and Mn2+ or antibody activation. *J. Biol. Chem.* 276:48670–48678.

15. Cochran, A.G. 2001. Protein-protein interfaces: mimics and inhibitors. *Curr. Opin. Chem. Biol.* 5:654–659.

16. Cook, N.D. 1996. Scintillation proximity assay: a versatile high-throughput screening technology. *Drug Discov. Today* 1:287.

17. Cronan, J.E. 1990. Biotinylation of proteins *in vivo*: a posttranslational modification to label; purify; and study proteins. *J. Biol. Chem.* 265:10327–10333.

18. Kellar, K.L., Kalwar, R.R., Dubois, K.A., Crouse, D., Chafin, W.D., Kane, B.E. 2001. Multiplexed fluorescent bead-based immunoassays for quantitation of human cytokines in serum and culture supernatants. *Cytometry* 45:27–36.

19. Daugherty, P.S., Chen, G., Iverson, B.L., Georgiou, G. 2000. Quantitative analysis of the effect of the mutation frequency on the affinity maturation of single chain Fv antibodies. *Proc. Natl. Acad. Sci. USA* 97:2029–2034.

20. Drouvalakis, K.A., Neeson, P.J., Buchanan, R.R.C. 1999. Detection of anti-phosphatidylethanolamine antibodies using flow cytometry. *Cytometry* 36:46–51

21. Dunne J.F. 1991. Time window analysis and sorting. *Cytometry* 12:597–601.

22. Edwards, B.S., Curry, M.S., Tsuji, K., Brown, D., Larson, R.S., Sklar, L.A. 2000. Expression of P-selectin at low site density promotes selective attachment of eosinophils over neutrophils. *J. Immunol. Methods* 165:404–410.

23. Edwards, B.S., Kuckuck, F.W., Prossnitz, E.R., Okun, A., Ransom, J.T., Sklar, L.A. 2001.

Plug flow cytometry extends analytical capabilities in cell adhesion and receptor pharmacology. *Cytometry* 43:211–216.

24. Fay, S.P., Posner, R.G., Swann, W.N., Sklar, L.A. 1991. Real-time analysis of the assembly of ligand, receptor, and G protein by quantitative fluorescence flow cytometry. *Biochemistry* 30:5066–5075.

25. Fivash, M., Towler, E.M., Fisher, R.J. 1998. BIAcore for macromolecular interaction. *Curr. Opin. Biotechnol.* 9:97–101.

26. Francisco, J.A., Campbell, R., Iverson, B.L., Georgiou, G. 1993. Production and fluorescence-activated cell sorting of *Escherichia-coli* expressing a functional antibody fragment on the external surface. *Proc. Natl. Acad. Sci. USA* 90:10444–10448.

27. Fu, A.Y., Chou, H., Spence, C., Arnold, F.H., Quake, S.R. 2002. An integrated microfabricated cell sorter. *Anal. Chem.* 74:2451–2457.

28. Gary, R., Park, M.S., Nolan, J.P., Cornelius, H.L., Kozyreva, O.G., et al. 1999. A novel role in DNA metabolism for the binding of Fen1/Rad27 to PCNA and implications for genetic risk. *Mol. Cell. Biol.* 19:5373–5382.

29. Gauci, M.R, Vesey, G., Narai, J., Veal, D., Williams, K.L., Piper, J.A. 1996. Observation of single-cell fluorescence spectra in laser flow cytometry. *Cytometry* 25:388–393.

30. Gilbert, G.E, Arena, AA.1995. Phosphatidylethanolamine induces high-affinity binding-sites for factor-VIII on membranes containing phosphatidyl-L-serine. *J. Biol. Chem.* 270:18500–18505.

31. Goldstein, B., Coombs, D., He, X.Y, Pineda, A.R., Wofsy, C. 1999. The influence of transport on the kinetics of binding to surface receptors: application to cells and BIAcore. *J. Mol. Recog.* 12:293–299.

32. Graves, S.W., Habbersett, R.C., Nolan, J.P. 2001. A dynamic inline sample thermoregulation unit for flow cytometry. *Cytometry* 43:23–30.

33. Graves, S.W., Habbersett, R.C., Nolan, J.P. 2002. Dynamic thermoregulation of the sample in flow cytometry. In *Current Protocols in Cytometry*, ed. J.P. Robinson, pp. 1.18.1. New York: Wiley.

34. Graves, S.W., Nolan, J.P., Jett, J.H., Martin, J.C., Sklar, L.A. 2002. Nozzle design parameters and their effects on rapid sample delivery in flow cytometry. *Cytometry* 47:127–138.

35. Gray, M.L., Hoffman, R.A., Hansen, W.P. 1983. A new method for cell-volume measurement based on volume exclusion of a fluorescent dye. *Cytometry* 3:428–434.

36. Griffin, B.A., Adams, S.R., Jones, J., Tsien, R.Y. 2000. Fluorescent labeling of recombinant proteins in living cells with FlAsH. *Methods Enzymol.* 327:565–578.

37. Griffin, B.A., Adams, S.R., Tsien, R.Y. 1998. Specific covalent labeling of recombinant protein molecules inside live cells. *Science* 281:269–272.

38. Gurskaya, N.G., Fradkov, A.F., Terskikh, A., Matz, M.V., Labas, Y.A., et al. 2001. GFP-like chromoproteins as a source of far-red fluorescent proteins. *FEBS Lett.* 507:16–20.

39. Haugland, R.P. 2001. *Handbook of Fluorescent Probes and Research Products*. Eugene, OR: Molecular Probes.

40. Hermanson, G.T. 1996. *Bioconjugate Techniques*. San Diego, CA: Academic Press.

41. Hidaka, H., Hidaka, E., Tozuka, M., Nakayama, J., Katsuyama, T., Fidge, N. 1999. The identification of specific high density lipoprotein(3) binding sites on human blood monocytes using fluorescence-labeled ligand. *J. Lipid Res.* 40:1131–1139.

42. Hlavacek, W.S., Perelson, A.S., Sulzer, B., Bold, J., Paar, J., et al. 1999. Quantifying aggregation of IgE-Fc epsilon RI by multivalent antigen. *Biophys. J.* 76:2421–2431.

43. Hlavacek, W.S., Posner, R.G., Perelson, A.S. 1999. Steric effects on multivalent ligand-receptor binding: exclusion of ligand sites by bound cell surface receptors. *Biophys. J.* 76:3031–3043.

44. Hoffman, R.A. 2001. Standardization and quantitation in flow cytometry. *Methods Cell Biol.*, ed. Z. Darzynkiewicz, H.A. Crissman, J.P. Robinson, pp. 299–340. San Diego, CA: Academic Press.

45. Holler, P.D., Holman, P.O., Shusta, E.V., O'Herrin, S., Wittrup, K.D., Kranz, D.M. 2000. In vitro evolution of a T cell receptor with high affinity for peptide/MHC. *Proc. Natl. Acad. Sci. USA* 97:5387–5392.

46. Holmes, K.L., Lantz, L.M. 2001. Protein labeling with fluorescent probes. *Methods Cell Biol.* 63:185–204.
47. Iannone, M.A., Consler, T.G., Pearce, K.H., Stimmel, J.B., Parks, D.J., Gray, J.G. 2001. Multiplexed molecular interactions of nuclear receptors using fluorescent microspheres. *Cytometry* 44:326–337.
48. Jameson, D.M., Sawyer, W.H. 1995. Fluorescence anisotropy applied to biomolecular interactions. *Methods Enzymol.* 246:283–300.
49. Johnson, K.A. 1998. Advances in transient-state-kinetics. *Curr. Opin. Biotechnol.* 9:87–9.
50. Kachel, V., Glossner, E., Schneider, H. 1982. A new flow cytometric transducer for fast sample throughput and time resolved kinetic-studies of biological cells and other particles. *Cytometry* 3:202–212.
51. Kelley, K.A. 1989. Sample station modification providing online reagent addition and reduced sample transit-time for flow cytometers. *Cytometry* 10:796–800.
52. Kettman, J.R., Davies, T., Chandler, D., Oliver, K.G., Fulton R.J. 1998. Classification and properties of 64 multiplexed microsphere sets. *Cytometry* 33:234–243.
53. Kieke, M.C., Shusta, E.V., Boder, E.T., Teyton, L., Wittrup, K.D., Kranz, D.M. 1999. Selection of functional T cell receptor mutants from a yeast surface-display library. *Proc. Natl. Acad. Sci. USA* 96:5651–5656.
54. Kieke, M.C., Sundberg, E., Shusta, E.V., Mariuzza, R.A., Wittrup, K.D., Kranz, D.M. 2001. High affinity T cell receptors from yeast display libraries block T cell activation by superantigens. *J. Mol. Biol.* 307:1305–1315.
55. Kuckuck, F.W., Edwards, B.S., Sklar, L.A. 2001. High throughput flow cytometry. *Cytometry* 44:83–90.
56. Lauer, S., Goldstein, B., Nolan, R.L., Nolan, J.P. 2002. Analysis of cholera toxin-ganglioside interactions by flow cytometry. *Biochemistry* 41:1742–1751.
57. Lauer, S.A., Nolan, J.P. 2002. Development and characterization of Ni-NTA-bearing microspheres. *Cytometry* 48:136–145.
58. Lindberg, W., Scampavia, L.D., Ruzicka, J., Christian, G.D. 1994. Fast kinetic measurements and online dilution by flow-injection cytometry. *Cytometry* 16:324–330.
59. Lorincz, M., Roederer, M., Diwu, Z., Herzenberg, L.A., Nolan, G.P. 1996. Enzyme-generated intracellular fluorescence for single-cell reporter gene analysis utilizing *Escherichia coli*—lucuronidase. *Cytometry* 24:321–329.
60 Lorincz, M.C., Parente, M.K., Roederer, M., Nolan, G.P., Diwu, Z.J., Martin, D.I.K., Herzenberg, L.A., Wolfe, J.H. 1999. Single cell analysis and selection of living retrovirus vector-corrected mucopolysaccharidosis VII cells using a fluorescence-activated cell sorting-based assay for mammalian beta-glucuronidase enzymatic activity. *J. Biol. Chem.* 274:657–665.
61. Lund-Johansen, F., Davis, K., Bishop, J., Malefyt, R.D. 2000. Flow cytometric analysis of immunoprecipitates: high-throughput analysis of protein phosphorylation and protein-protein interactions. *Cytometry* 39:250–259.
62. Martin, J.C., Swartzendruber, D.E. 1980. Time: a new parameter for kinetic measurements in flow cytometry. *Science* 207:199–201.
63. Matko, J., Edidin, M. 1997. Energy transfer methods for detecting molecular clusters on cell surfaces. *Methods Enzymol.* 278:444–462.
64. McDonnell, J.M. 2001. Surface plasmon resonance: towards an understanding of the mechanisms of biological molecular recognition. *Curr. Opin. Chem. Biol.* 5:572–577.
65. McHugh, T.M. 1995. Flow microsphere immunoassay for the quantitative and simultaneous detection of multiple soluble analytes. In *Methods in Cell Biology*, ed. Z. Darzynkiewicz, J.P. Robinson, H.A. Crissman, pp. 575–595. San Diego: Academic Press.
66. Mirzabekov, T., Kontos, H., Farzan, M., Marasco, W., Sodroski, J. 2000. Paramagnetic proteoliposomes containing a pure, native, and oriented seven-transmembrane segment protein, CCR5. *Nat. Biotechnol.* 18:649–654.
67. Murphy, R.F. 1990. Ligand binding, endocytosis and processing. In *Flow Cytometry and Sorting*, ed. M.R. Melamed, T. Lindo, M.L. Mendelsohn, pp. 355–366. New York: Wiley-Liss.
68. Nolan, J.P., Chambers, J.D., Sklar, L.A. 1998. Cytometric approaches to the study of re-

ceptors. In *Phagocyte Function: A Guide for Research and Clinical Evaluation*, ed. J.P. Robinson, G.F. Babcock, pp. 1946. New York: Wiley-Liss.

69. Nolan, J.P., Lauer, S., Prossnitz, E.R., Sklar, L.A. 1999. Flow cytometry: a versatile tool for all phases of drug discovery. *Drug Discov. Today* 4:173–180.

70. Nolan, J.P., Posner, R.G., Martin, J.C., Habbersett, R., Sklar, L.A. 1995. A rapid mix flow cytometer with subsecond kinetic resolution. *Cytometry* 21:223–229.

71. Nolan, J.P., Shen, B.H., Park, M.S., Sklar, L.A. 1996. Kinetic analysis of human flap endonuclease-1 by flow cytometry. *Biochemistry* 35:11668–11676.

72. Nolan, J.P., Sklar, L.A. 1998. The emergence of flow cytometry for sensitive, real-time measurements of molecular interactions. *Nat. Biotechnol.* 16:633–638.

73. Nolan, J.P., Sklar, L.A. 2002. Suspension array technology: evolution of the flat-array paradigm. *Trends Biotechnol.* 20:9–12.

74. Olsen, M.J., Stephens, D., Griffiths, D., Daugherty, P., Georgiou, G., Iverson B.L. 2000. Function-based isolation of novel enzymes from a large library. *Nat. Biotechnol.* 18:1071–1074.

75. Omann, G.M., Coppersmith, W., Finney, D.A., Sklar, L.A. 1985. A convenient online device for reagent addition, sample mixing, and temperature control of cell-suspensions in flow-cytometry. *Cytometry* 6:69–73.

76. Patel, S., Harris, A., G O.B., Cook N.D., Taylor C.W. 1996. Kinetic analysis of inositol trisphosphate binding to pure inositol trisphosphate receptors using scintillation proximity assay. *Biochem. Biophys. Res. Commun.* 221:821–825.

77. Patel, S.S., Wong, I., Johnson, K.A. 1991. Pre-steady-state kinetic analysis of processive DNA-replication including complete characterization of an exonuclease-deficient mutant. *Biochemistry* 30:511–525.

78. Perez, O.D., Nolan, G.P., 2002. Simultaneous measurement of multiple active kinase states using polychromatic flow cytometry. *Nat. Biotechnol.* 20:155–162.

79. Randolph, J.B., Waggoner A.S., 1997. Stability, specificity and fluorescence brightness of multiply-labeled fluorescent DNA probes. *Nucleic Acids Res.* 25:2923–2929.

80. Reed, D.S., Smoll, J., Gibbs, P., Little, S.F. 2002. Mapping of antibody responses to the protective antigen of *Bacillus anthracis* by flow cytometric analysis. *Cytometry* 49:1–7.

81. Ringquist, S., Parma, D. 1998. Anti-L-selectin oligonucleotide ligands recognize CD62L-positive leukocytes: binding affinity and specificity of univalent and bivalent ligands. *Cytometry* 33:394–405.

82. Roederer, M., Bowser R., Murphy R.F. 1987. Kinetics and temperature dependence of exposure of endocytosed material to proteolytic enzymes and low pH: evidence for a maturation model for the formation of lysosomes. *J. Cell Physiol.* 131:200–209.

83. Scampavia, L.D., Blankenstein, G., Ruzicka, J., Christian, G.D. 1995. A coaxial jet mixer for rapid kinetic-analysis in flow-injection and flow injection cytometry. *Anal. Chem.* 67:2743–2749.

84. Schuck, P. 1997. Use of surface plasmon resonance to probe the equilibrium and dynamic aspects of interactions between biological macromolecules. *Ann. Rev. Biophys. Biomol. Struct.* 26:541–566.

85. Seamer, L.C., Kuckuck, F., Sklar, L.A. 1999. Sheath fluid control to permit stable flow in rapid mix flow cytometry. *Cytometry* 35:75–79.

86. Selvin, P.R. 2000. The renaissance of fluorescence resonance energy transfer. *Nat. Struct. Biol.* 7:730–734.

87. Shapiro, H.M. 1995. *Practical Flow Cytometry*. New York: Wiley-Liss.

88. Shen, B.H., Nolan, J.P., Sklar, L.A., Park, M.S. 1997. Functional analysis of point mutations in human flap endonuclease-1 active site. *Nucleic Acids Res.* 25:3332–3338.

89. Sklar, L.A., Edwards, B.S., Graves, S.W., Nolan, J.P., Prossnitz, E.R. 2002. Flow cytometric analysis of ligand-receptor interactions and molecular assemblies. *Ann. Rev. Biophys. Biomol. Struct.* 31:97–119.

90. Sklar, L.A., Vilven, J., Lynam, E., Neldon, D., Bennett, T.A., Prossnitz, E. 2000. Solubilization and display of G protein-coupled receptors on beads for real-time fluorescence and flow cytometric analysis. *Biotechniques* 28:976–985.

91. Sohn, G., Sautter, C. 1991. R-phycoerythrin as a fluorescent label for immunolocalization of bound atrazine residues. *J. Histochem. Cytochem.* 39:921–926.

92. Song, X.D., Shi, J., Swanson, B. 2000. Flow cytometry-based biosensor for detection of multivalent proteins. *Analytical Biochemistry* 284:35–41.
93. Steen, H.B. 1990. characteristics of flow cytometers. In *Flow Cytometry and Sorting*, ed. M.R. Melamed, T. Lindo, M.L. Mendelsohn, pp. 11. New York: Wiley.
94. St-Pierre, Y., Desrosiers, M., Tremblay, P., Esteve, P.O., Opdenakker, G. 1996. Flow cytometric analysis of gelatinase B (MMP-9) activity using immobilized fluorescent substrate on microspheres. *Cytometry* 25:374–380.
95. Stuart, M.C.A, Reutelingspeger, C.P.M., Frederik, P.M. 1998. Binding of annexin v to bilayers with various phospholipid compositions using glass beads in a flow cytometer. *Cytometry* 33:414–419.
96. Van Regenmortel, M.H.V. 2001. Analysing structure-function relationships with biosensors. *Cell. Mol. Life Sci.* 58:794–800.
97. VanAntwerp, J.J., Wittrup, K.D. 2000. Fine affinity discrimination by yeast surface display and flow cytometry. *Biotechnol. Prog.* 16:31–37.
98. Waller, A., Pipkorn, D., Sutton, K.L., Linderman, J.J., Omann, G.M. 2001. Validation of flow cytometric competitive binding protocols and characterization of fluorescently labeled ligands. *Cytometry* 45:102–114.
99. Watson, J.V. and Dive, C. 1994. Enzyme kinetics. In *Methods Cell Biol., Flow Cytometry Part A*, ed. L. Wilson, P. Matsudeira, pp. 469–507. London: Academic Press.
100. Zimmer, M. 2002. Green fluorescent protein (GFP): applications, structure, and related photophysical behavior. *Chem. Rev.* 102:759–781.

Part II

BIOLOGICAL
APPLICATIONS

10

Multiparameter Analysis:
Application to
Vaccine Analysis

MARIO ROEDERER AND STEPHEN C. DE ROSA

Overview

Multicolor Technology

Fluorescence-based flow cytometry was introduced in the late 1960s and is now used extensively both in basic research and in the clinic. Flow cytometry allows not only for the rapid multiparametric analysis of cells on a cell-by-cell basis but also for the viable separation, or sorting, of highly purified populations of cells. In this chapter, we will discuss only the analysis aspects.

The earliest flow cytometry experiments had three parameters: one fluorescence measurement and two scattered light signals. An early "one-color" experiment successfully separated antibody-secreting B cells from mouse splenocytes (5). This and other early studies quickly demonstrated the usefulness of this technology in immunological studies.

However, measurement of only one fluorescence was a limitation. By adding detectors collecting light in specified wavelength ranges, multiple fluorescence measurements could be made simultaneously. By 1984, four-color fluorescence experiments could be routinely performed, at least in the most sophisticated flow cytometry laboratories (4), but it took another 10 years before most laboratories could perform routine three-color experiments. One reason for this delay is that it took some time to recognize the need for measuring multiple parameters in addressing questions that explored the complexity of the immune system. Another reason was that it was not until the late 1980s that four-color benchtop instrumentation became available. The AIDS epidemic also had a major effect on the expansion of flow cytometry into the research community, as early in the epidemic, the enumeration of CD4 was found to serve as a surrogate marker for disease progression. During this period, we were examining a number of functionally-important T cell subsets in HIV-infected adults and children, including naïve and memory, using three-color flow cytometry. These studies demonstrated clearly for the first time the loss of both CD4 and CD8 naïve T cells during HIV disease progression (11, 13). This loss had not been previously recognized either because appropriate combinations of reagents were not used or because the studies were limited to two colors.

Having demonstrated that multiple markers used in combination could lead to clinically relevant findings that were previously missed, we wondered how many other important subsets could be detected by measuring additional parameters. Therefore, working in the Herzenberg laboratory at Stanford University and in collaboration with the flow cytometry engineering group at Stanford, we undertook a development effort that culminated in our current instrument, capable of measuring 12 fluorescence parameters (14 total) simultaneously (3). The challenges involved in this effort have been discussed elsewhere (2).

Measuring Antigen-Specific Responses

The 1990s also saw an explosion in assays to identify antigen-specific T cells. These are T cells that respond to a given peptide from an immunogen presented in the context of Major Histocompatibility (MHC) class I (for CD8 T cells) or class II (for CD4 T cells). Two different forms of such assays are currently employed: stimulation-based assays that identify antigen-specific T cells on the basis of the upregulation of activation markers or the expression of cytokines in response to the stimulation (6, 10), and fluorescent, peptide-loaded MHC-multimers that selectively bind to T cells that are specific for such a peptide (1).

The stimulation assays were initially done with whole antigens. Purified protein antigens were included in short-term incubations of T cells (in the presence of antigen-presenting cells, AgPC). The AgPC digest the protein and present immunogenic peptides on MHC; T cells recognizing the peptides would be triggered. The triggering phenomenon is usually measured after 6 hr by either the upregulation of an early activation marker on the cell surface (such as CD69) or the production of cytokines (typically, γIFN). The disadvantage of this type of assay is that it requires functional AgPC (which are often severely depleted by cryopreservation, rendering the assay less useful on archival samples). The other aspect, measuring a response such as an activation marker or cytokines, is both an advantage and a disadvantage: If the appropriate response marker is not measured, then the cells could be missed even though they are present; however, the specific kind of response (e.g., Th1 vs. Th2) can be selectively measured (9).

More recently, it was discovered that the responses could be accurately and easily measured by using pools of peptides (15 amino acids in length) derived from the immunogen of interest. Pools are made so as to overlap by 11 amino acids, so that every T cell epitope is presumably present in the pool. This approach has a significant advantage in that it does not require AgPC function, and therefore can be used on cryopreserved samples. However, the cost of the peptide pools can be quite high, and it would be virtually impossible to make useful pools from complex virions such as cytomegalovirus that contain hundreds of proteins. The use of overlapping peptide pools is becoming the dominant assay for measuring responses to vaccination, where only one or very few proteins are given as immunogens.

The other technology for identifying antigen-specific T cells is the use of peptide-loaded MHC multimers (commonly referred to as "tetramers"). These multimers are synthesized in vitro to contain only a single peptide; they will bind to T cells that recognize that peptide, presented by the synthesized MHC molecule. This technology can never provide a broad analysis of the immune response to even a single protein, as it would be virtually impossible to make sufficient tetramers from all immunodominant peptides for each of the MHC alleles in a given test subject, not to mention the need to synthesize multimers from a sufficient number of MHC alleles so that all subjects in a trial could be tested. Nonetheless, it has found great utility because the responses to proteins are often limited to one or two peptides for a given MHC allele, and measurement of these responses in persons who express that MHC haplotype provides an indication of the immune response to the holoantigen.

The primary advantage of the tetramer approach is that it does not require stimulation of the T cells. Hence, there is no need to decide on which response is going to be quantitated; indeed, even nonresponsive (anergic) T cells can be identified by this technology (7). These aspects alone make the tetramer approach an important adjunct to virtually any experiments measuring antigen-specific responses.

A Confluence of Technologies: Application to Vaccine Development

The contemporaneous development of the high-end multicolor technology and the assays to measure antigen-specific T cells is fortunate. Because antigen-specific T cells are often extremely rare (typically <0.1% of lymphocytes), it is difficult to perform repetitive assays on these cells given the limitation of sample volume. Hence, it is important to perform as many distinct measurements simultaneously as possible. The unique capability of flow cytometry to make as many as 14 measurements on each cell confers the ability to simultaneously determine both phenotypic and functional characteristics of extremely rare cells.

The goal of vaccination is to create a functional cellular or humoral response to a given immunogen. Even with successful cellular responses, the frequency of antigen-specific T cells can still be quite low. During the iterative process of vaccine development, however (and, e.g., during dose-escalation studies), initial T cell responses are likely to be much lower than those finally achieved. Decisions on which vaccine approaches to take forward for optimization and eventual phase III clinical trials must be made on the comparison of what is likely to be exceedingly small responses to determine incremental benefits. In addition, during development phases, it may not be clear what the desired T cell response may be—thus, the decision of which cytokines, for example, are secreted becomes a key parameter for success. The ability to measure multiple different cytokines simultaneously ameliorates the need to decide in advance what the surrogate of vaccine efficacy should be. Together, these constraints illustrate that multicolor flow cytometric analysis of immunogenicity will be crucial for fast and economically viable vaccine development efforts.

The Requirement for More Colors

Although the use of three- or four-color technology has been very successful in research in immunology, it is now clear that four-color technology is often insufficient. Figure 10.1 illustrates a phenotypic division of the peripheral lymphocyte compartment in humans. Based on several years of phenotyping experiments, combined with functional analysis, we have a much better idea of the level of complexity in the immune system. First of all, it is important to recognize that there are more than just CD4 and CD8 T cells—there are probably at least seven distinct lineages (as distinct as CD4 vs. CD8 T cells). Within each of these lineages, naïve and multiple types of memory T cells can be identified, all with distinct functional characteristics.

Figure 10.1 ignores a much more recent and effective way of discriminating memory subsets, using the distinct expression of chemokine receptors [e.g., (8)]. We are only now beginning to evaluate the coordinate expression of the markers shown in figure 10.1 with these receptors to understand whether or not this adds an additional layer of subsetting or simply reinforces the functional distinctiveness of the subsets shown. In any case, we are probably not close to achieving one of our eventual goals: to phenotypically identify functionally homogeneous subsets of lymphocytes. Once this goal is achieved, fewer parameters will need to be measured. At present, we need to correlate phenotype with functional assays. Once this is documented, simple phenotypic analysis will be sufficient to determine functional repertoire.

Figure 10.2 illustrates an example of measuring multiple cytokines in stimulated peripheral T cells in a multicolor setting. These data show the potential complexity of antigen-specific responses (i.e., not only can antigen-specific T cells make a variety of different cytokines but it is possible that their response is heterogeneous). The fact that heterogeneity exists means that it is possible that there may be differential clinical outcomes; for example, perhaps an efficacious vaccination is accompanied by the generation of T cells with a particular cytokine expression profile. Measuring only single cytokines may not sufficiently resolve this heterogeneity to identify the subset that is pertinent. Note that in this experiment, not only were four different cytokines evaluated simultaneously (γIFN, TNFα, IL2, and MIP1β) but sufficient phenotyping was performed to uniquely identify lineage (CD3, CD4, and CD8) and differentiation stages (CD11a, CD45RO, and CD57). In addition, viability was assessed by the exclusion of ethidium monoazide bromide to eliminate artefact introduced by nonspecific binding of reagents by dead cells.

Implementation of Multicolor Analysis

Fundamentally, there are four classes of hurdles to overcome before successfully implementing high-end multicolor flow cytometric analysis. These are hardware, chemistry, software, and experimental optimization. These issues have been addressed in detail elsewhere (2) and will only be reviewed briefly here. In particular, there are now commercial solutions for some of these problems, and it is no longer necessary to have everything "home brewed."

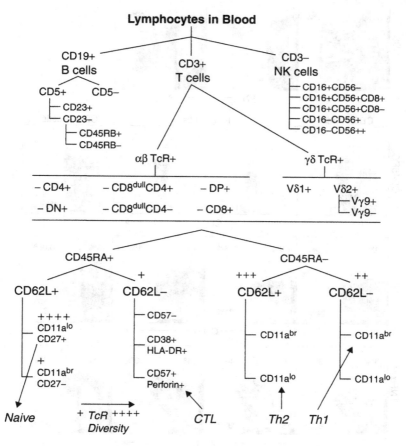

Figure 10.1. The complexity of the peripheral human immune system: A hierarchical representation of phenotypically and functionally distinct lymphocytes found in adult peripheral blood. Although B and NK cells can be divided into a few distinct subsets each, it is the T cell compartment that shows the richest degree of heterogeneity. Within CD3+ lymphocytes, as many as eight distinct lineages (in the same way that CD4 and CD8 T cells are considered lineages) can be identified. Each of these lineages has phenotypically definable subsets related by differentiation (e.g., naïve vs. memory). In fact, within each of the eight lineages listed above, as many as a dozen different memory subsets (and one naïve subset) can be found. Each of these subsets has different functional properties; for example, different proliferative capacity, cytokine profiles, apoptotic susceptibility, and so forth. Some identified functions (Th1/Th2 polarization, that is, the propensity to produce IL4, IL5, IL13 (Th2), or γIFN (Th1); cytolytic (CTL) activity; and naïve cells that have not encountered cognated antigen, are shown, as well as T cell receptor Vβ chain diversity for some subsets to illustrate the degree of clonality.

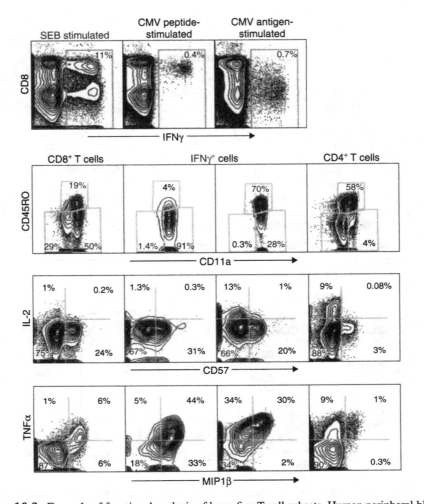

Figure 10.2. Example of functional analysis of hyperfine T cell subsets. Human peripheral blood mononuclear cells were stimulated for 6 hr with the superantigen SEB (which stimulates a relatively large [10%–30%] fraction of all T cells, depending on which Vβ gene they express), total cytomegalovirus (CMV) antigen, or a single immunodominant peptide from a CMV protein recognized by CD8 T cells in the presence of monensin to inhibit secretion of newly synthesized cytokines. Cells were then stained for surface antigens (CD3, CD4, CD8, CD11a, CD45RO, and CD57) in the presence of ethidium monoazide bromide (viability exclusion dye). After fixation and permeabilization, cells were further stained for the cytokines IL2, γIFN, TNFα, and MIP1β. Cells were gated for live T lymphocytes (not shown). The top row shows the profile of γIFN versus CD8 for SEB stimulated cells, CMV peptide stimulated cells, or CMV antigen stimulated cells. SEB will stimulate both CD4 and CD8 cells; hence, γIFN+ cells are found as both CD8+ (CD4−, not shown) and CD8− (CD4+). The CMV peptide is specifically recognized by CD8 T cells and stimulates a subset of CD8 T cells specific for this peptide, presented by MHC. The response to CMV protein in this assay is principally by CD4 T cells because of the requirement for antigen uptake and presentation. The bottom three rows show the phenotype and cytokine profile for γIFN+ cells among (from left to right): SEB-stimulated CD8 T cells, CMV-peptide stimulated T cells (all CD8+CD4−), CMV-antigen-stimulated T cells (all CD4+CD8−), and SEB-stimulated CD4 T cells. Although the phenotype and cytokine profiles of the CMV-specific T cells are restricted compared to the complete repertoire revealed by SEB, they are still heterogeneous.

Hardware

The extension of standard flow cytometric emission and excitation optics to collect up to 12 fluorescence and two scatter parameters originally involved, solving several engineering problems. However, at present, this represents perhaps the least complex hurdle, because the solutions originally devised are readily implemented on existing benches. Existing cytometers can be outfitted with three different lasers, providing the distinct excitation beams needed for 12-color cytometry. Note that it is possible to achieve 10-color detection with just a 488-nm argon laser and a 595-nm dye laser, and that it is possible to perform nine-color detection with the more typical combination of a 488-nm argon laser and a 633-nm HeNe laser.

The more difficult problem is arranging the emission optics to provide optimal detection of each fluorochrome while maintaining a relatively easily focused emission path. In general, going from four- or five-color emission optics to 12-color optics relies on the placement of additional photomultiplier tube (PMT) blocks in series with the existing optics. The selection of filters to block most of the undesired fluorescence emission is absolutely crucial. Even small differences can translate to several-fold differences in the efficiency of collected light. At this point, many fluorescence signals are highly signal limited [i.e., photon-counting statistics become the major source of measurement error; the effect of this limitation on data is significant (12)].

Although future instrumentation will use cheaper and smaller solid-state lasers, the problem of overlapping emission is unlikely to be solved with current-generation fluorochromes. Hence, emission optics are by far the most important focus of optimization in this technology.

Chemistry

At this time, there are seven distinct colors of conjugated antibodies that are commercially available (fluorescein, phycoerythrin [PE], Texas Red–phycoerythrin [TRPE], Cy5PE, Cy5.5PerCP, Cy7PE, allophycocyanin [APC], and Cy7APC). A few reagents can be found for virtually all of these colors; most others are available in only a much more limited set. For researchers who endeavor to perform routine six- or seven-color experiments, or for those who wish to measure more colors, it is necessary to conjugate antibodies in the laboratory. Detailed protocols for these conjugations can be accessed over the Internet (www.drmr.com/abcon/).

For generating reagents not commercially available, the primary need is obtaining purified unconjugated antibodies, which can either be purchased directly from manufacturers (as bulk concentrated reagent without exogenous protein) or isolated from available hybridomas. The other principal need is for the reactive fluorescent reagents. Fluorescein, Cascade Blue, Alexa 430, and Alexa 595 can be purchased, and the manufacturer's kits can be used for the conjugations. When doing a relatively large number of conjugations, the reactive dyes can be purchased in bulk.

The phycobiliproteins (PE and APC) and their tandems (e.g., Cy5PE) are somewhat more complex. Typically, a relatively large quantity of the phycobiliproteins should be purchased (bulk quantity discounts are significant). To make the reactive tandems, reactive Cyanine dyes (Cy5, Cy5.5, or Cy7) are also commercially available; it is likely that other resonance energy acceptors suitable for use with PE and APC will soon be available. Making reactive PE and APC is straightforward; the protocol is on the Internet (www.drmr.com/abcon/). Making reactive tandems requires some more effort, though, as each tandem must be optimized for your system. However, once optimized, the chemistry can be scaled up to the point at which a large quantity of tandem can be produced—sufficient to conjugate, for example, dozens of antibodies at 1 mg each. The advantage of making a large lot of tandem is that each antibody synthesized will have identical emission spectra and will therefore use the same compensation settings.

Using the available protocols, once the reactive dyes have been properly prepared (they are stable for months to years in the refrigerator), conjugation becomes very simple and quick: A typical researcher can prepare a dozen conjugates in a single day. Nonetheless, there is a significant investment in terms of reagents and optimization time required to get to this point. A reasonable estimate would be approximately US$25,000 in reactive dyes and 2 months of time. (This does not include the cost of obtaining or producing the purified antibodies.) However, the eventual cost of fluorescent antibodies is less than 10% that of commercially available sources.

Data Analysis and Presentation

By far, the analysis and presentation of the data resulting from six or more color experiments is the most difficult hurdle that researchers must deal with. Unfortunately, this is a problem that will likely persist for the foreseeable future; powerful "expert" systems that can automate the analysis of immunophenotyping data do not exist.

One of the most difficult processes in multicolor analyses is that of spectral compensation. It is beyond the scope of this chapter to address the problems inherent with multicolor compensation. There are a number of resources, both printed and online, that serve to explain compensation in detail [e.g., (12, 18) and www.drmr.com/compensation/].

In the multicolor world, staining controls are more important than ever. Rather than focusing on isotype controls, which have numerous inherent problems, we focus on staining controls that include all reagents except one. The difference in distributions between such a control and the fully stained sample is the only way to appropriately identify expression patterns. These types of controls are referred to as *FMO* (fluorescence minus one) controls; their use is described in detail elsewhere (12).

Nonetheless, the most time-consuming and complex part of this technology is the primary analysis of the data. A typical 12-color experiment can have upward of 100 distinct gates (or regions) applied to one- or two-dimensional regions. Thus, subsets such as those shown in figure 10.2 must be identified using a series of gates. Figure 10.3 shows an example of this gating approach, as well as a representation of such gating in the user interface.

However, even otherwise-simple statistical analyses become complex. The most common statistic employed, a "percentage of" value that connotes the representation of a subset, is no longer a single value. In a complex phenotyping analysis, it may be quite relevant whether a given subset is expressed as a percentage of, for example, CD4 T cells, total T cells, or total lymphocytes, or, conversely, that the representation is expressed as the percentage of the given subset that are CD4 T cells, total T cells, or lymphocytes. Fundamentally, the amount of information in these data files is enormous, with researchers forced to summarize them using a few numbers or, at most, one or two graphs.

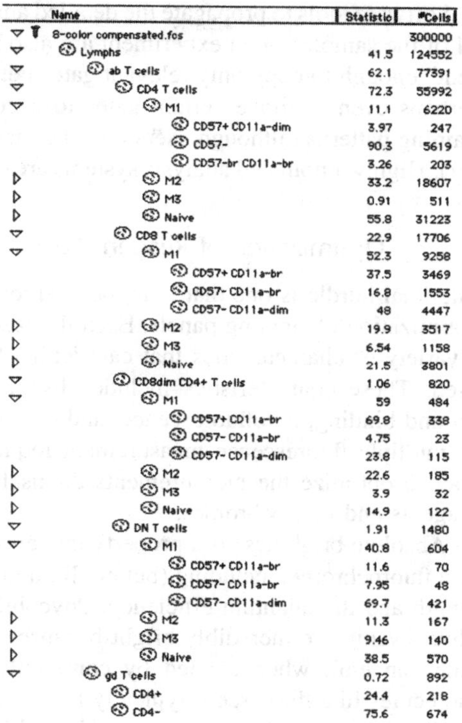

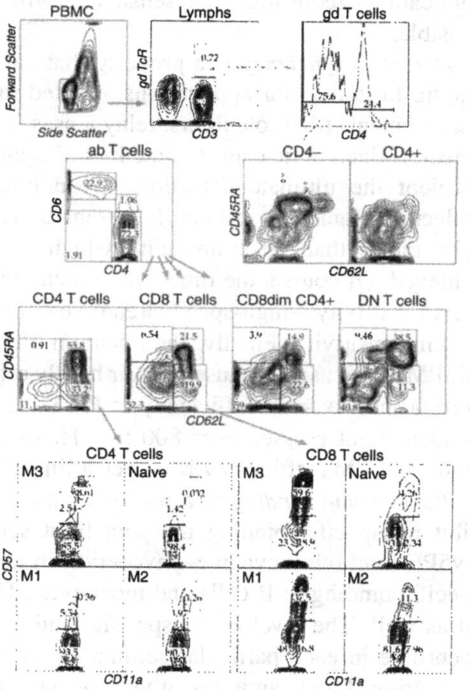

Figure 10.3. Identifying hyperfine subsets by progressive gating. (*A*) Implementation of the representation of a hierarchical gating structure to represent T cell subsets (similar to the conceptual diagram in Figure 10.1), using FlowJo analysis software (Tree Star, Inc., San Carlos, CA; www.flowjo.com). Each subset is indicated by an indentation level; the "statistic" shows the frequency within the "parent" population of the given gate. (*B*) Actual gates used to create this hierarchical representation. For example, the peripheral blood mononuclear cells sample was gated on forward- and side-scattered light for lymphocytes; only lymphocytes are shown in the next gate. Further subsetting is shown as finer subsets are identified. Note that it is not always possible to use the same "regions" or gates for different subsets, even if they are drawn on the same parameters. For example, gates identifying naïve and different memory subsets (M1, M2, and M3) must be different for CD4 T cells than for other T cell lineages because of the different level of expression of CD45RA. The hierarchical representation on the left abstracts the subset identification from the actual gates needed to identify the cells. This abstraction makes it easier for researchers to create statistical and graphical reports, for example, by requesting graphics or statistics on "CD4 naïve T cells" without having to ensure that all of the necessary gates are applied.

Name	Statistic	#Cells
8-color compensated.fcs		300000
Lymphs	41.5	124552
ab T cells	62.1	77391
CD4 T cells	72.3	55992
M1	11.1	6220
CD57+ CD11a-dim	3.97	247
CD57-	90.3	5619
CD57-br CD11a-br	3.26	203
M2	33.2	18607
M3	0.91	511
Naive	55.8	31223
CD8 T cells	22.9	17706
M1	52.3	9258
CD57+ CD11a-br	37.5	3469
CD57- CD11a-br	16.8	1553
CD57- CD11a-dim	48	4447
M2	19.9	3517
M3	6.54	1158
Naive	21.5	3801
CD8dim CD4+ T cells	1.06	820
M1	59	484
CD57+ CD11a-br	69.8	338
CD57- CD11a-br	4.75	23
CD57- CD11a-dim	23.8	115
M2	22.6	185
M3	3.9	32
Naive	14.9	122
DN T cells	1.91	1480
M1	40.8	604
CD57+ CD11a-br	11.6	70
CD57- CD11a-br	7.95	48
CD57- CD11a-dim	69.7	421
M2	11.3	167
M3	9.46	140
Naive	38.5	570
gd T cells	0.72	892
CD4+	24.4	218
CD4-	75.6	674

Software tools to propagate the detailed analysis performed on a single sample across all of the samples in an experiment are also necessary. These tools should be "intelligent" enough to copy only relevant gates between tubes that have different stains, and perhaps even to move certain gates to accommodate sample-to-sample variation in staining patterns (although users must be made aware of such processes when they occur). Highly automated analysis systems are crucial to the success of these experiments.

Optimization of Reagent Panels

The final hurdle is one that must be overcome for each new experiment: the iterative optimization of staining panels. Each fluorescence measurement has associated with it a variety of characteristics that can detract from the sensitivity with which it can be used. These characteristics include absolute brightness, detection efficiency, background binding, autofluorescence, and compensation requirements. Optimal detection of multiple fluorescence measurement requires that the instrument be designed in a way to optimize the measurements themselves, following the appropriate pairing of reagents and fluorochromes.

Absolute brightness is a property that encompasses a number of inherent properties of a fluorochrome, including (but not limited to) the absorbance at the excitation wavelength and the quantum efficiency. Phycobiliproteins such as phycoerythrin and allophycocyanin are incredibly bright because of the number of photons that each molecule can emit when excited by commonly used lasers. In contrast, small organic molecules like fluorescein typically have an order of magnitude less emission. Thus, applications requiring high sensitivity will employ the phycobiliproteins whenever possible.

Detection efficiency is a property that encompasses the emission optics and wavelength. In multicolor applications, emitted light typically has to traverse multiple optical elements (dichroic filters, relay lenses, and interference filters) before reaching the detector. Each optic causes some loss of light; thus, the more optical elements, the less efficient the ultimate detection. In addition, the numerical aperture of the light-collecting element is critical; high-sensitivity applications will require the use of flow cells rather than jet-in-air, with which a much higher numerical aperture can be achieved. Of course, the time over which the cells are interrogated proportionately affects sensitivity—high-speed stream velocity will be associated with a commensurate loss in sensitivity. Finally, the quantum efficiency of the detector is critical. Most of all, PMTs are used because of their highly efficient conversion of incident photons into current; however, the efficiency of PMTs drops off dramatically as the wavelength of incident light crosses over 800 nm. Hence, for far-red fluorochromes (like the Cy7 tandems), PMT efficiency becomes a limiting factor.

Background binding is a property inherent to fluorochromes. Some molecules exhibit nonspecific binding that can limit sensitivity of measurements. For example, Cy5PE (and other cyanine–phycoerythrin tandems) can exhibit a high degree of nonspecific binding to B cells and monocytes. Other examples of nonspecific binding exist as well: The level of nonspecific binding must always be evaluated for each fluorochrome in each particular setting. In the case of Cy5PE, sensitivity is greatest for measuring T cell antigens, given the lack of nonspecific binding of the reagent to T cells.

Autofluorescence is a property of the cells. All cells are autofluorescent (in that they have naturally fluorescing components in the cytoplasm). The spectrum of the emitted light is very similar to that of fluorescein (i.e., highest in the green and lowest in the far red). Therefore, all other aspects being equal, sensitivity for fluorescence detection is much greater in the far red than in the green, as there is much lower background signal from the cells themselves. Note that some correction for autofluorescence can be performed to increase sensitivity (16).

Finally, *compensation* requirements have the most significant effect on the sensitivity of any given fluorochrome. The effect of compensation on sensitivity is beyond the scope of this chapter; see (12) for a detailed discussion. Suffice it to note that the more compensation that is required, the greater the effect on deteriorating sensitivity. Hence, some fluorescence channels (e.g., allophycocyanin), which can be used with exquisite sensitivity in the absence of other fluorescence measurements, become far less useful because of the number of other fluorochromes that emit in the same detector channel.

Given these many different effects on sensitivity, it is perhaps not surprising that it is nearly impossible to predict in advance which channels will provide the greatest sensitivity for any given instrument, or even in any given combination of fluorochrome reagents. Instead, an iterative approach to trying out multiple different pairs of reagent:fluorochrome combinations is required. Each combination is evaluated, and the optimal combination of reagents is generated. Although experience helps considerably in choosing initial combinations, we find that we still require multiple experiments to optimize the combinations.

Future Directions

Multicolor flow cytometry measuring greater than six or eight colors is still in its infancy. A number of issues need to be addressed before this technology becomes as prevalent as three to four colors are now. These issues include increased automation in instrumentation and software. Current instruments require considerable expertise to operate, and even after instrument designs are improved, data analysis will be a limiting factor. This is currently the most significant time constraint. Therefore, we will need new software tools capable of organizing the analyses and creating databases for the results. New tools that can aid in the identification and comparisons of populations in multiple dimensions are also needed. We have published algorithms for comparing multidimensional distributions (14, 15, 17) that at this point can only be considered to be first steps in this direction.

References

1. Altman, J.D., Moss, P.A., Goulder, P.J., Barouch, D.H., McHeyzer-Williams, M.G., Bell, J.I., McMichael, A.J., Davis, M.M. 1996. Phenotypic analysis of antigen-specific T lymphocytes. *Science* 274:94–96.
2. Baumgarth, N., Roederer, M. 2000. A practical approach to multicolor flow cytometry for immunophenotyping. *J. Immunol. Meth.* 243:77–97.
3. De Rosa, S.C., Herzenberg, L.A., Roederer, M. 2001. 11-color, 13-parameter flow cytome-

try: identification of human naive T cells by phenotype, function, and T-cell receptor diversity. *Nat. Med.* 7:245–248.

4. Hardy, R.R., Hayakawa, K., Parks, D.R., Herzenbergm L.A. 1984. Murine B cell differentiation lineages. *J Exp Med* 159:1169–1188.

5. Hulett, H.R., Bonner, W.A., Barrett, J., Herzenberg, L.A. 1969. Cell sorting: automated separation of mammalian cells as a function of intracellular fluorescence. *Science* 166:747–749.

6. Kern, F., Surel, I.P., Brock, C., Freistedt, B., Radtke, H., Scheffold, A., Blasczyk, R., Reinke, P., Schneider-Mergener, J., Radbruch, A., et al. 1998. T-cell epitope mapping by flow cytometry. *Nat. Med.* 4:975–978.

7. Lee, P.P., Yee, C., Savage, P.A., Fong, L., Brockstedt, D., Weber, J.S., Johnson, D., Swetter, S., Thompson, J., Greenberg, P.D., et al. 1999. Characterization of circulating T cells specific for tumor-associated antigens in melanoma patients. *Nat. Med.* 5:677–685.

8. Liao, F., Rabin, R.L., Smith, C.S., Sharma, G., Nutman, T.B., Farber, J.M. 1999. CC-chemokine receptor 6 is expressed on diverse memory subsets of T cells and determines responsiveness to macrophage inflammatory protein 3 alpha. *J. Immunol.* 162:186–194.

9. Mitra, D.K., Rosa, S.C.D., Luk, A., Balamurugan, A., Khaitan, B.K., Tung, J., Mehra, N.K., Terr, A.I., O'Garra, A., Herzenberg, L.A., Herzenberg, L.A., Roederer, M. 1999. Differential representations of memory T cell subsets are characteristic of polarized immunity in leprosy and atopic diseases. *Intl. Immunol.* 11:1801–1810.

10. Pitcher, C.J., Quittner, C., Peterson, D.M., Connors, M., Koup, R.A., Maino, V.C., Picker, L.J. 1999. HIV-1-specific CD4+ T cells are detectable in most individuals with active HIV-1 infection, but decline with prolonged viral suppression. *Nat. Med.* 5:518–525.

11. Rabin, R.L., Roederer, M., Maldonado, Y., Petru, A., Herzenberg, L.A. 1995. Altered representation of naive and memory CD8 T cell subsets in HIV− infected children. *J. Clin. Invest.* 95:2054–2060.

12. Roederer, M. 2001. Spectral compensation for flow cytometry: visualization artifacts, limitations and caveats. *Cytometry* 45:194–205.

13. Roederer, M., Dubs, J.G., Anderson, M.T., Raju, P.A., Herzenberg, L.A. 1995. CD8 naive T cell counts decrease progressively in HIV-infected adults. *J. Clin. Invest.* 95:2061–2066.

14. Roederer, M., Hardy, R.R. 2001. Frequency difference gating: a multivariate method for identifying subsets that differ between samples. *Cytometry* 45:56–64.

15. Roederer, M., Moore, W., Treister, A.S., Hardy, R.R., Herzenberg, L.A. 2001. Probability binning comparison: a metric for quantitating multivariate distribution differences. *Cytometry* 45:47–55.

16. Roederer, M., Murphy, R.F. 1986. Cell-by-cell autofluorescence correction for low signal-to-noise systems: application to epidermal growth factor endocytosis by 3T3 fibroblasts. *Cytometry* 7:558–565.

17. Roederer, M., Treister, A., Moore, W., Herzenberg, L.A. 2001. Probability binning comparison: a metric for quantitating univariate distribution differences. *Cytometry* 45:37–46.

18. Stewart, C.C., Stewart, S.J. 1999. Four color compensation. *Cytometry* 38:161–175.

11

FACS-Based High-Throughput Functional Screening of Genetic Libraries for Drug Target Discovery

ERLINA PALI, MARK POWELL, ESTEBAN MASUDA,
YASUMICHI HITOSHI, SACHA HOLLAND,
MARK K. BENNETT, SUSAN M. MOLINEAUX,
DONALD G. PAYAN, AND JAMES B. LORENS

Introduction

A primary aim of functional genomics in pharmaceutical applications is to identify genes whose function is critical to maintaining a disease state and to determining whether therapeutic modulation of this function results in a beneficial clinical response. However, although many genomic approaches can identify disease-associated genes, lengthy follow-up studies are usually required to determine which genes are functionally important and are causally linked to a given disease. In contrast, retrovirally mediated functional genetic screening approaches enable rapid identification of physiologically relevant targets (22). Genetic screens are designed to detect functional changes that result in changes in cellular function that correlate with disease amelioration. Retroviruses possess unique properties that allow delivery of complex libraries of potential genetic effectors to a variety of cell types. These effectors can perturb specific interactions required to achieve a complete functional response and establish a direct relationship with a cellular function. Functional screens are employed to select for cells endowed with a desired genetic effector–induced change in phenotype. Identification of a genetic effector that causes an altered cellular phenotype that correlates with clinical benefit can explicate critical signaling components suitable for therapeutic intervention. Flow cytometry represents a uniquely powerful methodology to monitor complex multiparametric changes of individual cells in large populations. In conjunction with recent advances in retroviral expression systems, the sensitivity and speed of flow cytometry enables a highly efficient functional screening of complex libraries in a wide range of cell-based assays. In this chapter, we discuss the process of functional genetic screening and show specific examples of its implementation. We focus particularly on the critical parameters involved in the design and execution of functional genetic screening approaches based on FACS (fluorescence activated cell sorter).

Retroviral Library Technology

Retroviruses provide a powerful method of introducing genes into mammalian cells in an efficient and stable manner. Recent advances in retroviral vector technology and packaging systems have extended their application to allow efficient and stable delivery of highly diverse libraries encoding various types of genetic effectors, including cDNAs, peptides, and ribozymes, into a broad range of cell types (22). Retroviral vectors can accommodate up to 7 kb of exogenous sequence and are compatible with a wide array of genetic elements. In particular, cotranslational reporter systems such as internal ribosome entry sites and conditional expression systems impart significant flexibility in screen design (J.B. Lorens, unpublished results).

Infectious retroviruses are generated by transient transfection of retroviral packaging cells (figure 11.1). The high efficiency of this transient transfection step is critical to maintaining the diversity of retroviral libraries. Most cell types grown in cell culture can be infected with viral vector systems derived from the Moloney murine leukemia virus (MMLV) (32). The viral genome integrates into a target cell chromosome during cell division, requiring cell proliferation during the infection step. Alternative lentiviral systems support infection of nondividing cells such as neurons (2). Retroviral transduction of a dominant genetic effector library produces a population of individually "mutagenized" cells, where specific cellular functions are interrupted by a dominantly acting protein, peptide, or RNA molecule (figure 11.2). The diversity of available effectors allows the perturbation of virtually any cellular process. These libraries can be derived from endogenous sequences (e.g., cDNA) or generated from random sequences (e.g., random peptides). Those cells exhibiting an altered phenotype in a disease-relevant assay are selected, and the genetic effector sequence causing the change is recovered. The isolated genetic effector is then recloned and retested in naïve cells to confirm that the phenotypic change is a result of perturbations exerted by the effector molecule and not of a fortuitous mutation in the isolated cell. Conditional expression vectors are particularly useful in genetic library screening schemes to distinguish library member–derived effects from host cell compensations. The ability to regulate viral transcription provides the opportunity to directly correlate genetic effector expression with the target phenotype. This approach may be used to select against heritable background events that are otherwise indistinguishable from a screening hit (figure 11.3; see following discussion).

Overview of Flow Cytometry–Based Functional Genomic Screening

Flow cytometry provides a powerful and versatile tool for the execution of functional genomic screens. Recent applications have expanded beyond traditional polychromatic immunophenotyping to include direct measurements of cell function, such as cell viability, apoptosis, activation, proliferation, and signal transduction, as well as other in-

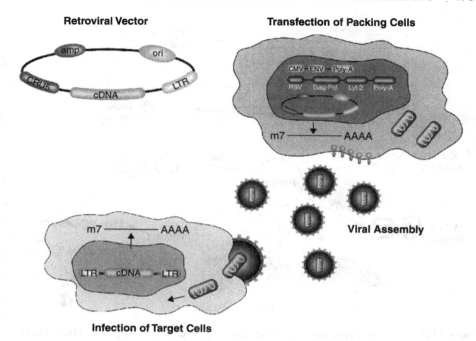

Figure 11.1. Retroviral vector system. Retroviral delivery systems require two main components: a transmission vector and a packaging cell line. The retroviral vector itself contains genomic sequences required for transcription, packaging, reverse transcription, and integration of the viral genome. The packaging cell provides the viral structural proteins required to convert the retroviral vector RNA into a mature infectious viral particle and Pol-derived enzymes for reverse transcription and integration. To generate infectious retroviral vector particles, a retroviral vector plasmid is transfected into Phoenix 293T-based packaging cells. The retroviral RNA genome is transcribed from the plasmid, transported to the cytoplasm, and assembled into a viral particle. Mature virions bud from the plasma membrane, and infectious virus recovered from the packaging cell media is used to infect a desired target cell population. Following entry into the cell retroviral vector, genomic RNA is reverse transcribed and integrated into the host cell genome. This stable integration ensures that all progeny cells continue to express the encoded RNA/peptide/protein. A wide variety of cell types may be efficiently transduced using this approach (32). Retroviral vectors can accommodate various genetic features such as regulatable expression cassettes and internal ribosome entry sites (see Figure 11.3).

tracellular processes. High-throughput cytometric applications facilitate quantitative multiparametric analyses and sorting at single-cell resolution, resulting in the recovery of highly enriched target cell populations. This attribute is particularly useful in the context of functional genetic screening. An example of a FACS-based retroviral screen is shown in figure 11.4. A retroviral library is transduced into a population of cells at an infection rate that ensures a single viral integration per cell. Hence, individual cells express a single genetic effector. The transduced population is then treated with the desired stimulus, and the response is measured using a FACS-based readout, often in conjunction with a non–flow cytometric selection parameter. Cells are sorted on the basis of gating schemes that reflect a desired change in the response phenotype. The

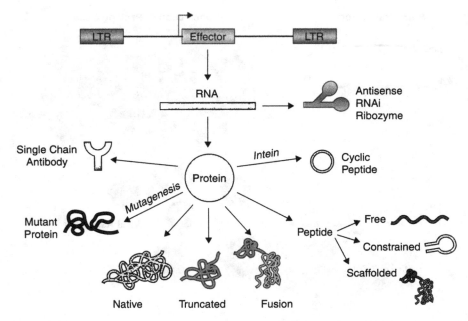

Figure 11.2. Retroviral generation of genetic effectors. Several types of genetic effectors can be efficiently delivered and expressed by retroviral vectors. RNA, protein, and peptide effectors are readily expressed by retroviral vectors. Libraries of genetic effectors can be constructed from genomic sources or as random oligonucleotides. RNA effector libraries such as random hairpin ribozymes have been used in library screening approaches (1, 19). Libraries of cDNAs that encode full-length, truncated, fusion, or mutagenized proteins have been screened to reveal protein function. Mutagenesis of a cDNA encoding a known pathway regulator by error-prone polymerase chain reaction (18), random fragmentation (10), or DNA shuffling (31) can be used to rapidly map functionally important domains of the molecule or to impart novel characteristics. Screens with full-length cDNA libraries can be used to identify the gene defect in a mutant cell line by restoring the original phenotype (33). Other screening approaches have used full-length or fragmented retroviral cDNA libraries to isolate genes encoding dominant effectors of cellular phenotypes (11, 27). Recently, we (14) reported a FACS-based functional cDNA screen to identify inhibitors of antigen-receptor induced B-cell activation, using cell-surface CD69 upregulation as a marker (see figure 11.4). Green fluorescent protein (GFP)–fusion protein libraries have been used to identify proteins with novel subcellular localization patterns (23, 26). Libraries expressing GFP fused to cDNA-derived domains have been successfully used to identify novel regulators of angiogenesis (J.B. Lorens, unpublished data). Random peptides, constrained or displayed by a larger protein scaffold, can probe cellular responses following specific biological or disease stimuli (34). Random peptides may be displayed on GFP scaffolds (17, 26) as surface-exposed loops or terminal extentions. Recently, it has been demonstrated that cyclic peptides may be expressed intracellularly, using a novel protein-splicing approach (17). This exciting technology allows the identification of highly potent regulators of cellular function.

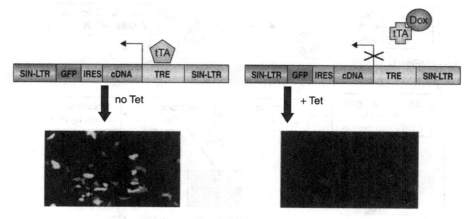

Figure 11.3. Regulation of genetic effector expression from retroviral vectors. Conditional expression of retrovirally delivered effectors provides a method to distinguish cells displaying a desired phenotype resulting from the expression of a genetic effector from those caused by heritable mutations. Tetracycline-regulated expression systems enable effector expression to be switched off (Tet-off system) and the wild-type phenotype to be restored. A target cell line that constitutively expresses the tetracycline transactivator (tTA; e.g., from an integrated retroviral vector) is infected with a retroviral vector encoding a library of effector elements whose expression is driven by a tetracycline-regulated promoter that comprises a tetracycline regulatory element (TRE). Transcription from the TRE is dependent on tTA binding to seven recognition sites upstream of a basal promoter. This vector also carries a self-inactivating mutation in the long terminal repeat (SIN-LTR) that removes the viral enhancer/promoter activity on viral integration, obviating transcriptional interference. On addition of tetracycline to the culture media, tTA binding to the TRE is inhibited and library expression is blocked. In this example, a lung carcinoma cell line (A549.tTA) is infected with a retroviral construct that coexpresses a cDNA and green fluorescent protein (GFP) from the viral vector transcript via an internal ribosome entry site. On the addition of tetracycline to the cell culture medium, cDNA and GFP expression are repressed. The loss of GFP fluorescence serves as a surrogate marker of cDNA expression.

selection is then repeated until a sufficient enrichment is achieved. The enriched population of cells is single-cell cloned and further analyzed to verify that the phenotype is dependent on expression of the genetic effector. The encoded library effector sequence is elucidated by polymerase chain reaction and reintroduced into naïve cells to verify the effect.

FACS-Based Assays for Functional Genomic Screening

Genetic screens distinguish between two states: response versus nonresponse, quiescence versus activation, arrest versus growth, and so forth. A clear definition of each state facilitates the development of selective parameters to discriminate affected cells in a large population. For example, a resting population of cells activated by a specific cytokine will undergo multiple measurable changes, including appearance of new cell surface markers and changes in growth rate, size/granularity, and exo- or endocytosis. Table 11.1 lists FACS-based parameters for cellular properties and functions that can

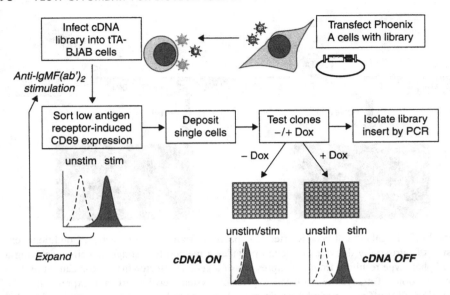

Figure 11.4. Overview of a cell surface marker screen: cDNA screen for inhibitors of B cell receptor (BCR) signaling. Changes in cell surface marker levels often correlate with cell signaling following a specific stimulus. For example, antigen-induced transcriptional upregulation and surface expression of CD69 upregulation is a prominent early-activation marker in both B and T lymphocytes. A FACS-based genetic screen for inhibitors of B cell activation was developed using CD69 as a marker. A BJAB cell line that expresses the tTA was infected with a cDNA library. Cells were stimulated with anti-IgM, stained with an APC-conjugated anti-CD69, and then sorted for nonresponders. These cells were expanded, and this process was repeated several times until enrichment of nonresponding cells was seen in the population. Single cells were deposited, expanded, and then grown in the presence and absence of tetracycline (doxycycline, *dox*) to extinguish cDNA expression. Positive clones were picked on the basis of their CD69 expression profile in the presence and absence of doxycycline: In the absence of doxycycline, with cDNA expression on, anti-IgM-induced CD69 expression should be repressed, and in the presence of dox, with the cDNA turned off, induced CD69 expression should be restored. Finally, library inserts were recovered from positive clones by reverse transcriptase polymerase chain reaction, using vector primers.

be used to monitor effector-induced changes in cellular physiology. Several examples of functional genomic screens using FACS-based readouts are listed in table 11.2. The following section discusses the application of these assays for the development and implementation of a functional genomic screen.

Changes in Cell Surface Phenotype

Changes in a cell surface marker level are a convenient and straightforward method to monitor a specific change in a phenotypic response. Surface proteins are often differentially regulated in response to disease-relevant activation signals. For example, the inflammatory cytokine TNFα elicits a signal transduction cascade in epithelial cells that leads to transcriptional activation of the ICAM-1 gene. The subsequent robust in-

Table 11.1. Flow cytometric assays can be used to assess many cellular functions

FACS-Based Assay	Parameters
Cell Surface Immunofluorescence	*Surface Protein Expression*
Immunofluoresence staining using specific antibodies and multiple fluorescent dyes define particular phenotypic subsets and activation antigens. Ligand–receptor interactions can be done in a quantitative manner using fluorescently labeled ligands.	1. Leukocyte differentiation antigens 2. Adhesion molecules 3. Receptors
Cellular Function Assays	*Cellular Function*
Viable cells are stained with specific fluorescent probes. In response to activation or changes in cellular biochemistry, these probes will undergo shifts in emission spectrum or intensity, which can be measured by FACS.	1. Viability/apoptosis 2. Exocytosis 3. Oxidative burst 4. Calcium flux 5. Proliferation 6. Cell cycle distribution 7. Mitochondrial potential 8. Intracellular pH
Intracellular Staining	*Intracellular Components, e.g.,*
After permeabilization, conjugated antibodies can be used to stain intracellular proteins.	1. Cytokines 2. Nucleoprotein 3. Phosphoproteins 4. Kinases
Flow Enzymology	*Enzymes, e.g.,*
Fluorescent substrates when taken up in viable cells allow characterization based on enzymatic activity.	1. Esterases 2. Glucosidases 3. Aminopeptidase 4. Peroxidases
Secretion Capture	*Cell Secretions*
Cells are encapsulated in agarose drops, within which their secreted products are captured and measured using a specific fluorescent antibody reporter.	1. Immunoglobulins 2. Cytokines
Gene Reporters	*Reporter Molecules*
Screening lines can be created by fusion of gene reporters that can be detected by FACS.	1. Autofluoresecent proteins GFP, YFP, RFP, CFP 2. Beta-Galactosidase 3. Cell surface proteins, e.g. CD2, NGFR, Thy1.2
Fluorescence Resonance Energy Transfer	*Protein–Protein Interactions*
Proximity and orientation of proteins can be established utilizing the spectral properties of appropriate pairs of dyes or autofluorescent proteins.	1. FITC to RITC conjugates for Cell surface antigen proximity 2. CFP to YFP or BFP to GFP for Intracellular proteins

Table 11.2. FACS-Based Genomic Screens

Screening Cell Line	Screen	Reference
T Cell Line	Redirected TCR (random peptide in TCR)	16
CHO	HDL binding (mutagenized scavenger receptor)	9
CHO	Measles virus vector	8
CHO	CHOP promoter activation	24
Neural tumor	IRES activity (random oligo vector)	25
Neuroblastoma	Promoter activity (random oligo in vector)	5
tTA-A549	ICAM-1 down-regulation	M. Powell, unpub. data
tTA-Jurkat N	CD69 down-regulation	3
tTA-BJAB	CD69 down-regulation	14
tTA-BJAB	Inhibition of Epsilon response to IL4	17
tTA-BMMC	Inhibition of exocytosis	S.J. Holland, A. Rossi, M. Bennett, unpub. data
tTA-A549	Cell cycle arrest	13
HUVEC	Surface protein changes	J.B. Lorens, unpub. data

Several examples of successful FACS-based genetic screens have been conducted. These encompass a wide range of cell types and selection parameters, indicative of the broad application of this approach. tTA, tetracycline transactivator.

duction of surface ICAM-1 levels is thus a marker for many prior molecular interactions in the cell. Expression of a genetic inhibitor of TNFα signal transduction, IκB, blocks the transcriptional activation of ICAM-1 as well as surface-level induction. Hence, IκB can be used to demonstrate the feasibility and relevance of ICAM-1 as a surface readout (figure 11.5). Using IκB as a positive control, we established conditions for a FACS-based functional genetic screen to identify novel regulators of TNFα-signaling. This approach yielded new interactions in TNFα-mediated signaling, providing fresh insight and novel targets for therapeutic intervention (M. Powell, unpublished data). CD69 is an early surface marker induced in both T and B cells on receptor activation. Using this parameter, we isolated cDNA inhibitors of T cell receptor signaling (3) and antigen receptor-induced signaling in the B cells (12). An overview of the B cell screening approach is depicted in figure 11.4. Cell surface changes are particularly useful for monitoring changes in primary cells that are not amenable to engineered reporters (see following). Phenotypic changes in cell surface protein levels are easily monitored by FACS analysis, and gates that define a desired cell state are used for sort selections. Several novel genetic regulators of angiogenesis were identified in FACS-based selection schemes that used changes in surface protein levels on primary human vascular endothelial cells (J.B. Lorens, unpublished data).

Changes in Cellular Functions

Some changes in cellular behavior and biochemistry can be directly probed by specific fluorescent dyes that bind selected macromolecules. Cell proliferation can be monitored using a wide variety of cell-tracker dyes, which stain either intracellular or membrane components. With each cell division, the dye is diluted between the daughter cells. Hence, cells affected by cell cycle inhibitors remain bright, while proliferating cells become dimmer. Arrested cells can thus be enriched by sorting the highest fluo-

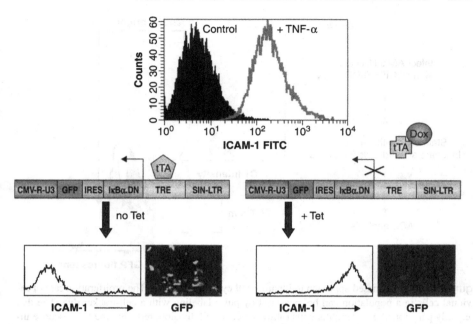

Figure 11.5. Optimization of screening parameters using a regulated positive control. To optimize screening and sorting parameters for a FACS-based screen, it is desirable to use a known genetic effector as a positive control. For example, in a screen aimed at identifying inhibitors of TNFα signaling (M. Powell, unpublished data), a dominant negative form of IκBα (IκBα.DN) was used as a known inhibitor of TNFα-induced NF-κB activation (32). The readout for TNFα-induced signaling was cell surface expression of ICAM-1 in the A549 (lung epithelial) cell line (a). A549.tTA cells were infected (95% efficiency) with the IκBα.DN–green fluorescent protein (GFP)–expressing vector, and on TNFα treatment, these cells fail to express ICAM-1. On the addition of tetracycline, which causes shutoff of IκBα.DN, and coexpressed GFP, ICAM-1 expression is restored. To optimize flow-sorting parameters for a phenotypic screen, IκBα.DN-GFP expressing cells can be spiked into and subsequently rescued from a large population of unaffected cells.

rescent population (figure 11.6). Demo et al. (4) developed a novel FACS-based assay for exocytosis using fluorescently labeled annexin V, which binds specifically to vesicles exposed on the cell surface after stimulation. This assay was used to identify a novel regulator of mast cell exocytosis (S.J. Holland, A. Rossi, and M. Bennett, unpublished data). Dynamic events such calcium flux and oxidative burst can also be used. Changes in the secretion of cytokines or immunoglobulins (29) of single cells can be monitored and used as a sorting parameter.

Naturally Fluorescent Protein Reporters

The isolation of FACS optimized variants of green fluorescent protein (GFP) has made this molecule an ideal probe for studying cellular components and processes. A distinct advantage of using fluorescent proteins such as GFP is the ability to rapidly detect expression in viable cells without the need for sample staining or the addition of

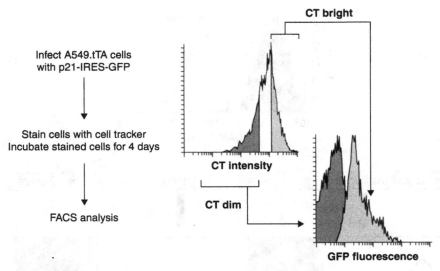

Figure 11.6. FACS-based approach to identify cell cycle modulators. The proliferation rate of individual cells in a population can be measured by pulse labeling with a cell-tracker dye. The dye is equally partitioned into daughter cells following each cell division, reducing the fluorescence intensity by about half. Arrested cells maintain the higher undiluted fluorescence and thus can be distinguished from proliferating cells. An example is shown here in which a lung carcinoma cell line (A549.tTA) was infected with a tetracycline-regulatable vector expressing the cyclin-dependent kinase inhibitor p21. Expression of p21 in these cells results in a potent arrest in the G1 phase of the cell cycle (21). The infected cells were demarcated by the cotranslation of a green fluorescent protein marker. The infected population was pulse labeled with cell-tracker dye and cultured for 4 days. The cells were then detached and analyzed by FACS. Gating of the cell-tracker fluorescence demonstrated a correlation between high cell-tracker mean fluorescence and green fluorescent protein positivity, indicative of a cell cycle arresting activity of the encoded p21 gene. This approach has been adopted to isolate novel cell cycle regulators in genetic library screens (13).

fluorochromes. GFP can be used as a reporter for the expression of various genetic effectors or as part of a reporter system, which can be easily monitored using any cytometer with a 488-nm argon ion laser. Stable cell lines can be generated that express reporter systems with a specific promoter that drives GFP expression on treatment with a selected stimulus. Although stable transfection techniques are commonly employed, these reporter constructs can be efficiently delivered to virtually any cell type by retroviral vector transduction (figure 11.7). Kinsella et al. (17) used an interleukin 4 responsive promoter construct that drives expression of a dual reporter system comprising GFP and a toxin receptor. B cells were transduced with this vector construct, and a cell line was isolated. Cells failing to respond to interleukin 4 stimulation were insensitive to the toxin and expressed only basal levels of GFP. Random cyclic peptide inhibitors of the interleukin 4 pathway were isolated in a screen that used a combination of GFP-gated sorting and toxin treatment. Peelle (unpublished data) isolated an optimized red fluorescent protein (RFP) by random mutagenesis and repeated FACS sorting. This red variant can be measured simultaneously with GFP with minimal spec-

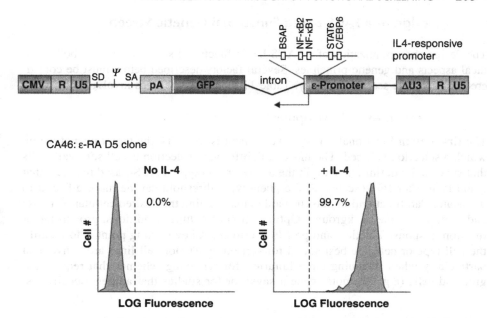

Figure 11.7. Reporter gene construct. Reporter constructs are powerful tools in developing cell-based screens for assaying various aspects of cell physiology. Traditionally, a promoter that responds predictably to a given stimulus is coupled to a convenient readout in a plasmid vector. Stable transfectants are generated and a subclone is identified with the proper response profile. A critical limitation to this approach is the transfection step. Many cell types are recalcitrant to transfection, and as the frequency of stable integration of plasmid concatamers is only 1 in 10^6, reporter cell lines may not be obtainable. To obviate the requirement of the transfection step and increase the repertoire of cell types compatible with a reporter gene assay approach, we employed self-inactivating retroviral vectors to stably introduce promoter constructs. As retroviral infection results in stable integration into the target cell genome, the frequency of potential cell lines equals the infection rate, which is a drastic improvement over the transfection-based approach. An example is shown here: The CA46 B-cell line was transduced with a retroviral vector carrying the interleukin 4–responsive germline e-promoter driving expression of green fluorescent protein (GFP). An optimal screening cell line was isolated by a sequential sorting approach. Infected CA46 cells treated with interleukin 4 were sorted for high GFP fluorescence. Following a rest period, cells with low basal expression were sorted. This population was restimulated, and GFP-expressing cells were single cell cloned. One of these clones (D5) demonstrated optimal kinetic and quantitative response characteristics. Using a dual laser system, we successfully screened a blue flourescent protein–scaffolded random peptide library for a regulated inhibition of interleukin 4–induced B-cell class switching response using an epsilon-promoter-based GFP reporter (17).

tral overlap. Combinations of different fluorescent proteins provide opportunities for using multiple reporters to enhance the specificity of the readout and the selection process. Several variants, including yellow, red, cyan, and blue, are available and can be used in multiparameter reporter systems and for studying protein–protein interactions based on fluorescence resonance energy transfer between an appropriately matched pair.

Design of a FACS-Based Functional Genetic Screen

The design and optimization of a FACS-based functional screen comprises both technical aspects and genetic principles. Several factors described below must be considered to maximize screen efficiency. Table 11.3 outlines the steps involved.

Selection Assay Development

The first step in functional screen development is to define the target phenotype on which a selection is based. The most straightforward selection is cell survival: Cells that survive, by definition, exhibit the desired phenotype. FACS-based readouts that quantify marker fluorescence as the phenotypic threshold can be more difficult to delineate. Careful attention must be paid to maximizing the desired cellular response and minimizing the background. Optimization of culture conditions, growth factor titration, response kinetics, and profile is critical. A key starting point is to identify the cell type or cell line best suited for screening. Tumor cell lines are convenient particularly when screening large libraries, for screening schemes that require engineered cells (e.g., promoter-gene assays), or for studies that address specific as-

Table 11.3. Optimization of Library Screen Sorting Process Involves
Several Factors

Selection Assay

For screening line:
    Select cell type or clone with good response and low background
Maximize response:
    Optimize culture conditions, growth factors, and time course
Optimize staining conditions

Library

Select the appropriate library
Determine number of cells required to cover library complexity

Screen Strategy

Choose an adequate pre-selection approach
Establish phenotypic criteria for hit identification
Optimize the sort strategy:
    enrich followed by purity sorting;
    mixed mode or iterative purity sorting
Delineate phenotypic sort gate
Determine efficiency of the enrichment process by spiking positive and negative controls
Calculate background
Isolate the target cell by single cell cloning

Hit Confirmation and Validation

Rescue and transfer DNA to naïve responding cells
Reproduce phenotypic response
Evaluate in counter assays

---

pects of malignant cells. A problem with many transformed cells is stability: Tumor cell lines are often unstable, leading to phenotypic drift. Hence, screening lines that have been optimized by FACS should be archived and cells with low passage should be used for screens. The screening line should maintain a stable profile for the duration of the selection procedure. Alternatively, primary cells may be used. Although more challenging to maintain, primary cells often are much more stable and predictable than their transformed counterparts and are less prone to confounding mutations.

The target phenotypic state is best defined by expressing a positive control genetic inhibitor of the pathway. This can be a known negative pathway regulator (e.g., IκB as a positive control for TNFα-signaling; figure 11.3), a dominant negative mutant, or antisense RNA/RNAi of a known pathway member. In lieu of genetic positive controls, unresponsive cell mutants or simply unstimulated cells can provide information on the correct gate. Pilot sorts should be conducted to optimize staining conditions and assess the efficiency of recovery and percentage background.

Retroviral infection conditions should be assessed with standard control viruses. GFP-expressing viruses allow a rapid FACS assessment of infection efficiency. Importantly, initial infection optimization should include a dilution of viral stocks to determine the multiplicity of infection. The maximal infection rate varies among different cell types. Highly infectible cell types such as fibroblasts can reach 100% transduction efficiency. In these cells, the optimal library infection rate is 30%–40%, which by Poisson statistics, predicts that on average there will be a single viral integration event per cell. Other cell types may exhibit significantly lower maximal transduction rates as a result of a variety of factors such as proliferation rate and viral receptor density. Importantly, in some cases only a subpopulation of cells may be infectible at any given time. Hence, although only 30% of cells are infected, many may be superinfected. This is evidenced in GFP-titration studies by a gradual increase in mean fluorescence without an accompanying increase in infection rate. In cases in which the infection rate is particularly low, overexpression of the MMLV receptor can boost infection rates (12).

Another important factor to consider in the overall design of a genetic screen is library size. In general, cDNA libraries have a complexity on the order of 10^6–10^7, and random peptide libraries are at least 10^9. Statistically, three times the library complexity must be screened to attain a 95% confidence level that every library member has been tested. Thus, different types of libraries will require different numbers of cells in the screen. The execution of the functional assay and the selection process will be greatly affected by the enormous number of cells involved. In FACS-based approaches, it is often prudent to preenrich the cell population before sorting to reduce the sorting time. The combination of different selection schemes can greatly enhance the overall efficiency of the screen.

In any genetic screen, a critical parameter is the background. Three major types of background are encountered: sporadic, epigenetic, and heritable. Sporadic background events are noninheritable and are the result of inefficiencies in the selection procedure. For example, wild-type cells that fail to stain with an antibody or that contaminate the sorted sample as a result of fluidic bursts. Epigenetic background events are cells that transiently show the desired phenotype as a result of nonheritable changes in cell physiology. For example, some cells in a population may be in a phase of the cell cycle in which they are unable to respond to the activation stimulus. Heritable background is a result of a mutation in the cell that blocks the response. This type of background is more frequent in genetically unstable cell lines than in primary cells. Importantly, heritable background is indistinguishable from library encoded effectors and will coenrich with them. The frequency of these background events must be determined. This is accomplished by mock selecting a population and reanalyzing it. For example, a cell population activated with a cytokine and monitored for the induction of a cell surface marker is gated for nonresponders (cell surface marker low; figure 11.8). The sorted population is subsequently restained and reanalyzed to determine the sporadic background. The inheritable and noninheritable background is determined by growing the sorted cells and reanalyzing for a change in the frequency of nonresponders. The relationship between the frequency and nature of the background events and the frequency of effector hits in the library determines the enrichment attainable per sort.

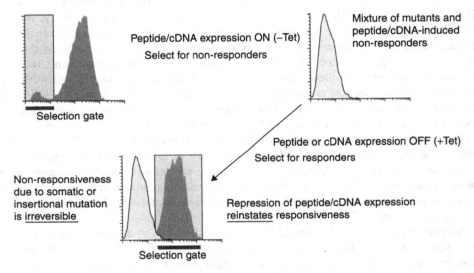

Figure 11.8. Conditional expression enhances selection efficiency. In any genetic selection strategy, the enrichment of the desired phenotype is restricted by the background. Sporadic or epigenetic background events can often be efficiently removed by iterative selections. However, heritable background will efficiently coenrich with a library-encoded effector. The use of conditional expression vectors is a powerful method to enhance overall screening efficiency by distinguishing heritable and library effector-dependent phenotypes. For example, in a selection strategy designed to isolate inhibitors of a given stimulus (e.g., cytokine activation of a cell surface marker), the sorted (low fluorescence) cells will comprise sporadic, epigenetic, and heritable background events and the desired library effector. Subsequent low-fluorescence gate sorts can effectively enrich a nonresponsive population, which is now a mixture of heritable mutant background events and the desired library effector. To distinguish between these two phenotypes, the library encoded effector expression is repressed by the addition of tetracycline (see Figure 11.3) reinstating responsiveness; in contrast, mutant cells remain unresponsive. Thus, sorting on a subsequent high-fluorescence gate (responsive) effectively enriches for the library-encoded response inhibitor.

Functional Screening Strategy Development

When the target phenotype confers resistance to chemotherapeutics (27, 35) or Fas-induced apoptosis (12), a straightforward selection method that highlights a survival or growth advantage can be powerful. This principle is readily applied to gene reporter selection schemes that express proteins that confer susceptibility to cytotoxic agents. Using a selective ablation of wild-type cells, resistant cells can then be readily enriched more than a thousand-fold (17). Faster growing cells can be identified in the FACS using a cell-tracker dye.

Selections based on changes in cell surface phenotype can be improved by presorting the cell population using magnetic beads conjugated with anti-idiotype or antifluorophore antibodies (www.miltenyibiotec.com). Both positive and negative selection strategies may be conducted, and the procedure is generally well tolerated by different cell types. Enrichment factors vary depending on the cell type, antibodies, and protocol used; however, enrichments of from 100-fold to 1000-fold are achievable, greatly reducing the number of cells before a subsequent FACS step. The magnetic beads generally do not interfere with analysis of the flow cytometric detection parameters, nor do they affect cell viability. A surface marker gene may also be included as part of a bicistronic gene reporter construct enabling a MACS preenrichment step (J.B. Lorens, unpublished results).

The large number of cells required for a functional genomic library screen imparts particular importance to the sorting process. We use MoFlo cell sorters (www.dakocytomation.com) originally designed at Lawrence Livermore for high-speed sorting of chromosomes (28). Our approach emphasizes overall system efficiency rather than maximum speed. This involves meticulous instrument maintenance, calibrations, and sort quality control. Using 100-μ nozzle tips avoids frequent disruptions resulting from clogging or obstructions. Sorting is often performed using two sorters simultaneously, and large libraries are processed in fractions to minimize cell settling. Both total sort duration and operator endurance are considered in the overall planning. Process monitoring involves using the SortMaster for maintaining accurate drop delay. In addition, critical parameters such as scatter and mean fluorescence are carefully monitored for drift. Adjustments and recalibrations are performed whenever necessary. At the onset of a genetic library screen, the frequency of desired events is very low. Hence, the first selection is especially critical. When using cell sorting as the primary selection approach, an initial enrichment is conducted to recover an impure collection of rare cells within a reasonable time frame. This is best accomplished by an enrichment sort.

The anticoincidence circuitry is disabled, and drops containing cells that fit the sort criteria will be deflected, regardless of the presence of accompanying unwanted events. An enrichment factor of from 5- to 15-fold can be attained, depending on how the sort region resolves from the rest of the population. The enriched fraction can either be rerun through the purity sort on the same day or put back in culture for expansion. An alternative is to use the mixed-mode sorting to obtain both purity and maximum recovery at the same time. Although purified events are being collected from the left deflection, the aborted but desired events are deflected to the right and are collected rather than going to waste. Only the latter fraction requires resorting, thus saving time.

For selection based on phenotypes exhibiting down-modulation or loss of function, iterative purity sorting using less than 1% low expression as a cutoff may be performed, allowing growth and expansion between sorts. It is, however, always important to limit the growth of cells between selections. Often the desired phenotype has a growth disadvantage relative to wild-type cells. Hence, enrichment can be compromised and clones lost if wild-type background cells are allowed to overgrow the culture.

The sort strategy should include quality control measures to assess the performance of the enrichment process and verify that recovery of a rare target cell is achievable. Ideally, a known positive control is monitored in parallel with the screen. When possible, a positive control can be spiked into the library cell population and coenriched. Various GFP variants are convenient markers for following cell enrichment through the functional screen process (figure 11.9).

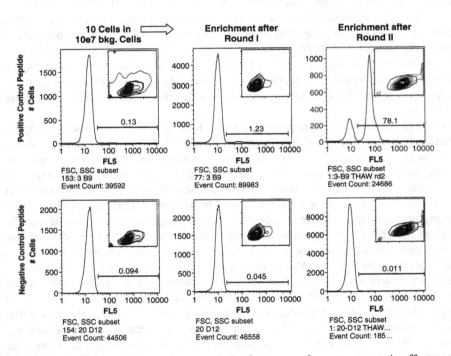

Figure 11.9. Spiking experiment. In general, the frequency of penetrant genetic effectors in any given library is very low, often on the order of 1 in 10^6. Hence, the efficiency of a given screening approach is important to evaluate with proper controls before embarking. One effective strategy to ascertain the overall screening efficiency of a given genetic screen is to demonstrate that small numbers of spiked control cells in a population can be recovered. An example is shown here, where 10 cells expressing a blue fluorescent protein–scaffolded peptide that blocks the epsilon-promoter response to interleukin 4 are doped into 10^7 wild-type cells. The enrichment of the BFP-peptide is measured by FACS (in FL5) following successive enrichment steps. As shown above, the positive control peptide is enriched approximately 10^4-fold (to ~1% of the population) in the first selection procedure. A second selection step results in a 1:4 ratio of background-to-control cells; notably this background comprises heritable mutants. A negative control peptide shows no enrichment. Using this scheme, Kinsella et al. (17) isolated peptide effectors of interleukin 4 signal transduction.

Verifying Isolated Clones

Once a cell with the desired phenotype has been isolated, the effect must be demonstrated to depend on the encoded genetic effector. In cases in which the library member is conditionally expressed, inheritable background may be distinguished from a library effector-dependent effect by repressing viral expression: A response dependent only on expression of the genetic effector will demonstrate a regulatable phenotype; mutant background cell phenotypes are not regulatable. This principle is readily applied both as a selection parameter and to validate a sorted clone (e.g., figure 11.5). Single cells are sorted from populations into microtiter plates and expanded for analysis. Depending on the type and complexity of the library, we have sorted up to 40,000 clones for this validation step. Each clone is analyzed by FACS using the same functional readout employed for the enrichment step.

This screening process is facilitated by available sample handling devices. Applying established hit criteria for the particular screen, the target clones or "hits" are identified. The overall frequency of identifiable genetic effectors is generally between one in from 10^5 to 10^7. Finally, the encoded sequences are isolated from cell clones by polymerase chain reaction. Alternatively, strategies that take advantage of the transferable nature of retroviruses may be used (11, 27).

Future Directions

Continuing advances in the utility of retroviral vector technology together with the rapid development of flow instrumentation herald great opportunities for function-based gene discovery. New genetic library approaches such as intracellular cyclic peptide production and interference RNA expression are expanding the repertoire of available genetic effectors. Implementation of more exotic flow technologies, such as "straight ahead" sorting (20), or high-speed photodamage cell sorting (15) could enhance both rare target cell purity and recovery in a functional screen. Plug flow coupling–based valve technology and continuous rapid multiwell sampling (6) for endpoint readouts will serve to increase the throughput of single-cell clone analysis, facilitating larger screens (7). Novel flow-based receptor-ligand (30) and protein–protein interaction screening applications can be used to develop functional genetic screening assays to identify specific modulators of cell-signaling complexes. Bead-based procedures that quantify soluble analyte levels would facilitate a screen for effectors that engender a distinct secretion molecule profile. An integration of flow cytometry and high-content cell imaging can be used to sharpen the selection criteria and incorporate novel cell morphological changes. Recent advances in microfluidics and microchip technology should facilitate a more integrated screen process at the single-cell level.

References

1. Beger, C., Pierce, L.N., Kruger, M., Marcusson, E.G., Robbins, J.M., Welch, P., Welch, P.J.,Welte, K. King, M.C., Barber, J.R., Wong-Staal, F. 2001. Identification of Id4 as a regulator of BRCA1 expression by using a ribozyme-library-based inverse genomics approach. *Proc Natl Acad Sci USA.* (98):130–135.

2. Chinnasamy, D., Chinnasamy, N., Enriquez, M.J., Otsu, M., Morgan, R.A., Candotti, F. 2000. Lentiviral-mediated gene transfer into human lymphocytes: role of HIV-1 accessory proteins. *Blood.* 96(4):1309–1316.
3. Chu, P., Pardo, J., Zhao, H., Li, C., Pali, E., Shen, M., Qu, K., Yu, S., Huang, B., Yu, P., Masuda, E., Molineaux, S., Kolbinger, F., Aversa, G., de Vries, J., Payan, D., Liao, X.C. 2003. Systemic identification of regulatory proteins critical for T-cell activation. *J. Biol.* 2(3): 21.
4. Demo, S.D., Masuda, E., Rossi, A.B., Throndset, B.T., Gerard, A.L., Chan, E.H., Armstrong, R.J., Fox, M.P., Lorens J.B., Payan, D.G., et al. 1999. Quantitative measurement of mast cell degranulation using a novel flow cytometric annexin-V binding assay. *Cytometry* 36:340–348.
5. Edelman, G.M., Meech, R., Owens, G.C., Jones, F.S. 2000. Synthetic promoter elements obtained by nucleotide sequence variation and selection for activity. *Proc Natl Acad Sci USA* 97:3038–3043.
6. Edwards, B.S., Kuckuck, F., Sklar, L.A. 1999. Plug flow cytometry: An automated coupling device for rapid sequential flow cytometric sample analysis. *Cytometry* 37:156–159.
7. Edwards, B.S., Kuckuck, F., Prossnitz, E., Ransom, J., Sklar, L. 2001. HTPS Flow cytometry: A novel platform for automated high throughput drug discovery and characterization. *J Biomol Screening* 6:83–90.
8. Erlenhoefer, C., Wurzer, W.J., Loffler, S., Schneider-Schaulies, S., ter Meulen, V., Schenider-Shcaulies, J. 2001. CD150(SLAM) is a receptor for measles virus but is not involved in viral contact-mediated proliferation inhibition. *J Virol.* 75:4499–4505.
9. Gu, X., Lawrence, R., Kroeger, M. 2000. Dissociation of the high density lipoprotein and low density lipoprotein binding activities of murine scavenger receptor class B type I (msR-B1) using retrovirus library-based activity dissection. *J Biol Chem* 275:9120–9130.
10. Gudkov, A.V., Roninson, I.B. 1999 Functional approaches to gene isolated in mammalian cells. *Science* 285:299a.
11. Hannon, G.J., Sun, P., Carnero, A., Xie, L.Y., Maestro, R., Conklin, D.S., Beach, D. 1999. MaRX: An approach to genetics in mammalian cells. *Science* 283:1129–1130.
12. Hitoshi, Y., Lorens, J., Kitada, S.I., Fisher, J., LaBarge, M., Ring, H.Z., Francke, U., Reed, J.C., Kinoshita, S., Nolan, G.P. 1998. Toso, a cell surface, specific regulator of Fas-induced apoptosis in T cells. *Immunity* 4:461–471.
13. Hitoshi, Y., Gururaja, T., Pearsall, D., Lang, W., Sharma, P., Huang, B., Catalano, S., McLaughlin, J., Pali, E., Vialard, J., Janicot, M., Wouters, W., Luyten, W., Bennet, M.K., Anderson, D.C., Lorens, J.B., Bogenberger, J., Demo, S. 2003. Cellular localization and antiproliferative effect of peptides discovered from a fuctional screen of a retrovirally derived random peptide library. *Chem. Biol.* 10:975–987.
14. Holland, S.J., Liao, X.C., Mendenhall, M.K., Zhou, X., Pardo, J., Chu, P., Spencer, C., Fu, A., Sheng, N., Yu, P., et al. 2001. Functional cloning of Src-like adapter protein-2 (SLAP-2), a novel inhibitor of antigen receptor signaling. *J Exp Med.* 194(9):1263–1276.
15. Keij, J.F., Groenewegen, A.C., Visser, J. 1994. High-speed photodamage cell sorting: An evaluation of the ZAPPER prototype. *Methods Cell Biol.* 42:371–386.
16. Kessels, H.W., van den Boom, M.D., Spits, H., Hooijberg, E., Schumacher, T.N. 2000. Changing T cell specificity by retroviral T cell receptor display. *Proc Natl Acad Sci USA* 97:14578–14583.
17. Kinsella, T.M., Ohashi, C.T, Harder, A.G., Yam, G.C., Li, W., Peelle, B., Pali, E.S., Bennett, M.K., Molineaux, S.M., Anderson, D.A., et al. 2002. Retrovirally delivered random cyclic peptide libraries yield inhibitors of IL-4 signaling in human B cells. *J Biol Chem.* 277:37512–8.
18. Kitamura, T. 1998. New experimental approaches in retrovirus-mediated expression screening. *Int J Hematol.* 4:351–359.
19. Kruger, M., Beger, C., Li, Q.X., Welch, P.J., Tritz, R., Leavitt, M., Barber, J.R., Wong-Staal, F. 2000. Identification of Id4 s a regulator of BRCA1 expression by using a ribozyme-library-based inverse genomics approach. *Proc Natl Acad Sci USA.* 97:8566–8571.
20. Leary, J. 1994. Strategies for rare cell detection and isolation. *Methods Cell Biol.* 42:331–358.
21. Lorens, J.B., Bennett, M.K., Pearsall, D.M., Throndset, W.R., Rossi, A.B., Armstrong, R.J.,

Fox, M.B., Chan, E.H., Luo, Y., Masuda, E., et al. 2000. Retroviral delivery of peptide modulators of cellular functions. *Mol Ther.* 1(5):438–437.

22. Lorens, J.B., Sousa, C., Bennett, M.K., Molineaux, S., Payan, D.G. 2001. The use of retroviruses as pharmaceutical tools for target discovery and validation in the field of functional genomics. *Curr Opin Biotechnol.* 12:613–621.

23. Misawa, K., Nosaka, T., Morita, S., Kaneko, A., Nakahata, T., Asano, S., Kitamoura, T. 2000. A method to identify cDNAs based on localization of green fluorescent protein fusion products. *Proc Natl Acad Sci USA* 97:3062–3066.

24. Novoa, I., Zeng, H., Harding, H.P., Ron, D. 2001. Feedback inhibition of the unfolded protein response by GADD34-mediated dephosphorylation of elF2a. *J Cell Biol.* 153:1011–1022.

25. Owens, G.C., Chappell, S.A., Mauro, V. P., Edelman, G.M. 2001. Identification of two short internal ribosome entry sites selected from libraries of random oligonucleotides. *Proc Natl Acad Sci USA* 98:1471–1476.

26. Peelle, B., Lorens, J., Li, W., Bogenberger, J., Payan, D.G., Anderson, D.C. 2001. Intracellular protein scaffold-mediated display of random peptide libraries for phenotypic screens in mammalian cells. *Chem Biol.* 5:521–534.

27. Perez, O.D., Kinoshita, S., Hitoshi, Y., Payan, D.G., Kitamura, T., Nolan, G.P., Lorens, J.B. 2002. Activation of the PKB/AKT pathway by ICAM-2. *Immunity* 16(1):51–65.

28. Peters, D., Branscomb, E., Dean, P., Pinkel, D., van dilla M., Gray, J.W. 1985. The LLNL high-speed sorter: design features, operational characteristics, and biologic utility. *Cytometry* 6:290–301.

29. Powell, K.T., Weaver, J.C. 1990. Gel microdroplets and flow cytometry: rapid determination of antibody secretion by individual cells within a cell population. *Biol Technol.* 8:333–337.

30. Sklar, L.A., Edwards, B.S., Graves, S.W., Nolan, J.P., Prossnitz, E.R. 2002. Flow cytometric analysis of ligand-receptor interactions and molecular assemblies. *Ann Rev Biophys Biomol Struct.* 31:97–119.

31. Stemmer, W.P. 1994. iDNA shuffling by random fragmentation and reassembly: in vitro recombination for molecular evolution. *Proc Natl Acad Sci USA* 91:10747–10751.

32. Swift, S., Lorens, J., Achacoso, P., Nolan, G.P. 1999. Rapid production of retroviruses for efficient gene delivery to mammalian cells using 293T cell based systems. *Curr Protocols Immunol.* 10:14–29.

33. Titus, S.A., Moran, R.G., 2000. Retrovirally mediated complementation of the glyB phenotype. Cloning of a human gene encoding the carrier for entry of folates into mitochondria. *J Biol Chem.* 275:36811–36817.

34. Traenckner, E.B., Pahl, H.L., Henkel, T., Schmidt, K.N., Wilk, S., Baeuerle, P.A. 1995. Phosphorylation of human I kappa B-alpha on serines 32 and 36 controls I kappa B-alpha proteolysis and NF-kappa B activation in response to diverse stimuli. *EMBO J.* 14:2876–2883.

35. Xu, X., Leo C., Jang, Y., Chan, E., Padilla, D., Huang, B.C., Lin, T., Gururaja, T., Hitoshi, Y., Lorens, J.B., et al. 2001. Dominant effector genetics in mammalian cells. *Nat Genet.* 27:23–29.

12

Applications of Flow Cytometry in Protein Engineering

GEORGE GEORGIOU, BARRETT R. HARVEY,
KARL E. GRISWOLD, AND BRENT L. IVERSON

Introduction

In recent years, the application of evolutionary methods for protein engineering has created tremendous optimism regarding our ability to generate proteins with tailored functional properties such as ligand binding, improved stability, allostery, and catalytic activity (1, 28, 39, 48, 50). The power of directed protein evolution lies in its simplicity: First, a gene encoding a polypeptide is subjected to mutagenesis, and the resulting ensemble of mutated genes is expressed in a suitable cellular host. Second, the population of expressed proteins is subjected to a screening process. Often, multiple rounds of screening are required to isolate the rare clones within the population that can satisfy the functional screen. Third, DNA is isolated from the enriched clones and subjected to additional rounds of mutagenesis and screening under increasingly stringent conditions. This iterative process is repeated several times until either little functional improvement is observed between sequential rounds or proteins that satisfy the chosen criteria have been generated.

There is a plethora of methods for generating an ensemble of mutated genes. Specifically, sequence diversity can be created by random mutagenesis, typically accomplished using error-prone polymerase chain reaction techniques (5); by homologous in vitro recombination (10, 43); or by nonhomologous recombination. The latter involves two families of methods collectively known as incremental truncation for the creation of hybrid enzymes (25, 35) and sequence-homology independent protein recombination (41). Regardless of the means for generating sequence diversity, the next and by far the more technically challenging step in directed evolution is the screening of the resulting library of protein-expressing cells to isolate those that are expressing a protein variant that exhibits the desired function.

It is fair to say that evolutionary protein design has been hampered by limitations in screening technologies. The quantitative determination of protein function for each and every clone in a library in a high-throughput fashion is a difficult and technically demanding task. In broad terms, there are four general strategies suitable for the screening of combinatorial protein libraries: phage display; biological assays that include selections (19) and assays that use reporter enzymes [e.g., two-hybrid-like techniques for detecting interacting proteins (36)]; single-well assays using high-density microtiter well plates; and flow cytometry (FC) methods. Each of these methods has a different set of advantages and shortcomings. Phage display is a relatively straightforward and powerful technique for the isolation of ligand-binding proteins and has been used with considerable success in applications such as the in vitro isolation of antibodies to desired antigens and the isolation of bioactive peptides (28, 40)

The two kinds of biological assays useful for protein library screening purposes, selection and two-hybrid system methods, have been used successfully for the isolation of proteins with novel properties (1, 16, 34) and for elucidating the protein–protein interaction network in the cell, respectively (52). Although phage display and biological assays satisfy the requirement for high throughput, they neither are quantitative nor provide information about the function of all the clones in a library. Selections in which cell viability is tied to some enzymatic activity are complicated by false positives that arise when biological mechanisms override the function provided by a cloned protein (53). Finally, microtiter well plate assays of clonal populations are limited by low throughput, and therefore, only small libraries of low complexity (e.g., encoding only one to two amino acid replacements per gene) can be realistically screened in this fashion.

By comparison, FC offers numerous advantages for the screening of combinatorial protein libraries:

1. It is a truly high-throughput screening technique. As many as 1×10^9 cells/hr can be processed with state-of-the-art research instrumentation, and somewhat lower but comparable rates are attainable with top-of-the-line commercial instruments. The isolation of rare clones represented within a heterogeneous population at frequencies as low as from $1:10^6$ to $1:10^7$ has been demonstrated (11, 24).

2. Quantitative assays can be performed on large populations, with single-cell resolution. In other words, FC analysis provides the opportunity to examine the distribution of protein functions within a library and to determine the fraction of clones having an activity of interest.

3. FC is well suited to the isolation of polypeptides that bind to target molecules. Ligand-binding equilibria and dissociation kinetics can be readily determined by FC of whole cells (11).

4. FC can be readily used to select clones on the basis of catalytic activity (33). An increasing number of fluorescent single-cell assays suitable for library screening purposes are being developed.

5. Multiple quantitative parameters for each cell can be analyzed simultaneously, including fluorescence signals of various wavelengths as well as forward and 90° light scattering. Recently, Perez and Nolan demonstrated the simultaneous flow cytometric detection of 13 different cellular parameters (37). The ability to carry out multiple assays on each cell is important for minimizing false positives and for the facile selection of variants having a desired new specificity or other property of interest. For example, mutations that yield a high signal in a functional as-

say as a result of an increase in expression level rather than a bona fide change in specific activity can be easily detected.

Of course, FC has its own shortcomings. First of all, it is applicable only when a fluorescent assay suitable for single-cell analysis is available. Although there is an ever-increasing gamut of fluorescent probes, there are many protein functions, particularly in enzymatic catalysis, that are not yet amenable to fluorescent assays. Second, there is the erroneous perception that FC requires considerable technical sophistication. Although this is still true for high-end instruments such as the Cytomation MoFlo, most current bench-top sorters are, in fact, quite easy to use. The number of protein engineering labs that use FC for library screening purposes has been increasing rapidly in recent years, and screening has attracted considerable commercial interest. There is no doubt that FC will play a key role in the development of the next generation of binding proteins and engineered enzymes.

Flow Cytometric Isolation of Binding Proteins

The most straightforward application of FC in combinatorial library screening is the isolation of cells expressing proteins that bind to a fluorescent ligand. In this case, the experimental design is simple. A pool of genes encoding a diverse set of polypeptides is created by one of the methods outlined in the introduction. The genes are transformed and expressed in a host, followed by incubation with a fluorescent ligand. Cells expressing binding proteins become fluorescent in proportion to their ligand affinity/expression level, and those clones that exhibit a desired fluorescence profile are isolated by sorting (figure 12.1).

The size of the protein library from which ligand binders are to be isolated can vary considerably, depending on the nature of the experiment. As a rough guide, the number of independent sequences in protein libraries cloned into *Escherichia coli* is usually somewhere between 10^5 and 10^8. Following transformation, the pool of transformed cells is grown for a few generations to amplify the number of clones that express the same sequence. Thus, typically, each clone in the library is represented at least 3–10 times to minimize the probability that a desirable, functional clone will be lost from the library during sorting.

Sorting by FC of cells expressing ligand-binding proteins is challenging for a number of reasons. First, the hosts used for the vast majority of combinatorial library screening experiments are bacteria or yeast, which, because of their smaller size relative to mammalian cells, are more difficult to detect and sort (see following). Second, the sorted populations are usually very large, whereas the binding clones are rare (11). Third, the fluorescence intensity of the labeled cells is often not very high, resulting in a low signal-to-noise ratio. For example, it is not uncommon for the positive control (i.e., clones having high ligand affinity) to exhibit only 5-fold higher fluorescence than the background of unlabeled cells. Needless to say, the low signal-to-noise ratio complicates the isolation of rare clones. Finally, not all cells in a population express a recombinant protein at the same level (42). As a result, clones expressing the same polypeptide exhibit a distribution of fluorescent labeling intensities as a result of variations in protein levels, cell sizes, and effects related to the mode of protein display, such as the details of expression induction (13). An example of this is the fluorescence

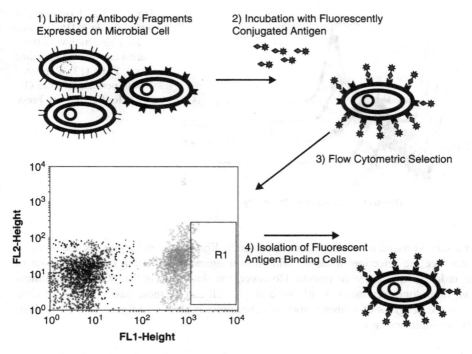

1) Library of Antibody Fragments Expressed on Microbial Cell

2) Incubation with Fluorescently Conjugated Antigen

3) Flow Cytometric Selection

4) Isolation of Fluorescent Antigen Binding Cells

Figure 12.1. Schematic of protein engineering selection strategy.

distribution of *E. coli* expressing the 26–10 antidigoxin scFv induced for various time intervals (figure 12.2).

The heterogeneity of protein expression (which is typically quantified on the basis of the coefficients of variation [CVs] of the distribution), low signal-to-noise ratio, and the often small difference in ligand affinity between the desired clones and the background population set a limit on the degree to which positive cells can be enriched. In most cases, the enrichment ratio (number of positive events in the postsort/number of positive events in the presorts) is not sufficiently high to allow the isolation of desired ligand-binding cells in a single step. For this reason, multiple rounds of sorting are required to isolate high-affinity clones, particularly when such clones are very rare.

With mechanical sorters, such as the Becton-Dickinson FACSCalibur, the sort solution is very dilute and cannot be directly applied back onto the instrument for resorting. Therefore, the sorted cells must be either concentrated by filtration/centrifugation or mixed with growth media and allowed to grow for several generations, as needed, to amplify the population and achieve the desired cell density. Cell amplification is usually preferred because the representation of the positive clones in the sorted population is increased, in turn, allowing for better enrichment during the next sorting round. Usually three to four rounds of sorting and regrowth yield a cell population dominated by the desired clones.

In droplet deflection sorters, the cells are collected in a small volume of fluid, and therefore the sort solution can be resorted immediately without the need for regrowth. Sorts can be made into a sip tube and the collected cells put directly back on the FC

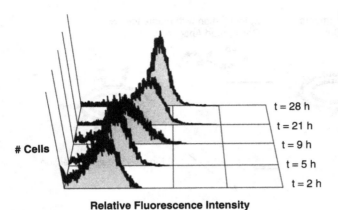

Figure 12.2. Expression of scFv(dig) in JM109/pB30D after induction with 0.6% arabinose for the indicated time. From (13). Reprinted with permission from Oxford University Press.

for a second round of sorting. This expedites the library screening process and can also encourage the selection of clones with slower ligand dissociation rates as a result of the reduced time between resorts. However, the downside is that without amplification, rare clones represented only by a single cell can be more easily lost. An example of the progress of a library sorting experiment is shown in figure 12.3 (B.R. Harvey, unpublished results).

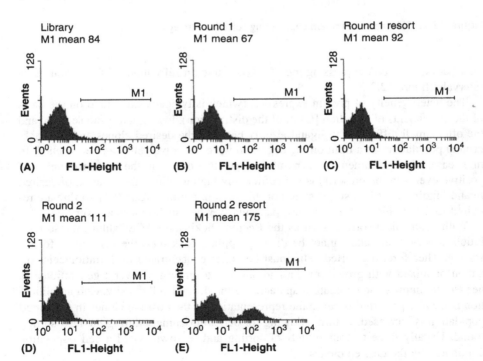

Figure 12.3. Selection of high-affinity scFvs to Methamphetamine-fluorescein conjugate via two rounds of sorting/re-sorting on a droplet deflection flow cytometer. (A) Naive library gate M1 mean 84.03. (B) Round 1 sort, gate M1 mean 66.86. (C) Round 1 re-sort, gate M1 mean 91.59. (D) Round 2 sort, gate M1 mean 111.18. (E) Round 2 resort, gate M1 mean 175.48.

The high level expression of heterologous proteins is taxing for the cell and generally results in slower growth rate or even cell death. If the synthesis of the expressed protein is not repressed during the amplification step, then the desired clones can be at a substantial growth disadvantage compared to cells that have lost the ability to produce the protein. This can result in deenrichment of the positive clones during regrowth following a successful sort. Fine-tuning the expression system and the induction parameters, namely, the concentration of inducer and time before the cells are harvested, can sometimes circumvent these problems (13). Other methods include using a repressor to inhibit protein expression or growing sorted cells on plates rather than in liquid media to spatially reduce competition between cells.

Perhaps the greatest advantage of sorting as a screening technology is that it is highly quantitative. It is straightforward to adjust the labeling conditions to achieve the isolation of clones expressing proteins with specified ligand affinity. Selections for ligand affinity can be performed either by equilibrating the library with a concentration of the probe equal to the desired K_D (i.e., under thermodynamic control) or by labeling with a concentration of ligand above the K_D, followed by addition of competitor for a specified period of time (i.e., under kinetic control). The latter strategy effectively selects for protein variants that exhibit a desired range of rate constants for ligand dissociation. For most protein–ligand interactions, the rate of association does not change significantly from clone to clone, so selecting for a specified dissociation rate is effectively a selection for the corresponding equilibrium constant. Another useful feature of FC is the ability to determine directly the protein:ligand dissociation rate constant, using whole cells and without the need for purifying the protein. In fact, the kinetics of ligand dissociation as determined by FC are in good agreement with measurements using in vitro assays such as Surface Plasmon Resonance (3, 11).

Quantitative analysis by FC is also useful in library experiments because the relative activity of each and every library member can be determined under a given set of conditions. Quantification greatly facilitates optimization of library and screening parameters. This is an important issue, as almost by definition, one never knows what to expect from a given library a priori. All too often, directed evolution/library experiments are run simply as searches for "needles in haystacks," with no systematic way to optimize preparation of the "haystack" or search conditions. The quantitative features of FC screening allow optimization of both. A recent example that underscores the value of quantitative FC library screening used a set of surface-expressed scFv libraries produced with increasing mutational error rates (12). A logarithmic decrease in the number of active clones was observed by FC with increasing average library mutation frequency, except at very high mutagenesis rates (>2%–3% at the nucleotide level). In these highly mutated libraries, approximately two orders of magnitude more active clones were present compared to the number expected based on extrapolation of the lower–mutagenesis rate libraries. In addition, the most active clones in these highly mutated libraries had greater levels of gain-of-function than the most active clones in libraries with lower overall mutagenesis rates. This experiment has now been repeated independently with similar results (A. Taylor, R. Loo, G. Georgiou, B. Iverson, unpublished results). This insight, namely, that relatively high error rates can produce surprisingly fertile libraries, would have been difficult to identify without the ability to quantitatively screen entire libraries.

Because most fluorescent ligands cannot permeate into the cytoplasm, library screening applications require the use of an expression system that renders the protein of in-

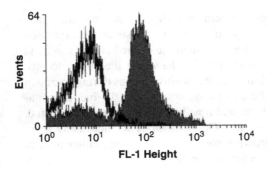

Figure 12.4. Anti-digoxin scFv vs. anti-atrizine scFv surfaced displayed using Lpp-pOmpA and labeled with a 200nM concentration of bodipy-digoxin. Surface displayed anti-digoxin 26-10scFv (solid) mean FL 111.28. Surface displayed anti-atrizine mean FL 11.33 (open).

terest accessible to the extracellular fluid. The expression of protein libraries in a ligand-accessible format is generally referred to as protein display, even though in some instances the protein is not technically anchored on the cell surface (see PECS). Systems for protein display in microorganisms, namely *E. coli* and yeast, have been reviewed recently (6, 16, 50) and will only be briefly outlined below.

Protein display hinges on the fusion of the polypeptide of interest to a surface targeting/anchoring sequence derived from an authentic protein of the host cell. The first example of a microbial protein display system suitable for library screening applications was developed by Georgiou and coworkers in the early 1990s (15). Proteins were displayed on *E. coli* by using the first nine amino acids of Braun's lipoprotein Lpp fused to a transmembrane domain of the outer membrane protein OmpA (Lpp-OmpA). The resulting Lpp-OmpA chimera contained a lipophilic anchor, derived from Lpp, that attached to the interior of the *E. coli* outer membrane, with the membrane spanning regions of OmpA allowing for display of a C-terminal fusion to the outside of the *E. coli* cell (14). This system has been used for isolating high-affinity antibodies to various antigens (figure 12.4) and for the screening of trypsin inhibitor II from *Ecballium elaterium* libraries (8). A more recent display system makes use of the N-terminal region of Intimin EaeA, a member of the bacterial adhesion family from the pathogenic *E. coli* 0157:H7. Intimin EaeA has been used successfully for the display of several proteins including trypsin inhibitor II (35aa), Interleukin 4 (128aa), the Bence-Jones protein REIv(108aa; 49), and for the identification of linear peptide Mab epitopes from peptide gene fragments screened by FC (9). Two other systems have been used for the display of folded proteins in Gram-negative bacteria (as opposed to short peptides, which can be readily accommodated within surface-exposed loops of outer membrane proteins): The Ice nucleation proteins from various *Pseudomonas* sp. have been used successfully to display heterologous proteins in *E. coli* (6, 20, 21). Another system, termed autodisplay, relies on the special properties of outer membrane proteins (autotransporters) of Gram-negative bacteria in which the C-terminal domain forms a pore that serves as a conduit for the export of the N-terminal part onto the cell surface. Autotransporter proteins include the *E. coli* adhesin, (AIDA-I; 22) and the IgA1 protease from *Neisseria gonorrhoeae* (23). For protein engineering applications, the native N-terminal domain is replaced with a heterologous protein (27). Thus, the protein is fused at its C terminus rather than the N terminus, as is the case with all other display techniques. However, so far, the utility of autodisplay for combinatorial protein library screening applications has not been demonstrated.

Saccharomyces cerevisiae has also been used extensively for the isolation of ligand-binding polypeptides by FC. Proteins are displayed on the surface of yeast as C-terminal fusions to the Aga2 mating adhesion receptor (50). Yeast offers a number of advantages as a host for protein engineering purposes. Being a eukaryotic organism, it has a more complex protein biosynthetic machinery that performs certain posttranslational modifications not carried out by bacteria. In yeast, most surface-displayed polypeptides are correctly folded, as unfolded molecules are largely eliminated by the quality-control mechanisms in the endoplasmic reticulum. Preventing unfolded proteins from being displayed on the cell surface is desirable in protein screening applications because exposed hydrophobic regions in unfolded molecules can bind fluorescent probes nonspecifically, giving rise to false-positive FC signals. Finally, the larger size of yeast as compared to bacteria results in greater side and forward scatter and makes it considerably easier to gate on the cells and distinguish them from debris. So far, yeast display has been used primarily for engineering immunological proteins, including single-chain antibodies and T cell receptors. In one, rather amazing, study three rounds of mutagenesis and flow cytometric screening under increasingly stringent conditions were used to improve the binding affinity of the antifluorescein 4-4-20 antibody by over four orders of magnitude (3). The main limitation of yeast display is that the efficiency of DNA transformation is considerably lower than that of *E. coli*, and therefore, it is much more difficult to construct very large libraries in that organism. Large libraries are desirable because they provide the means for exploring a larger portion of protein sequence space, which in turn increases the probability of isolating polypeptides with unique properties.

Even without display on the cell surface, chemical treatments may be employed to render proteins localized in secretory compartments accessible to fluorescent ligands. This is the basis for a technique called PECS, periplasmic expression with cytometric screening. PECS is a "display-less" method whereby the protein of interest is secreted into the periplasm of the host *E. coli* cell while the outer membrane is made more permeable to allow access of fluorescent ligands. Chen et al. (7) identified a combination of strain and incubation conditions that allowed fluorescent molecules up to about 10 kDa to equilibrate into the periplasm without significant loss of cell viability or leakage of proteins from the periplasmic space. An important benefit of this system is that there is no fusion partner to complicate the presentation of the protein. In addition, selected proteins are expressed directly in soluble form as needed for subsequent characterization and production purposes. Finally, the periplasmic expression format is directly compatible with the use of many preexisting libraries constructed for phage display. Effectively, any library that has been made for display in filamentous bacteriophage can also be screened by PECS (7).

Phage display itself can clearly benefit from the real-time, quantitative, and multiparameter features of flow cytometric screening. However, with dimensions of 1000 by 10–15 nm, filamentous bacteriophage (the most common phage used for display) are extremely small, making detection by scattered light difficult. It has been shown that detection of individual virus particles, although none yet as small as filamentous phage, can be identified by light scattering on specially designed flow cytometers (18). An alternative for detection may be to fluorescently label the phage, using fluorescence as the event trigger. For example, the T4 head region can be labeled with GFP (30), and viruses with genomes as small as 7.4 kb have been labeled with nucleic dyes for detection with flow cytometers (4). Two-color sorting could then be used to isolate clones that bind to a ligand labeled with a different dye. Advances in cytometer design could easily make the use of filamentous phage/phagemid as FC library display platforms much more common in the future.

Flow Cytometric Screening of Enzyme Libraries

Unlike the isolation of ligand binding polypeptides by FC, the screening of libraries for enzymatic activities is still in its infancy. The limited application of FC in enzyme engineering derives from several factors, but perhaps the most significant barriers are the synthetic effort required in producing enzyme probes and the fact that unique probes are generally necessary for each enzymatic activity to be engineered. Probes useful for FC enzyme assays must exhibit cytometer-compatible spectral characteristics, adequate solubility, and good cell accessibility and retention characteristics, and they must, of course, exhibit a change in fluorescence in response to the enzymatic reaction (either because they serve as substrates in the reaction or because they are sensitive to a reaction product; e.g., the generation of hydronium ions by carboxyl hydrolases).

First and foremost, a probe should possess spectral properties that are amenable to FC analysis. It must have an excitation wavelength that is compatible with the light sources of the available cell sorters; it should preferably possess a large extinction coefficient, quantum yield, and stokes shift; it should exhibit minimal photobleaching/self quenching; and it should be only modestly influenced by environmental factors such as ionic strength or pH.

Because FC assays are most useful when performed on viable cells, probes should be soluble in aqueous solutions with minimal doping of organic solvents. Many enzyme probes are only soluble in high–organic content solutions, which are incompatible with maintaining cells in a viable state. One class of probe molecules whose utility is limited by solubility is the carboxyl esters of Elf 97 alcohol, marketed by Molecular Probes. The alcohol itself offers near ideal spectral qualities and precipitates in aqueous solutions as a result of intramolecular hydrogen bonding. The formation of insoluble aggregates on hydrolysis allows for excellent product retention following enzymatic transformation. However, esterifying the phenolic oxygen results in a highly hydrophobic molecule, which tends to have negligible solubility in aqueous solutions. Esters as small as the acetate (table 12.1, entry 1) have been found to precipitate on addition to aqueous solutions, resulting in highly fluorescent aggregates before hydrolysis (K.E. Griswold, unpublished data). On occasion, surfactants may be used to overcome solubility limitations. However, the complex behavior of micelle solutions can also lead to unanticipated complications such as inaccessibility of the probe and triggering on the micelles themselves.

Probes must also be capable of reaching the cellular locale of the enzyme to be assayed. This requirement is trivial if the enzyme is displayed on the cell surface. However, in many instances surface display is not an option because either the enzyme is not amenable to export onto the cell's surface or the catalytic activity is dependent on cofactors that are present in the cytoplasm (e.g., NADH, ATP, GSH, etc.). Thus, for cytoplasmic enzymes, it is necessary to use probes that can permeate though the membrane (or membranes, in the case of Gram-negative bacteria that have both an inner and outer membrane).

Another issue that must be addressed in probe design is the retention of the fluorescent product by the cell. Diffusion of the fluorophore can lead to several problems including loss of fluorescent signal for active clones, increased background fluorescence, and nonspecific labeling of inactive clones. Therefore, it is important that the chemical transformation of the probe by the enzyme of interest results in formation of a molecule that is quantitatively associated with the cell.

Cell retention can be accomplished by a variety of strategies. The enzymatic reaction can result in

1. Formation of a poorly soluble fluorescent product that precipitates within the cell. As mentioned above, Elf 97–derived enzyme probes are designed to generate the precipitating alcohol on enzymatic hydrolysis of the weakly fluorescent substrate. Although production of the ester substrates has been discontinued, other soluble Elf 97 probes are available including galactosidase, glucuronidase, phosphatase, and chitinase/N-acetylglucosaminidase substrates (2, 27, 45)
2. A molecule with high affinity for an intracellular component. Examples of this class of probe include 5-dodecanoylaminofluorescein di-β-D-galactopyranoside (29) and 2-dodecylresorufin β-D-galactopyranoside, both of which posses lipophilic tails, allowing for probe retention via high affinity for cellular membranes.
3. Covalent attachment of the fluorescent product to a membrane impermeable cellular component. For example, monochlorobimane readily permeates into the cytoplasm of rodent cells where glutathione S-transferase conjugates it to glutathione, resulting in the formation of a membrane-impermeable molecule that accumulates intracellularly (table 12.1, entry 2; 17). Monoclorobimane and other related compounds can be employed for the FC screening of glutathione S-transferase libraries in *E. coli* (K.E. Griswold, B.L. Iverson, and G. Georgiou, unpublished data).
4. Chemical transformation that converts a membrane permeable probe into a molecule that exhibits impaired diffusion across membranes and is therefore retained in the cytoplasm. This approach differs from 3, above, in that the fluorescent product is not conjugated to a cytoplasmic molecule. Zlokarnik et al. have designed a fluorescence resonance energy transfer (FRET) probe suitable for labeling Jurkat cells expressing recombinant β-lactamase as a reporter (54). Ratiometric measurement of green versus blue fluorescence allowed for exceptionally sensitive measurement of β-lactamase activity (<100 enzyme molecules per cell) with little interference from factors such as variable cell size and probe concentration. Functional cell sorting was demonstrated by isolation of positive clones from a 10^6-fold excess of negative control cells. Other probes do not require FRET for fluorescence quenching. Hydrolysis of 6-carboxyfluorescein diacetate by cytoplasmic esterases results in the formation of charged fluorescein monomers that are retained within the cytoplasm, resulting in selective cell labeling (table 12.1, entry 3; 44). In a similar manner, cells have been selectively labeled on the basis of β-D-galactosidase activity with the probes fluorescein di-β-D-galactopyranoside (32) and resorufin β-D-galactopyranoside (table 12.1, entry 4; 51). However, fluorescein and other negatively charged soluble probes can slowly leak from the cell, and thus the fluorescence intensity is time dependent.
5. Retention of the product via electrostatic interactions. Unlike the other approaches listed above, electrostatic product retention allows the quantitative capture of re-

Table 12.1 Representative target enzymes with substrates, products, and means of product retention

Entry	Target Enzyme	Substrate Structure	Product Structure	Means of Product Retention	Reference
1	Carboxyl hydrolase			Precipitating fluorophore	K.E. Griswold (unpublished data)
2	Glutathione-S-transferase			Covalent attachment	17
3	Carboxyl hydrolase			Membrane barrier	44
4	β-D-galactosidase			Membrane barrier	51

5 Protease

Electrostatic capture

34

H$_2$N-R-G-R-G-K-G-R-CONH$_2$

6 Protease

Electrostatic capture

J. Gam (unpublished data)

H$_2$N-R-G-R-G-K-G-R-CONH$_2$

E-E-G-R-G-R-G-K-CONH$_2$

action products on the cell surface. Coupled with the display of proteins on the cell surface, this technique circumvents the need for probe penetration into the cell. Electrostatic product retention is based on the observation that, in low–ionic strength buffers, fluorescent probes with a net charge of +3 or greater become tightly adsorbed onto the surface of *E. coli*, which is highly negatively charged because of the presence of the anionic lipopolysaccharide layer. The original surface-based screening scheme, developed by Olsen et al. (34), used FRET substrates consisting of a fluorophore (Fl), a positively charged moiety, the scissile bond, and a quenching fluorophore that acts as an intramolecular FRET partner (Q; table 12.1, entry 5). Enzymatic cleavage of the scissile bond separates the Fl and Q moieties, disrupting intramolecular FRET quenching. The product containing the quenching fluorophore Q has no net charge and presumably diffuses away from the cell, whereas the product containing the fluorescent group Fl, in this case BODIPY, remains cell associated. As a result, the cells become fluorescently labeled with intensity proportional to substrate turnover.

Electrostatic retention of FRET substrates was the first methodology applied to the FC screening of directed evolution enzyme libraries. This strategy was used for isolating mutants of OmpT exhibiting altered substrate selectivity. OmpT is a trypsin-like bacterial surface endopeptidase that exhibits high activity and specificity for cleaving between pairs of basic residues. Substrate 5 has an overall +3 charge and contains the BODIPY (Fl) and tetramethylrhodamine (Q) fluorophores on either side of an Arg–Arg sequence optimal for OmpT cleavage. Hydrolysis of substrate 5 by OmpT at the Arg–Arg peptide bond gives rise to a *C*-terminal product containing the BODIPY (Fl) fluorophore on a short peptide with a +3 overall charge (two Arg guanidinium groups as well as the unmasked N-terminal amine). In low–ionic strength buffers containing sucrose to avoid plasmolysis, ompT$^-$ cells incubated with substrate 5 display no fluorescence. In contrast, cells expressing the wild-type enzyme exhibit intense green fluorescence. This system has been shown to provide very efficient enrichment of cells expressing catalysts from negative control cells (>5000-fold per round of screening), to afford the rapid screening (less than an hour) of libraries containing millions of clones, and to result in the isolation of clones exhibiting dramatically increased activity (up to a 60-fold increase in the k_{cat}/K_m was obtained via the screening of a library with a single round of sorting; 34).

Recently, a simplified but more sensitive FC screening assay employing a substrate with a single fluorescent dye molecule (no FRET) was devised (table 12.1, entry 6; 33a). Zwitterionic peptides containing the Arg–Arg cleavage site and a total of three positive and three negative charges each are synthesized. A single Fl is conjugated via the lone Lys residue. Enzymatic hydrolysis generates a product having a net +3 charge. This product, which also contains the fluorescent moiety, becomes associated with the cell surface via electrostatic interactions. One advantage of the zwiterionic probes is that they are more easily synthesized than the FRET substrates. In addition, because they contain only one fluorophore, multiple substrates labeled with different dyes can readily be generated and used in tandem to isolate enzymes with desired selectivities (i.e., that are capable of hydrolyzing one substrate but not another). The ability to select for catalytic specificity using multiple fluorescent substrates in a single assay is one of the most important advantages of FC-based enzyme screening technology. With this approach, we have succeeded in engineering remarkable and unprecedented changes in the substrate specificity of hydrolytic enzymes (Varadarajan, Gam, Olsen, Georgiou, and Iverson, unpublished data).

Significantly, other directed enzyme evolution researchers are beginning to exploit the power of FACS for the alteration and enhancement of enzyme function including modification of Cre recombinase specificity (38). Control of GFP expression was used as a fluorescent signal for FACS screening.

Technical Considerations

Cell Viability

Although it is possible to sort dead cells and then recover the DNA encoding the protein of interest (B.R. Harvey, unpublished data), library screening is simplified if the cells are maintained in a viable state. Dead cells within the population can be selectively labeled with stains such as propidium iodide or oxonol, which produce an enhanced FL signal when the inner membrane is compromised, allowing the dye to intercalate within DNA (figure 12.5). In library screening applications, viability stains may be used to maximize the viability of cells by fine-tuning important parameters such as strength of promoter used, growth phase of culture, and induction time/temperature (13). In addition to fluorescent probes, other parameters useful for qualitatively assessing the viability of cells expressing protein libraries are the FSC and SSC profiles. These properties are characteristic of particular organisms and expression systems. A healthy culture has relatively small CVs for these parameters, and as viability decreases, the scatter signal becomes more diffuse.

Finally, the postsort handling of cells can be a critical factor in the viability of recovered clones. On mechanical sorters such as the Becton Dickinson FACS Calibur,

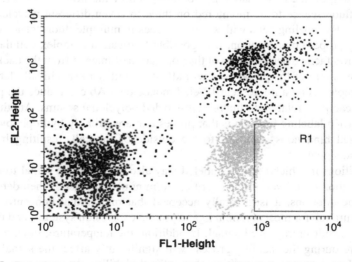

Figure 12.5. Dot plot representation of *Escherichia coli* labeled with ligand of interest in FL-1 and live/dead nucleic acid stain in FL-2. R1 region selects binders to ligand and excludes nonviable cellular events.

cells are sorted into 50-ml Falcon tubes of sheath fluid. In many cases, the viability of cells is adversely affected over time in common sheath fluids. This problem may be addressed by sorting into 25 ml 2× growth media, or immediately filtering the sorted sheath fluid through 0.22-μM filters and placing the filters on agar plates. Droplet deflection sorters such as the MoFlo by Cytomation have the advantage of small recovery volumes, making it possible to sort directly onto agar plates or 96-well plates with appropriate attachments (B.R. Harvey, K.E. Griswold, unpublished data).

Signal-to-Noise Ratio

Maximizing the signal-to-noise ratio is critical for sorting rare events. Typically, microbial populations are analyzed by triggering on forward- or side-scatter signals and carefully adjusting the threshold setting. Proper laser selection is important when triggering on small cells such as bacteria. With side scatter as the trigger, we have found that the low-power 635-nm red diode lasers do not produce an adequate signal for *E. coli* cultures. However, side-scatter trigger from a 488- or 350–360-nm ultraviolet laser typically results in an acceptable event rate. We have also found that, as with many mammalian cell sorting applications, a 488-nm forward-scatter trigger using a photo diode detector is sufficient to discriminate bacteria from background particulates (B.R. Harvey, K.E. Griswold, unpublished data). For some applications, it is preferable to fluorescently label the cells and use the resulting fluorescence, rather than light scatter, for triggering (46). For example, microbial cells can be labeled with fluorescent antibody conjugates specific for cellular surface markers. The resulting signal may be detected in fluorescence channel 1, which is designated as the trigger. The probe molecule used to screen the library is then detected in a second fluorescence channel. In fact, multiple fluorescent probes may be used to improve gating of the cell population (31). Depending on the exact hardware configuration of the flow cytometer, it is not unusual for fluorescence to be monitored on three to seven different wavelengths simultaneously. By labeling cells with specific probes in multiple fluorescence channels and applying proper compensation, it is possible to create a complex cellular fingerprint that allows facile discrimination of the population of interest from the background. However, the nature of the probe used will affect its ability to specifically label a target. For example, using a fluorescent-labeled monoclonal Ab can reduce the probability of cross reactivity relative to fluorescent-labeled polyclonal serums. Another factor to consider when labeling bacteria is that probes that are strongly hydrophobic or positively charged bind nonspecifically to the outside of the cell, resulting in loss of target-specific signal.

The conditions in which cells are labeled may also influence the signal-to-noise ratio. Concentration of a given probe is critical to minimizing noise. When determining optimal probe dilutions, it is generally accepted that micromolar concentrations are above the saturation for most common ligand protein expression systems, and that more dilute probes result in improved signal. In addition, the temperature, time, and washing procedure during the labeling process will significantly affect the signal-to-noise levels. As when determining conditions for optimal viability, these parameters should be evaluated on a case-by-case basis.

Conclusion

The use of FC in directed evolution strategies for the screening of protein libraries for desirable function is emerging as a promising method for protein engineering. By maintaining, via cellular scaffold, the genetic information of the displayed protein and by linking functionality of that protein to an increase in fluorescent signal, individual members of a library can be selected on the basis of improved activity. FC technology creates a high-throughput, real-time, multiparameter selection strategy for proteins of tailored function, a combination of parameters not found in other directed evolution schemes. Significantly, FC is a quantitative technique allowing the analysis of each and every library member. Rapid, whole-library quantitation allows for library preparation and screening optimization—opportunities not afforded by other screening approaches. With increased interest in protein engineering both in academia and for industrial applications, FC will undoubtedly play an expanding role in the selection and laboratory evolution of ligand-binding proteins and enzymes for decades to come.

Acknowledgments This work was supported by a Multidisciplinary University Research Initiatives grant (DAAD 19-99-1-0207) from the Army Research Office.

References

1. Arnold, F.H. 2001. Combinatorial and computational challenges for biocatalyst design. *Nature* 409(6817):253–257.
2. Baty, A.M.I., Eastburn, C.C., Diwu, Z., Techkarnjanaruk, S., Goodman, A.E., and Geesey, G.G. 2000. differentiation of chitinase-active and non-chitinase-active subpopulations of a marine bacterium during chitin degradation. *Appl. Environ. Microbiol.* 66(8):3566–3573.
3. Boder, E.T., and Wittrup K.D. 2000. Yeast surface display for directed evolution of protein expression, affinity, and stability. *Methods Enzymol.* 328:430–44
4. Brussaard, C.P., Marie, D., and Bratbak, G. 2000. Flow cytometric detection of viruses. *J. Virol. Methods* 85(1–2):175–182
5. Cadwell, R.C., and Joyce, G.F. 1994. Mutagenic PCR. *PCR Methods Appl.* 3(6):S136–S140.
6. Chen, G and Georgiou, G. 2002. Cell-surface display of heterologous proteins: from high-throughput screening to environmental applications. *Biotechnol. Bioeng.* 79(5):496–503.
7. Chen, G., Hayhurst, A., Thomas, J.G., Harvey, B.R., Iverson, B.L., and Georgiou, G. 2001. Isolation of high-affinity ligand-binding proteins by periplasmic expression with cytometric screening (PECS). *Nat. Biotechnol.* 19(6):537–542.
8. Christmann, A., Walter, K., Wentzel, A., Kratzner, R., and Kolmar, H. 1999. The cystine knot of a squash-type protease inhibitor as a structural scaffold for *Escherichia coli* cell surface display of conformationally constrained peptides. *Protein. Eng.* 12(9):797–806.
9. Christmann, A., Wentzel, A., Meyer, C., Meyers, G., and Kolmar, H. 2001. Epitope mapping and affinity purification of monospecific antibodies by *Escherichia coli* cell surface display of gene-derived random peptide libraries. *J. Immunol. Methods* 257(1–2):163–173.
10. Crameri, A., Raillard, S.A., Bermudez, E., and Stemmer, W.P. 1998. DNA shuffling of a family of genes from diverse species accelerates directed evolution. *Nature* 391(6664): 288–291.
11. Daugherty, P.S., Chen, G., Olsen, M.J., Iverson, B.L., and Georgiou, G. 1998. Antibody affinity maturation using bacterial surface display. *Protein Eng.* 11(9):825–832.
12. Daugherty, P.S., Iverson, B.L., and Georgiou, G. 2000. Flow cytometric screening of cell-based libraries. *J. Immunol. Methods* 243(1–2):211–227.

13. Daugherty, P.S., Olsen, M.J., Iverson, B.L., and Georgiou, G. 1999. Development of an optimized expression system for the screening of antibody libraries displayed on the *Escherichia coli* surface. *Protein Eng.* 12(7):613–621.

14. Earhart, C.F. 2000. Use of an Lpp-OmpA fusion vehicle for bacterial surface display. *Methods Enzymol.* 326:506–516.

15. Francisco, J.A., Campbell, R., Iverson, B.L., and G. Georgiou 1993. Production and fluorescence-activated cell sorting of *Escherichia coli* expressing a functional antibody fragment on the external surface. *Proc. Natl. Acad. Sci. USA* 90(22):10444–10448.

16. Georgiou, G., Stathopoulos, C., Daugherty, P.S., Nayak, A.R., Iverson, B.L., and Curtiss, R. 3rd. 1997. Display of heterologous proteins on the surface of microorganisms: from the screening of combinatorial libraries to live recombinant vaccines. *Nat. Biotechnol.* 15(1):29–34.

17. Hedley, D.W., and Chow, S. 1994. Evaluation of methods for measuring cellular glutathione content using flow cytometry. *Cytometry* 15(4):349–358.

18. Hercher, M., Mueller, W., and Shapiro, H.M. 1979. Detection and discrimination of individual viruses by flow cytometry. *J. Histochem. Cytochem.* 27(1):350–352.

19. Hilvert, D. 2000. Genetic selection as a tool in mechanistic enzymology and protein design. *Ernst Schering Res. Found Workshop* 32:253–268.

20. Jeong, H., Yoo, S., and Kim, E. 2001. Cell surface display of salmobin, a thrombin-like enzyme from Agkistrodon halys venom on *Escherichia coli* using ice nucleation protein. *Enzyme Microb. Technol.* 28(2–3):155–160.

21. Kwak, Y.D., Yoo, S.K., and Kim, E. J 1999. Cell surface display of human immunodeficiency virus type 1 gp120 on *Escherichia coli* by using ice nucleation protein. *Clin. Diagn. Lab Immunol.* 6(4):499–503.

22. Lattemann, C.T., Maurer, J., Gerland, E., and. Meyer, T.F. 2000. Autodisplay: functional display of active beta-lactamase on the surface of *Escherichia coli* by the AIDA-I autotransporter. *J. Bacteriol.* 182(13):3726–3733

23. Klauser, T., Pohlner, J., and Meyer, T.F. 1992. Selective extracellular release of cholera toxin B subunit by *Escherichia coli*: dissection of *Neisseria* Igaβ-mediated outer membrane transport. *EMBO J.* 11:2327–2335.

24. Leary, J.F. 1994. Strategies for rare cell detection and isolation. *Methods Cell Biol.* 42(Pt B):331–358.

25. Lutz, S., Ostermeier, M., Moore, G.L., Maranas, C.D., and Benkovic, S.J. 2001. Creating multiple-crossover DNA libraries independent of sequence identity. *Proc. Natl. Acad. Sci. USA* 98(20):11248–11253.

26. Matsumura, I., Olsen, M.J., and Ellington, A.D. 2001. Optimization of heterologous gene expression for in vitro evolution. *Biotechniques* 30(3):474–476.

27. Maurer, J., Jose, J., and Meyer, T.F. 1997. Autodisplay: one-component system for efficient surface display and release of soluble recombinant proteins from *Escherichia coli*. *J. Bacteriol.* 179(3):794–804.

28. Maynard, J., and Georgiou, G. 2000. Antibody engineering. *Annu. Rev. Biomed. Eng.* 2:339–376.

29. Miao, F., Todd, P., and Kompala, D.S. 1993. A single-cell assay of β-galactosidase in recombinant *Escherichia coli* using flow cytometry. *Biotechnol. Bioengin.* 42:708–715.

30. Mullaney, J.M., and Black, L.W. 1998. Activity of foreign proteins targeted within the bacteriophage T4 head and prohead: implications for packaged DNA structure. *J. Mol. Biol.* 283(5):913–929.

31. Nebe-von-Caron, G., Stephens, P.J., Hewitt, C.J., Powell, J.R., and Badley, R.A. 2000. Analysis of bacterial function by multi-colour fluorescence flow cytometry and single cell sorting. *J. Microbiol. Methods* 42:97–114.

32. Nolan, G.P., Fiering, S., Nicolas, J.F., and Herzenberg, L.A. 1988. Fluorescence-activated cell analysis and sorting of viable mammalian cells based on beta-D-galactosidase activity after transduction of *Escherichia coli* lacZ. *Proc. Natl. Acad. Sci. USA* 85:2603–2607.

33. Olsen, M., Iverson, B., and Georgiou, G. 2000. High-throughput screening of enzyme libraries. *Curr. Opin. Biotechnol.* 11(4):331–337.

33a. Olsen, M.J., Gam, J., Georgiou, G., Iverson, B.L. 2003. High-throughput FACS method for directed evolution of substrate specificity. *Methods Mol. Biol.* 230:329–342.

34. Olsen, M.J., Stephens, D., Griffiths, D., Daugherty, P., Georgiou, G., and Iverson, B.L. 2000. Function-based isolation of novel enzymes from a large library. *Nat. Biotechnol.* 18(10):1071–1074.

35. Ostermeier, M., Shim, J.H., and. Benkovic, S.J. 1999. A combinatorial approach to hybrid enzymes independent of DNA homology. *Nat Biotechnol* 17(12):1205–1209.

36. Pelletier, J., and Sidhu S. 2001. Mapping protein-protein interactions with combinatorial biology methods. *Curr. Opin. Biotechnol.* 12(4):340–347.

37. Perez, O.D., and Nolan G.P. 2002. Simultaneous measurement of multiple active kinase states using polychromatic flow cytometry. *Nat. Biotechnol.* 20(2):155–162.

38. Santoro, S.W., and Schultz P.G. 2002. Directed evolution of the site specificity of Cre recombinase. *Proc. Natl. Acad. Sci. USA* 99(7):4185–4190.

39. Schmidt-Dannert, C. 2001. Directed evolution of single proteins, metabolic pathways, and viruses. *Biochemistry* 40(44):13125–13136.

40. Sidhu, S.S. 2001. Engineering M13 for phage display. *Biomol. Eng.* 18(2):57–63.

41. Sieber, V., Martinez, C.A., and Arnold, F.H. 2001. Libraries of hybrid proteins from distantly related sequences. *Nat. Biotechnol.* 19(5):456–460.

42. Siegele, D.A., and Hu, J.C. 1997. Gene expression from plasmids containing the araBAD promoter at subsaturating inducer concentrations represents mixed populations. *Proc. Natl. Acad. Sci. USA* 94(15):8168–8172.

43. Stemmer, W.P. 1994. Rapid evolution of a protein in vitro by DNA shuffling. *Nature* 370(6488):389–391.

44. Tanaka, Y., Yamaguchi, N., and Nasu, M. 2000. Viability of *Escherichia coli* O157:H7 in natural river water determined by the use of flow cytometry. *J. App. Microbiol.* 88(2): 228–236.

45. Telford, W.G., Cox, W.G., Stiner, D., Singer, V.L., and Doty, S.B. 1999. Detection of endogenous alkaline phosphatase activity in intact cells by flow cytometry using the fluorogenic Elf-97 phosphatase substrate. *Cytometry* 37(4):314–319.

46. Veal, D.A., Deere, D., Ferrari, B., Piper, J., and Attfield, P.V. 2000. Fluorescence staining and flow cytometry for monitoring microbial cells. *J. Immunol. Methods* 243(1–2):191–210.

47. Veiga, E., de Lorenzo, V., and Fernandez, L.A. 1999. Probing secretion and translocation of a beta-autotransporter using a reporter single-chain Fv as a cognate passenger domain. *Mol. Microbiol.* 33(6):1232–1243.

48. Walsh, S.T., Lee, A.L., DeGrado, W.F., and Wand, A.J., 2001. Dynamics of a de novo designed three-helix bundle protein studied by 15N, 13C, and 2H NMR relaxation methods. *Biochem.* 40(32):9560–9569.

49. Wentzel, A., Christmann, A., Adams, T., and Kolmar, H. 2001. Display of passenger proteins on the surface of *Escherichia coli* K-12 by the enterohemorrhagic *E. coli* intimin EaeA. *J. Bacteriol.* 183(24):7273–7284.

50. Wittrup, K.D. 2001. Protein engineering by cell-surface display. *Curr. Opin. Biotechnol.* 12(4):395–399.

51. Wittrup, K.D., and Bailey, J.E. 1988. A single-cell assay of beta-galactosidase activity in *Saccharomyces cerevisiae*. *Cytometry* 9:394–404.

52. Xenarios, I., and Eisenberg, D. 2001. Protein interaction databases. *Curr. Opin. Biotechnol.* 12(4):334–339.

53. Yano, T., and Kagamiyama, H. 2001. Directed evolution of ampicillin-resistant activity from a functionally unrelated DNA fragment: a laboratory model of molecular evolution. *Proc. Natl. Acad. Sci. USA* 98(3):903–907.

54. Zlokarnik, G., Negulescu, P.A., Knapp, T.E., Mere, L., Burres, N., Feng, L., Whitney, M., Roemer, K., and Tsien, R.Y. 1998. Quantitation of transcription and clonal selection of single living cells with β-lactamase as reporter. *Science* 279:84–88.

13

Applications of Flow Cytometry in Animal Reproduction

DUANE L. GARNER AND GEORGE E. SEIDEL, JR.

Introduction

Animals use many highly specialized cells to carry out their reproductive functions. These cells include not only germ-line cells—spermatozoa/oocytes—but also a variety of supporting cells of the remaining organs of the reproductive system and their precursors, thereby making production of viable offspring possible. Reproductive processes, at least in nonhuman mammals, also require functional mammary glands for adequate nourishment of newborn offspring. Specific analyses or sorting of the specialized cells of the reproductive system can provide a means of monitoring and modifying reproductive processes in animals and man.

Overview

Male gametes are relatively small, haploid cells suspended in fluid secretions from the testes and accessory reproductive organs. Fully functional, mature spermatozoa are incapable of dividing; thus, these terminal gametes are readily quantified and characterized by flow cytometry. Many thousands to millions of spermatozoa can be readily analyzed and sorted. This situation differs markedly from most other kinds of cells, with which cell division and cell cycle differences can make interpretations of data more difficult. Many of the cells of the reproductive and endocrine systems of mammals have been studied by flow analyses.

Female gametes, however, are relatively large, often exceeding 100 μ in diameter, and therefore are not readily analyzed using flow cytometry. Although the oocytes themselves usually are not useful targets, some of the supporting cumulus cells surrounding developing oocytes can be analyzed and thereby provide useful information about that particular ovarian follicle. The functional status of these specialized cells is an indicator of the health of the follicle and of the likelihood that the associated oocyte can produce viable offspring.

Reproductive functions in animals are controlled by various widely distributed cellular components of the endocrine system including the hypothalamus, the pituitary gland, the gonads, and the placenta. Such regulatory cells can be isolated and cultured and have their in vitro function monitored, using flow cytometric analysis of their internal components and surface receptors.

The supporting cells of animal reproductive systems make possible the production of viable offspring. The functional status of these supporting cells can be evaluated by flow analyses, including the epithelial cervical cells of the female tract, and most accessory tissues, including the male accessory sex glands. Another reproductive organ whose cells can provide extremely useful information on reproductive processes is the placenta. Fetal cells, either procured by aspiration of placental fluids or from maternal blood, can provide information on the status of the fetus and on its genetic profile, including sex.

The mammary glands provide nourishment to mammalian offspring. The health status of this gland can be assessed readily using flow cytometry. Flow analyses can delineate the types of somatic cells in milk and thereby establish the health of this organ in producing milk for human consumption or for nursing offspring.

Spermatozoa

Flow cytometric analyses of spermatozoa can provide rapid and precise quantitative measurements of specific attributes of these gametes. Fluorescently labeling spermatozoa with organelle-specific stains provides a means for quantifying the functional status of particular organelles. Furthermore, flow cytometry can be used to measure these attributes rapidly in many thousands of individual spermatozoa from each sample. Flow analyses can generate data from spermatozoa at 300 to 500 cells/s, simultaneously gathering more than 10,000 bits of data for several separate parameters in approximately 1 min. Flow cytometric data are normally gathered using signal gates, so that aggregates and particles that differ in light-scatter properties from spermatozoa are excluded from analyses. Each of the various fluorescent stains or combinations of organelle-specific stains results in unique scatter plot patterns whereby the functional status of that proportion of individual spermatozoa can be quantified. The initial application of flow cytometry to spermatozoa was assessment of cellular damage resulting from irradiation or other environmental agents (116). Spermatozoa, which are particularly sensitive to environmental insults, readily provide a time-delayed barometer of the health of the male producing them (116). Some of the important applications of flow cytometry to sperm analyses are given below.

Sperm Number Determinations

The actual number and concentration of spermatozoa in semen can be assessed accurately by flow cytometry (22, 29). Flow analyses have been shown to be considerably more accurate than the commonly used hemocytometer counts recommended by the 1992 WHO guidelines (121) for human fertility clinics (22) and than the indirect spectrophotometric methods based on the degree of light scatter (17). The routine hemocytometric, Coulter counter, and spectrophotometric methods for assessing sperm numbers in bull semen leave much to be desired (17, 29, 109). The asymmetry of spermatozoa, however, can be problematic in attaining accurate counts with flow cytometry (22). Nevertheless, flow cytometric assessments of the number of fluorescently stained spermatozoa under field conditions indicated that flow analyses were superior in accuracy to preexisting spectrophotometric determinations of sperm concentration in medium-diluted bull semen and in counting spermatozoa in straws packaged for insemination (22, 29). Flow analyses have also been used to assess sperm count and viability of rat spermatozoa simultaneously (122).

Evaluation of Sperm Morphology

Although most sophisticated analyses of sperm morphology use image analysis, the morphological normalcy of spermatozoa has been examined using slit-scan flow cytometry (4). Stained spermatozoa were oriented to flow lengthwise through a thin laser beam, whereby fluorescence emission was sampled every 20 ns, digitized, and stored in a waveform recorder to provide profiles consisting of approximately 50 fluorescence measurements per profile. The resultant slit-scan flow cytometry profiles were highly correlated with the visually based assays of sperm morphology (4). The slit-scan flow cytometry sperm head shape profiles were measured using the distribution of acriflavine stained gametes from mice, rabbits, hamsters, and bulls to establish norms for the various species, so that assessments of irradiation-induced changes could be monitored (figure 13.1; 4). The frequencies of misshapen spermatozoa increased relative to the dose of X-irradiation. Slit-scan technology has been applied to sperm sexing to attain consistent high-quality resolution DNA analysis of X- and Y-chromosome-bearing spermatozoa (figure 13.2; 101).

Evaluation of Sperm Organelle Function

Sperm organelle function can be quantified using fluorescent staining and flow cytometric analyses. Staining can be either a single-membrane impermeant fluorophore or a combination of stains based on differing membrane permeabilities. Some particular applications of various supravital stains are provided here.

R123 and Ethidium Bromide

The initial estimates of sperm viability used rhodamine 123 (R123) to study mitochondrial function, and they simultaneously used ethidium bromide to distinguish spermatozoa with intact membranes from the dead, membrane-damaged spermatozoa (25). Fluorescent staining, especially when combinations of membrane-permeant and membrane-impermeant dyes are used, can simultaneously identify living and dead spermatozoa in semen (41), as opposed to determining the percentage of live spermatozoa by the subtraction of the percentage of dead spermatozoa from 100%.

Figure 13.1. Slit-scan profiles from spermatozoa of four mammalian species: mouse, rabbit, bull, and hamster. These are average profiles (solid line) consisting of 20 flow measurements, with 1 standard deviation from the averages (dotted lines). Distinct profiles relative to length and shape were noted for spermatozoa from each species, with the characteristic head shapes shown in inserts. [Reprinted from (4) with permission from John Wiley & Sons, Inc.]

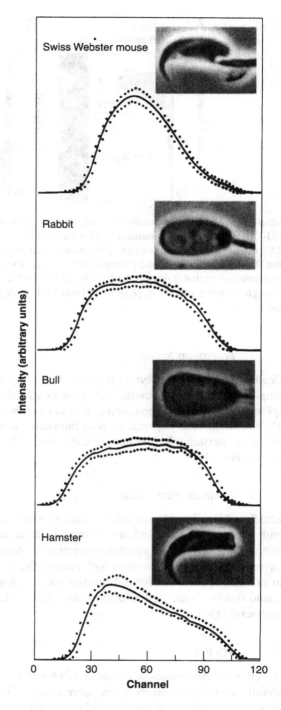

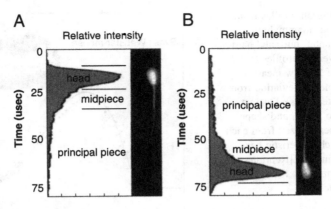

Figure 13.2. Slit-scan profiles of Dansyl Lysine–stained bovine spermatozoa excited at 351–364 nm, ultraviolet multiline and measured for relative intensity at a flow velocity of 0.65 m/s. Adjacent photomicrographs show the corresponding portions of stained spermatozoa. The head, midpiece, and principal piece sections were clearly identified. Profile A shows a spermatozoon that traversed the laser beam tail first, and B corresponds to a cell that passed through the beam head first. [Reprinted from (101) with permission from John Wiley & Sons, Inc.]

Propidium Iodide

Dead spermatozoa stain bright red with fluorescent dyes such as ethidium bromide or propidium iodide (PI) because such dyes, as with the colorimetric dye eosin, penetrate spermatozoa with damaged membranes but not those with intact membranes (25, 40, 41, 85). Membrane-impermeant phenanthridium compounds such as PI are very useful as counterstains for 488-nm excitable organelle-specific probes emitting at different wavelengths.

Ethidium Homodimer-1

Ethidium homodimer-1 and other related ethidium compounds are impermeant to cells with intact membranes and are sensitive to dead cell stains. They bind to DNA and oligonucleotides with a large fluorescence enhancement of about 30-fold (59), and they have been used in combination with carboxyfluorescein derivatives and merocyanine to examine sperm viability and function (9, 10). Another phenanthridium-based compound that has been used effectively as a dead cell stain for spermatozoa is ethidium monoazide (61).

YO-PRO

A series of membrane-impermeant DNA-specific cyanine dyes has been used to identify membrane-compromised spermatozoa. One of these particular membrane-impermeant dead cell stains, YO-PRO-1, which excites at 491 nm and emits at 509 nm, has been used in combination with stains, such as Merocyanine 540 to detect changes in fluidity of sperm membranes (58).

Fluorescein Diacetate Derivatives

The classical dead cell stain, PI, has been coupled with fluorescent probes that excite at 488 nm, such as carboxyfluorescein diacetate (CFDA) or its derivatives to provide differential staining combinations whereby dead spermatozoa are identified by the uptake of PI. Living spermatozoa stain bright green because the membrane-permeant CFDA is converted enzymatically by nonspecific esterases within the cytoplasm of gametes into a bright green, membrane-impermeant product (41). This stain combination has been used to assess the efficacy of cryopreserving bovine spermatozoa, using both egg yolk– and milk-based media (39), and to determine the effectiveness of new antibiotic combinations for preventing microbial growth in diluted bull semen (20, 21). The combination of CFDA and PI staining with flow cytometric analysis also has been used to assess the viability of dog spermatozoa (92).

SYBR-14

The double-staining approach for identifying the live and dead sperm populations was enhanced by using a combination of nucleic acid–specific fluorescent probes, one which was membrane permeant, while the other probe was not. The combination of SYBR-14, which stains the nuclei of membrane intact spermatozoa bright green, and PI, which stains the chromatin of dead or membrane-damaged spermatozoa red, is particularly useful in quantifying the live and dead sperm populations, using excitation at 488 nm (38, 40). Thus, the variability and time lag associated with the CFDA/PI staining has been overcome by using two stains, SYBR-14 and PI, that target the chromatin (40). The relative proportion of spermatozoa exhibiting green or red fluorescence can be rapidly quantified by flow cytometry (figure 13.3, 40). This stain combination can be readily standardized and calibrated using precisely prepared mixtures of living and killed spermatozoa (figure 13.3), and this combination of fluorophores has been used to determine the proportion of viable cells in samples of cryopreserved spermatozoa from bulls (11, 40), boars (38, 80), rams, rabbits, mice (38), stallions (37), and men (38). A major application is to assess sperm function following various treatments after in vitro storage of spermatozoa (47, 48). Furthermore, this differential stain combination, which is commercially available (Spermatozoa Viability Kit, Molecular Probes, Eugene, OR), has been successfully applied to spermatozoa from avian males (18, 19).

Merocyanine 540

Spermatozoa can be preloaded with merocyanine 540, a stain capable of detecting changes in the lipid architecture of the plasma membrane (58). This dye, which maximally excites at 555 nm and emits at 578 nm, binds preferentially to membranes with highly disordered lipids. Under conditions whereby spermatozoa are capacitated, an increase in the proportion of merocyanine 540 fluorescence-positive spermatozoa from 2%–5% to 30%–55% was noted (32). It was suggested that the observed increase in merocyanine 540 fluorescence indicated that the packing of plasma membrane phospholipids became disordered in capacitating spermatozoa (108). The combination of Merocyanine 540 and YO-PRO-1 can be used simultaneously with labeled peanut agglutinin to identify intra-acrosomal components (32).

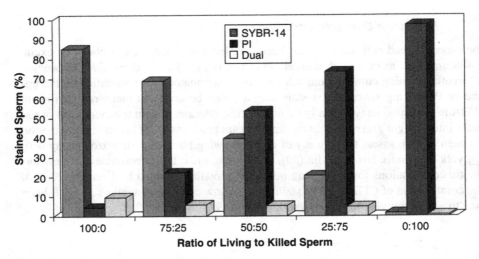

Figure 13.3. Shown are a standardization and calibration comparison of staining with the combination of SYBR-14 and propidium iodide (PI) for assessing viability of spermatozoa. Fresh ejaculates from two bulls were filtered through glass wool–Sephadex columns to remove the dead and damaged spermatozoa. Approximately one-half of the recovered filtered-living spermatozoa were killed by repeated freeze-thawing. Mixtures of living and killed spermatozoa prepared and stained with SYBR-14 and PI. The mixed samples were analyzed using flow cytometry to determine the proportions of living (SYBR-14-stained), killed (PI-stained), and dual-stained (both green and red fluorescence) spermatozoa. The correlation between the SYBR-14 stained spermatozoa and living sperm proportion was $r = 0.99$ ($P < .001$). [Reprinted from (40) with permission from the American Society of Andrology.]

Figure 13.4. Example of how the sperm chromatin structure assay detects damaged chromatin in stallion spermatozoa using acridine orange staining and flow cytometry to examine the structural stability of sperm nuclear chromatin. Shown are sperm chromatin structure assay results from analyzing semen samples of two stallions, one highly fertile (panels A, C, and E) and one with relatively low fertility and marginal semen quality (panels B, D, and F). Sperm samples were subjected to low pH–induced denaturation in situ before being stained with the dichromatic probe acridine orange (AO). Stained samples were examined using flow cytometry to determine the relative green and red fluorescence because binding of AO to double-stranded DNA emits green fluorescence, whereas AO bound to single-stranded DNA emits red fluorescence. Spermatozoa with uncompromised DNA fluoresce green from the bound AO, whereas those with denatured DNA emit red fluorescence. Analyses are expressed as the DNA fragmentation index (DFI), which is the ratio of red to total (red and green) fluorescence. The relative number of spermatozoa exhibiting red fluorescence (box 2 compared to box 3) is much greater for the marginal stallion (F, box 2) than for those from the fertile stallion (C, box 2). Figure provided by Dr. Donald Evenson.

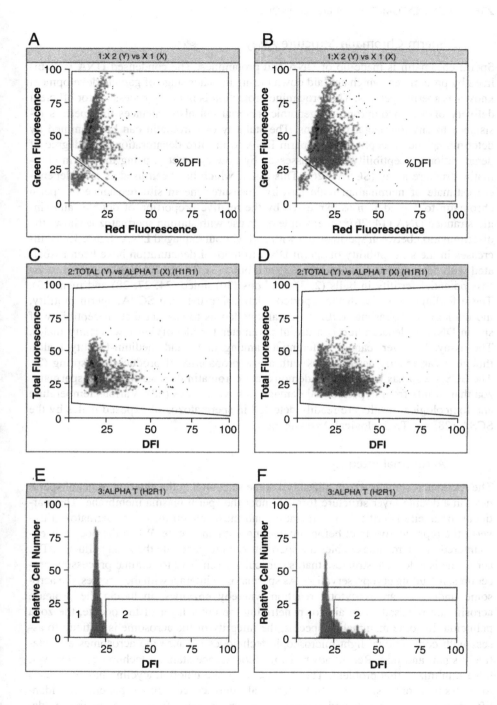

Sperm Chromatin Structure Assay

Sperm chromatin is made up of DNA and protamines. The condensed DNA is stabilized by protamines, which are laid down during the later stage of gamete development known as spermiogenesis. This protamine protection is not only necessary for efficient delivery of the uncompromised genomic material but also enhances the sperm's resistance to environmental mutagens. The stability of chromatin can be examined by determining the susceptibility of sperm DNA to in vitro denaturation. The degree of denaturation susceptibility can be assessed by flow cytometry, using the sperm chromatin structure assay (SCSA) (26). This assay, which has been used as an independent estimate of mammalian male fertility, measures the in situ resistance of sperm chromatin to degradation as expressed by the relative proportion of double- and single-stranded DNA (14, 26). An example of its use with stallion spermatozoa shows the discrimination between spermatozoa with and without damaged DNA (figure 13.4). Increases in the susceptibility of sperm DNA to thermal denaturation have been associated with decreased fertility in men (23) and bulls (3). The SCSA has been used to assess potential fertility in bulls (2, 3), stallions (30), mice (24, 27, 28) and men (23). These findings, along with the reported relationship between SCSA, sperm motility, morphology, and seasonal fertility in stallions led us to suspect that susceptibility of sperm DNA to denaturation is a useful parameter for identifying low fertility males. The assay, however, cannot differentiate among high- and medium-fertility males, though it can readily detect males with a low probability of producing offspring (2). The SCSA also has been used to demonstrate chromatin damage in bovine spermatozoa that results from hyperthermal damage to bovine testes (68). Various temperature and toxicological insults are readily detected in spermatozoa of exposed males by the SCSA (28; see Toxicological Assessment).

Acrosomal Integrity

The sperm acrosome is a specialized lysosome surrounding the anterior aspect of sperm head in a double-layer structure formed under the sperm plasma membrane. This particular organelle is critical to fertilization and must remain intact as spermatozoa traverse the reproductive tract before interacting with an oocyte. When the oocyte is encountered, the acrosome enables the spermatozoon to penetrate the zona pellucida. The acrosome is a delicate structure that is particularly sensitive to seminal processing procedures, including cryopreservation. As spermatozoa interact with the oocytes, the acrosome undergoes an exocytotic reaction whereby enzymes, including the protease acrosin, are released as an aid for penetrating the outer layer of the oocyte, the zona pellucida. In some mammalian species, the integrity of the acrosome is difficult to assess with conventional light-microscopic techniques because the acrosomes are relatively small and thin. Recent advances in fluorescence staining technology, however, have minimized this problem. The fluorescent probe chlortetracycline has been used to assess acrosomal status (118). Differential fluorescence staining patterns can identify spermatozoa that have undergone an acrosome-reaction from those undergoing degenerative changes. Interpretation of chlortetracycline staining patterns is, however, difficult and confusing. A more definitive approach for identifying acrosomal status is the use of the lectin, *pisum sativum* agglutinin (PSA), which binds to the outer acrosomal membrane and to the contents of the acrosome (31). This common pea lectin has been used to identify acrosomal damage accurately in stallion spermatozoa (31). The PSA is conjugated to fluorescein isothiocyanate (FITC) to yield a fluorescently labeled ligand, PSA-FITC. The PSA-FITC binds to permeabilized sperm acrosomes in a particular pattern, thereby revealing the presence or absence of the acrosomal contents. This reagent also has been used to monitor capacitation status as well as acrosomal integrity in spermatozoa (7, 84).

Labeled peanut (*Arachis hypogaea*) agglutinin (PNA) was bound exclusively to the outer acrosomal membrane (9). Labeled PNA actually stains sperm acrosomes from some species more brightly than PSA while minimizing nonspecific staining (50). PNA-labeled spermatozoa can be quantified rapidly by double staining with PI and then analyzing the stained spermatozoa by flow cytometry (113). Alexa 488-labeled PNA also can effectively reveal acrosomal status of spermatozoa and be readily quantified using flow cytometry (figure 13.5). This fluorescence assay can provide definitive data on the proportions of acrosomal-intact, acrosomal-damaged, and acrosomal-denuded spermatozoa (D.L. Garner, unpublished).

A novel approach to identifying and quantifying the level of acrosin, an inner acrosomal component, is using the irreversible biotinylated isocoumarin serine protease inhibitor, BIC (89). This "suicide" inhibitor binds to proacrosin/acrosin, thereby inactivating the enzyme on an equimolar basis. Quantification can be accomplished by

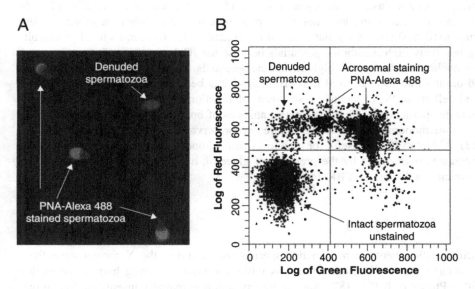

Figure 13.5. Fluorescence photomicrograph of bovine spermatozoa stained for acrosomal contents using Alexa 488-labeled peanut agglutinin and for DNA with propidium iodide (PI). (*A*) The flow histogram (*B*) is from analysis of a PNA-Alexa 488/PI-stained sample containing 50% live and 50% killed spermatozoa for the logs of green and red fluorescence. The histogram shows populations of spermatozoa with intact spermatozoa that resisted uptake of both stains, PI-positive spermatozoa retaining most of their acrosomal contents, and denuded spermatozoa that have shed the entire acrosome as a result of the freeze–thaw killing process (D.L. Garner, unpublished data).

secondary binding to fluorescein-conjugated avidin. Although a relationship between BIC staining and the fertility of the semen samples was not found, significant differences were evident among bulls in the amount of BIC bound to their spermatozoa (89).

Mitochondrial Function

The initial demonstration that the functional status of mitochondria in spermatozoa can be determined used rhodamine 123 (R123; 25). Although this membrane potential-dependent probe specifically identifies spermatozoa with functional mitochondria, some nonspecific staining has been encountered (113). The R123 has been used in combination with the membrane-impermeant nucleic acid probe, ethidium bromide, to differentiate between spermatozoa possessing functional mitochondria and gametes with compromised membrane integrity (25). The R123, which excites at 488 nm and emits at 515–575 nm, has specificity for functional mitochondria exhibiting a transmembrane potential, thereby identifying them as living cells. From flow cytometric examination of double-fluorescent stained stallion spermatozoa, the percentage of spermatozoa with optimally functioning mitochondria, as assessed with R123 fluorescence, was correlated ($r = 0.99$) with the percentage of viable spermatozoa as indicated by exclusion of PI (90). R123 staining, however, does not differentiate between sperm mitochondria exhibiting high membrane potential from those that possess mitochondria with relatively low membrane potentials, as does the probe 5,5′,6,6′-tetrachloro-1,1′,3,3′-tetraethylbenzimidazolyl-carbocyanine iodide (JC-1; 45). Not only does JC-1 identify mitochondria exhibiting low membrane potentials by the emission of green fluorescence (510–520-nm range), but it also differentiates these from mitochondria exhibiting relatively high membrane potentials because the JC-1 monomers form aggregates in mitochondria that exhibit high membrane potentials. These J aggregates emit a bright red-orange fluorescence (590 nm) (100). JC-1 has been combined with the classical dead cell stain, PI, to identify a spectrum of functional spermatozoa along with degenerate spermatozoa. Flow cytometric analysis of bull spermatozoa showed that the aggregate:monomer ratio differed before cryopreservation, but not afterward (46, 47, 113). Although JC-1 is useful for monitoring mitochondrial function, its ability to form aggregates is impaired by the presence of glycerol, a common cryoprotectant, or by the nucleic acid stain, SYBR-14 (46).

Sperm Sexing

Differentiation between mammalian spermatozoa carrying the X-chromosome from those carrying the Y-chromosome was initially established using human spermatids with a Phywe or ICP 22 (87). Such a flow system is essentially insensitive to cell orientation variability because of a coaxial measurement system (93). Early efforts to measure the DNA content of spermatozoa using orthogonal optics were only marginally successful because the highly condensed sperm nucleus makes stoichiometric fluorescent staining difficult, and the asymmetrical shape and high index of refraction of the nucleus make quantitation problematic (117). Fabrication of a specially built flow chamber that oriented the spermatozoa relative to the excitation beam minimized this problem (93). This orienting instrument used a beveled needle to hydrodynamically

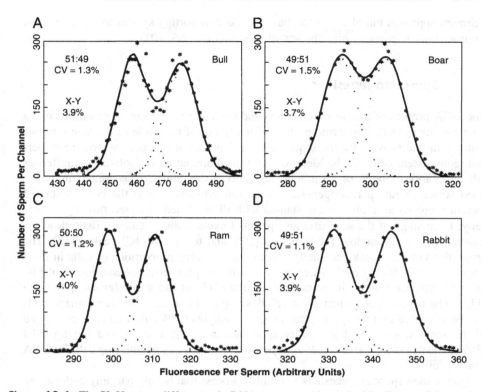

Figure 13.6. The X–Y sperm differences in DNA content as shown by histograms of flow cytometric analyses of 4'-6-diamine-2-phenylindole–stained sperm nuclei from ejaculates of the bull (*a*), boar (*b*), ram (*c*), and rabbit (*d*). The actual data points are shown for each channel by an asterisk (*), whereas computer-fitted curves are shown as a solid line ([2m]) and dots (. . .). Ratios of the areas of the two peaks are in the upper-left corner of each histogram, along with the coefficient of variation (CV) of the individual peaks. The ordinate values are arbitrary and do not reflect differences in DNA content. [Reprinted from (36) with permission from the Society for the Study of Reproduction.]

force each spermatozoon into a particular orientation as they passed through the flow cytometer, thereby allowing accurate measurement of the fluorescence emitted from the flat surface of the sperm nucleus (16, 33). The angular dependence of fluorescence from mouse sperm nuclei was less pronounced when measured from the flat surface than when measured from the edge of the spermatozoon (93). This system resulted from ongoing efforts at Lawrence Livermore National Laboratory to develop a sperm-based assessment of the effect of environmental insults, especially radiation, on mammalian males (49). Application of this innovation led to a system whereby the ratio of X- to Y-chromosome-bearing spermatozoa could be determined in semen samples of domestic livestock including bulls, boars, rams, and rabbits (figure 13.6; 36). Examination of cryopreserved spermatozoa using a variety of methods that purportedly enriched for either X or Y spermatozoa indicated that none of the methods altered the ratio from 50:50 (94). The ability to differentiate between X- and Y-chromosome-bearing spermatozoa has not only provided a means to determine the efficacy of any sperm enrichment approach but also was the basis of the flow-sorting system that is now used commercially to predetermine the sex of offspring (34, 66, 103).

Spermatogenesis

The DNA profiles of mouse spermatozoa and testicular nuclei can be examined using flow cytometry (83). Furthermore, the various stages of the cycle of the seminiferous epithelium that provide for the sequential development of rat spermatozoa from spermatogonial stem cells can be identified by transillumination of isolated seminiferous tubules. Various lengths of the tubular segments of the seminiferous epithelium at defined stages of rat spermatogenesis can be microdissected, and the cells contained therein prepared as a suspension, stained with PI, and then analyzed flow cytometrically. Each stage of the seminiferous epithelial cycle shows a characteristic flow cytometric pattern of haploid [1C], diploid [2C], and tetraploid [4C] peaks (115). The proportions of each peak accurately reflected the relative proportions of cells in most stages of the epithelial cycle when compared to morphometric measurements of histological preparations of the same stage of the cycle of the seminiferous epithelium (115). The products of human germ cell development and the testicular somatic cells can be separated into tetraploid primary spermatocytes (17%); diploid cells composed of the spermatogonia and secondary spermatocytes, Sertoli cells, and Leydig cells (36%); and the haploid spermatids and spermatozoa (45%) by quantifying cellular DNA content (60).

Mammalian spermatogenesis is a complex process that is only partially understood. Three-parameter flow cytometric analyses using PI to quantify DNA content, immunostaining for vimentin to identify somatic cells, and nonyl acridine orange to identify mitochondria enabled classification of 11 testicular cell subpopulations in the rat. In addition to the somatic cells, 10 subtypes of the germinal cells were identified using this stain combination (107). The subpopulations included residual bodies, spermatid development steps 17–18, spermatid development step 19, spermatids steps 1–5, spermatids steps 6–11, spermatids steps 12–16, secondary spermatocytes, preleptotene spermatocytes/spermatogonia, pachytene spermatocytes, and leptotene and zygotene spermatocytes (figure 13.7; 107). The ability to identify and quantify various stages of germ cell development provides a very powerful tool for evaluating spermatogenesis in both normal and perturbed situations.

Toxicological Assessments

Spermatozoa have been used to screen for possible adverse effects of chemicals on reproductive processes (106, 122). Various automated image assessments of sperm morphology (Automated Sperm Morphology Assay; 15) and motility (Computer Assisted Semen Analysis; 106), along with in vitro fertilization tests (5), have been used to detect reproductive damage in laboratory animals. Flow cytometric assessments of spermatozoa stained with organelle-specific fluorescent stains are useful in detecting very subtle changes in organelle function. The ability to assess simultaneously the functional status of several organelles including membrane integrity, mitochondrial func-

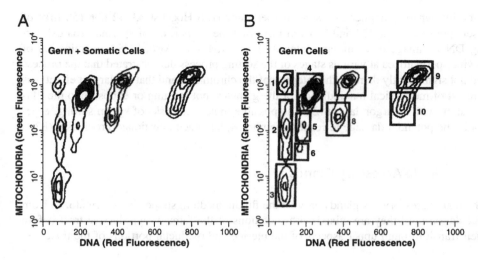

Figure 13.7. Flow cytometric assessment of spermatogenesis of rat testicular cells stained with antivimentin primary rabbit antibody and 7-amino-4-methycoumarin-3-acteic acid–conjugated secondary goat anti-rabbit IgG (blue), propidium iodide DNA-staining (red), and Nonyl-acridine orange (green). The defined gates (boxes 1–10) were used for sorting and identification. Contour density plot (*A*) represents DNA versus mitochondria staining of all testicular cells, whereas contour density plot (*B*) represents DNA versus mitochondrial staining of testicular germ cells after subtraction of the vimentin-positive somatic cells. The subhaploid peaks contained residual bodies (box 1) and spermatids (boxes 2 and 3). Boxes 4, 5, and 6 contained three different subpopulations from round spermatids (box 4) to elongated spermatids (box 6). The diploid populations contained the secondary spermatocytes and preleptotene spermatocytes (boxes 7 and 8) and spermatogonia (box 8). The tetraploid cells included the large pachytene spermatozoa (box 9) and smaller leptotene and zygotene spermatocytes (box 10). [Reprinted from (107) with permission from John Wiley & Sons, Inc.]

tion, and acrosomal status on many thousands of spermatozoa is especially useful in detecting toxicant-induced alterations in sperm function (52). The development of flow cytometric techniques to separate X and Y spermatozoa resulted from an effort to use spermatozoa for monitoring the effect of various toxicological insults on animal and human health (93, 117).

At one time, gossypol, a pigment from cottonseed, was used in China as an antimitotic agent to sterilize human males. Cottonseed is used widely as a protein supplement in cattle and sheep rations because ruminants were considered to be tolerant of dietary gossypol. Flow cytometric assessment of the reproductive toxicity of gossypol in prepubertal rams indicated that dietary gossypol levels of cottonseed could be detrimental to sperm development at high levels (71).

The process of staining, diluting, sorting, and concentrating mammalian spermatozoa to predetermine the sex of offspring imparts considerable stress to these gametes. Potential damage to spermatozoa at various steps in the flow-sorting process was assessed using SYBR-14/PI staining to determine which steps caused the most damage to the spermatozoa (44). The mechanical stresses occurring during sorting or postsorting centrifugation marginally increased the proportion of spermatozoa exhibiting changes in DNA integrity, whereas neither exposure of spermatozoa to Hoechst 33342 nor 150 mW of laser illumination at 352 and 364 nm increased the proportion of spermatozoa exhibiting DNA damage or membrane integrity losses (34, 44). Also, SCSA examination of bovine spermatozoa at various stages of the sorting process demonstrated that the process did not significantly damage the integrity of the chromatin, and that the largest effect was a result of mechanical manipulation of the gametes, not staining or exposure to the illumination with an argon laser (42). The possible mutagenic risks of sperm sorting, especially the potential damage to resulting embryos, had been questioned previously (82).

Male Accessory Glands

The male accessory sex glands provide the fluidic medium supporting spermatozoa from ejaculation until they are mixed with secretions of the female reproductive tract before their transit through upper aspects of the uterine and oviductal portions of the tract.

Prostate

There are many dozens of papers published annually in which flow cytometry is used to study prostate biology and pathology. By comparison, similar studies with other male accessory sex glands such as the seminal vesicles are rare. This obviously is related to human prostate cancer, which is the most common cancer of men in the United States, or the second most common after skin cancer in some populations. Only lung cancer causes more cancer deaths. Because of this situation, considerable funding is available for such studies. Much of this work concerns cell lines derived from prostate tumors (e.g., 69) or prostate tissue from animal models. Both of these options are very amenable to producing cells that can be studied with flow cytometry.

Surveying uses of flow cytometry for prostate studies range from the most basic to very applied clinical work. Perhaps as much as half of the recently published work falls into the following four categories: apoptosis, cell cycle regulation, growth factors, and prostate-specific antigen. Obviously there is considerable overlap even among these (e.g., apoptosis and cell cycle regulation; 54).

Apoptosis studies with prostate tissue recently have become prevalent (e.g., 54, 97, 110, 111). Although most of these studies have the ultimate goal of clinical application to regress tumors, considerable basic biology is being accomplished. Especially intriguing is the application of new tools such as ribozyme technology (110).

Examples of flow cytometer studies of cell cycle regulation of prostate cells include work of Lu and Epner (77), Kobayashi et al. (70), and Wartenberg et al. (120). Cell cycles of carcinogenic prostate cells usually are disregulated, and the above-cited studies serve to improve understanding of this phenomena; in some cases, the goal is simply to inhibit cell cycle progression, including induction of apoptosis (54, 73).

Much of cell cycle progression in prostate tumors is driven by growth factors (51, 88), although such progression also can be inhibited by growth factors such as transforming growth factor beta (6). Neutralizing the stimulators and enhancing the inhibitors of cell cycle progression are being studied routinely using flow cytometry (e.g., 64).

The fourth of the major categories mentioned earlier is prostate-specific antigen (PSA). Not only is this molecule used diagnostically by measuring its concentration in blood, but immunotherapy with anti-PSA antibodies is also being pursued on a large scale in certain patients with prostate cancer (8). Labeled anti-PSA plus flow cytometry provides a powerful tool for many kinds of studies. Production of prostate cell lines that continue to secrete PSA constitute excellent models for developing superior diagnostic approaches for identifying prostate tumors (74).

Ovarian Cells

Follicular Maturation Assessment

The production of mammalian female gametes—oogenesis—involves the cyclic degeneration of most ovarian follicles. This degenerative process, termed atresia, allows only one or a few follicles to mature and produce a secondary oocyte capable of achieving ovulation. Most ovarian follicles including their oocytes undergo cyclic degeneration, never reaching maturity. The major physiological regulation of follicular atresia is apoptosis, not only of the oocyte, but also of the granulosa cells that support the gamete during its final development and maturation. These subtle follicular events can be monitored by examining ploidy of dispersed granulosa cells aspirated from ovarian follicles at varying stages of development and atresia, using flow cytometry (56). Porcine follicles were classified as atretic or nonatretic on the basis of the percentage of cells at a particular stage of the cell cycle. Follicles exhibiting more than 10% A_0-stage cells were considered to be atretic, whereas follicles with less than 10% A_0 granulosa cells were nonatretic; the range was 0.2%–89% A_0 cells/follicle examined (56). These data were obtained from the DNA histograms of PI-stained granulosa cells in which the levels of DNA were used to determine the proportions of cells in the A_0, G_0/G_1, S, and G_2/M stages of the cell cycle (figure 13.8; 55), and they indicated that apoptosis-associated DNA fragmentation was a measure of follicular health. The importance of this approach is that it provides a means to determine oocyte health indirectly during in vitro manipulation while seeking to enhance the reproductive efficiency and genetic contribution of female mammals.

Oviduct, Uterus, and Cervix

Research on the tubular structures of the female reproductive system is particularly amenable to study with flow cytometry. The epithelia of these organs can be harvested readily, dissociated into single cells, cultured and passaged in vitro, and studied in a variety of ways. The interactions between the cells associated with these epithelia, such as lymphocytes and spermatozoa, also are readily studied with flow cytometry (78).

Oviduct

An example of studying the oviduct epithelial cells themselves is provided by the work of Tiemann et al. (114), who examined changes in the amount of platelet-activating-factor receptor through the estrous cycle, using an antibody to platelet-activating-factor receptor and flow cytometry. Thibodeaux et al. (112) used flow cytometry to

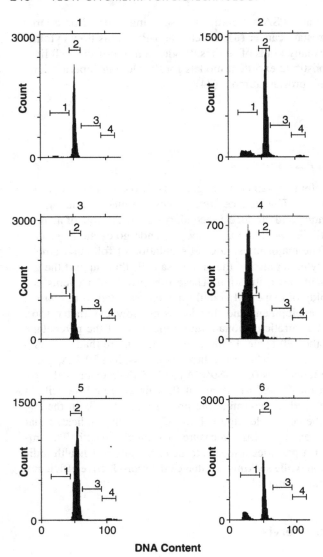

Figure 13.8. DNA histograms of granulosa cells from six Graafian follicles from a pig on day 7 postestrus. The percentage of cells containing subdiploid amounts of DNA (%A_0 cells) and the distribution of cells in the stages of the cell cycle were determined by flow cytometry of PI-stained nuclei of ethanol-fixed follicular cells. Region 1 represents subdiploid levels of DNA (%A_0 cells), region 2 represents G_0/G_1, region 3 the S stage, and region 4 the G_2/M stages of the cell cycle. The univariate histograms illustrate the proportions of granulosa cells in the A_0, G_0/G_1, S, and G_2/M stages in six follicles of varying maturity and states of atresia. [Reprinted from (55) with permission from the Society for the Study of Reproduction.]

monitor cell cycles of growing oviduct epithelial cells under various in vitro culture conditions. A very different approach is to assess the effects of oviduct secretions on sperm capacitation (78); Mahmoud and Parrish used a battery of lectins to monitor sperm surface changes with flow cytometry. These papers illustrate the variety of approaches for studying the oviduct, using flow cytometric procedures.

Uterus

Although flow cytometry has been used to study uterine cancer, this subject does not dominate the uterine literature in the same way that cancer dominates literature on the prostate and cervix. Of course, many important events occur in the uterus, including implantation, transport of spermatozoa, phagocytosis of nonfertilizing spermatozoon, development of the conceptus, and on occasion, pathology such as venereal disease, teratogenesis, preeclampsia, and so forth, all of which have been studied with flow cytometry. Monitoring cell cycles within the reproductive cycle is also done commonly with flow cytometry.

Rather than discussing a potpourri of studies on the above topics, we concentrate on one area of pregnancy biology: the fetus as an allograft. This issue has intrigued biological scientists for most of a century, and recent studies using flow cytometry are providing some explanations of how the maternal immune system tolerates this allograft. It has become clear that there are multiple mechanisms operating, and a sampling follows.

Peltier et al. (91) have shown that serpin, a progesterone-induced uterine serine protease inhibitor, was secreted by the ovine uterine epithelial cells in large quantities. Two-color flow cytometry was used to show that this protein inhibited concanavalin A-induced expression of CD25 in one type of T cells, but not in another type. Similar proteins are produced in cattle and pigs and probably many other species. This was but one finding in the complex action of this molecule, which inhibits lymphocyte proliferation under certain conditions.

A completely different mechanism of allograft protection is expression of the HLA-G major histocompatibility complex molecule on cytotrophoblast cells at the feto–maternal interface (102). This molecule appears to protect cells against natural killer cells by conferring immunological tolerance.

Still another mechanism was described by Zorzi et al. (123), who showed that CD95 ligand could trigger apoptosis in T cells that had recognized tissue of the conceptus. The variety of mechanisms that seem to be involved with protection of the fetal allograft and have been studied with flow cytometry is truly striking.

Cervix

As with the prostate gland, the use of flow cytometry for work with cells of the uterine cervix is dominated by cancer. There also is the special case of HeLa cells, which originated as a cervical cancer and since have become one of the most used cell lines for basic biological studies, as well as a model of cervical cancer (e.g., 105). Another important area of research involving flow cytometry with cervical cells is venereal disease, including the special case of HIV-1 virus; as one example, Prakash et al. (95) showed with flow cytometry techniques that cervical T lymphocytes are likely to be more susceptible to HIV-1 than circulating ones.

In many of these studies, flow cytometry was used primarily for analysis of the distribution of cells throughout the cell cycle, either for cells taken from the cervix (62) or in studies with cell lines (e.g., 76). However, more sophisticated uses also are common, such as monitoring apoptotic status via Annexin V binding (105) or detecting the oncoprotein E7 originating from human papilloma virus (13); this oncoprotein appears to be a major inducer of cervical cancer.

Another application that is gaining in importance is diagnostics. A simple example is prediagnostic analysis of samples of cervical cells (53) to determine whether they will be suitable for Pap smear analysis. At the other end of the diagnostic spectrum is analysis of tumor cells for treatment and prognostic purposes (12).

Fetal Cells in the Blood of Women

The hemochorial placentae of primates provide a means for obtaining fetal cells without intervention with the fetus itself. The presence of fetal cells in maternal blood is well documented, and the development of high-speed flow systems to analyze of these rare cells provides a means for determining the chromosomal makeup of the fetus as well as providing evidence of detrimental alleles. Adinolfi et al. (1) was one of the first groups to suggest that cytological analysis of fetal cells present in the maternal circulation might provide a means for prenatal diagnosis of chromosome and biochemical anomalies. Fetal conditions including sex; human leukocyte antigen and Rh blood types; trisomy 13, 18, and 21; triploidy, sickle cell anemia; and thalassemia have been identified using this approach (119). Although fetal cell isolation by flow cytometry, when combined with other purification techniques, could provide prenatal screening for diagnosis of common aneuploidies, further developments are needed before it can become a routine clinical test. The problem is magnified because only one in from 10^5 to 10^7 cells in maternal blood cells is of fetal origin (119), and adequate numbers of fetal hemopoietic progenitor cells may not circulate in the maternal blood before 16 weeks of gestation (63).

Mammary Gland

Breast Tumor Detection

Breast cancer is a major cause of premature death in women. This fact has generated considerable effort toward identifying metastatic tumor cells that originate from mammary tissue carcinomas (99). Aneuploidy levels have been identified, using simultaneous analyses of electronic nuclear volume and DNA content (72). A comparison of nuclei from normal human breast tissue and those from a primary breast tumor indicate that the incidence of aneuploidy is increased greatly in tumor cells (figure 13.9; 72).

Another intriguing application of flow analysis is the attempt to screen for breast cancer by examining the intracellular viscosity of the peripheral blood lymphocytes, using intracellular fluorescence polarization (96). This approach, although promising, has yielded variable results.

Cellular Components of Milk

Milk production, either for support of offspring or for human consumption, is dependent on healthy mammary glands. Inflammation of the mammary gland, commonly known as mastitis, can be caused by injury or infection with microorganisms and can severely decrease milk production. Somatic cell counts in milk are used as an indicator of mastitis, but such measurements alone lack the ability to detect subclinical mastitic changes and thus are not an optimal prognostic indicator of mammary gland health. A flow cytometric procedure was developed to diagnose mastitis before the development of overt clinical symptoms identified as high somatic cell counts (98). At least five or more populations of cells have been identified in mastitic milk using carboxymethylfluoroes-

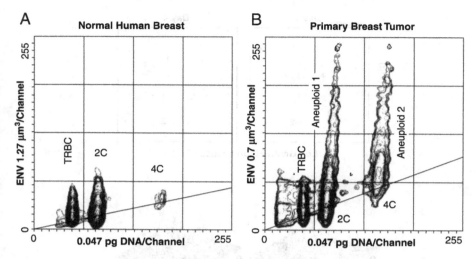

Figure 13.9. Flow cytometric comparison of normal (*A*) and cancerous (*B*) human breast tissue as assessed for nuclear packing efficiency and using trout red blood cells (TRBC) as internal controls. Nuclei from normal breast tissue and a primary breast tumor were prepared by mincing the tissue with crossed scalpels, staining for DNA content with DAPI, and filtering the result through 35–40-μm nylon mesh before analyses for electronic nuclear volume (ENV) and DNA content of nuclei. The aneuploidy populations exhibit elevated electronic nuclear volume values and are distinct from diploid populations 2C and 4C. [Reprinted from (72) with permission from John Wiley & Sons, Inc.]

cein diacetate. These populations included intact and degenerate neutrophils, lymphocytes, monocytes, and large activated macrophages containing vacuoles and phagocytosed particulates (figure 13.10; 98). This system could not only identify affected cows before overt symptoms developed but was also effective in providing prognostic information as to the efficacy of antibiotic treatments. A similar approach using the combination of SYBR Green I and PI has been used to detect subclinical mastitis in sheep (81). Flow cytometric assessment of the number and cell types in milk can provide significant improvements in understanding and treating mastitis not only in animals but also in women.

Flow Sorting

Sex-Sorting Spermatozoa

The sex of mammalian offspring can be predetermined by separating X- from Y-chromosome-bearing spermatozoa with a flow cytometer/cell sorter. This sexing process separates spermatozoa because X-chromosome-bearing spermatozoa contain about 4% more DNA than Y-chromosome-containing spermatozoa, as originally demonstrated in the late 1970s and early 1980s (36, 93). At present, spermatozoa are stained with a specific bisbenzimidazole DNA-binding dye, Hoechst 33342, to determine differences in DNA content before sorting (42, 65, 66, 67, 103). The fluorescence

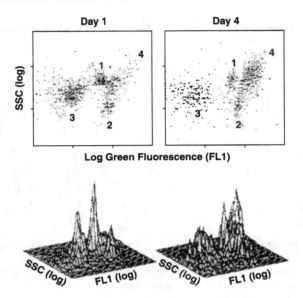

Figure 13.10. Flow cytometric analysis of the changes in somatic cell populations in bovine milk during mastitis. Bovine milk was collected daily from an infected quarter and stained with carboxy-4′-5′-dimethylfluorescien diacetate (CDMFA) and analyzed for the log of side scatter (SSC) and the log of green fluorescence. The top panels show samples collected 4 days apart, illustrating changes in cell populations. The cells in population 1 were identified as intact neutrophils, those in population 2 as lymphocytes, those in population 3 as disintegrating neutrophils, and those cells in population 4, which exhibited the greatest CDMFA staining and light-scatter properties, as a mixture of large lymphocytes, monocytes and large activated macrophages. Note the increases in the proportion of inflammatory cells after 4 days. [Reprinted from (98) with permission from John Wiley & Sons, Inc.]

signals emitted when Hoechst 33342–stained X and Y spermatozoa are illuminated with the 351- and 364-nm lines of an argon laser are accurately measured, so that the cells can be sorted (figure 13.11). The Hoechst 33342–stained X spermatozoa emit proportionally brighter fluorescence than Y spermatozoa because of their greater DNA content.

Many animals have been born from embryos produced from sex-sorted spermatozoa (104). The integrity of the DNA in sex-sorted bovine spermatozoa has been examined using the SCSA (42). These flow data agree with field trial results with Hoechst-stained, sex-sorted spermatozoa, in that reasonable levels of fertility have been attained (86, 104). The lack of an effect on the fertility of stained spermatozoa also has been demonstrated in pigs (35, 57, 65). Libbus et al. (75), however, reported that staining with Hoechst 33342 and ultraviolet-laser irradiating spermatozoa during flow sorting tended to increase the incidence of chromosome aberrations, as indicated by an increase in the number of DNA breaks and triradial and quadriradial exchanges between chromosomes that developed after the treated sperm nuclei were microinjected into hamster eggs. Nonetheless, several thousand calves have been born from sex-sorted living spermatozoa (103, 104). The flow sorting procedure using DNA content is the only verified method for selecting the sex of mammalian offspring (34, 103), however,

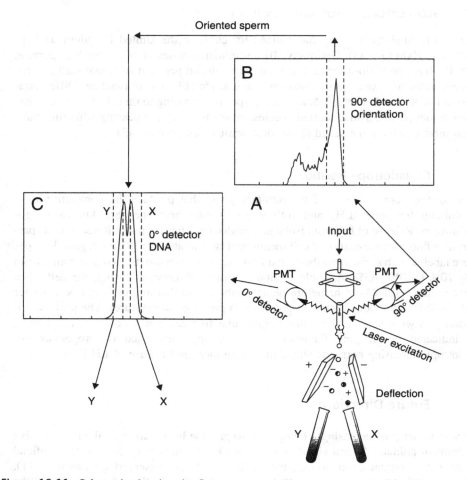

Figure 13.11. Schematic showing the flow cytometer/cell sorter system modified for sorting spermatozoa for DNA content to select live X- and Y-chromosome-bearing spermatozoa. Hoechst 33342–stained spermatozoa enter the system through an injection needle (sample input) and are passed into a flow cell that orients the sperm nucleus so that the DNA content of properly oriented spermatozoa can be accurately measured. The photomultiplier tube (PMT) at 90° detects the degree of orientation so that only properly oriented spermatozoa can be measured with the 0° detector. About 70% of the spermatozoa are sufficiently well oriented for DNA content to be measured accurately, as indicated by the spermatozoa selected in the highest peak in the 90° detector histogram, within the dotted lines (*b*). Those spermatozoa determined to be properly oriented are measured for DNA content with the 0° detector and show two overlapping bimodal peaks representing the X- and Y-sperm populations because X-chromosome-bearing spermatozoa contain about 4% more DNA than those with the Y-chromosome. The histogram is gated to eliminate those spermatozoa exhibiting overlapping fluorescence intensities so that only those spermatozoa within the paired dotted lines for the X- or Y-sperm populations are selected (*c*). As the stream exits the nozzle (*a*), it is vibrated at 80,000 cycle/s, so that droplets are formed. Those droplets containing the X spermatozoa are usually charged positively, and those with Y spermatozoa are negatively charged. The charged droplets are deflected into appropriate containers by oppositely charged plates so that the spermatozoa are directed into appropriate vessels. Those spermatozoa excluded by the gating process and those not properly oriented are disposed of as waste. [Adapted from (66) with permission from Elsevier Science.].

and this technology was commercialized for cattle in the United Kingdom on September 1, 2000 (34, 43). Optimally, 10 insemination doses of sexed bovine sperm at 2×10^6 live sperm/dose for each sex can be produced per hour of actual sorting. Production rates of sexed sperm, however, are considerably less in practice. Differentiation of the X- and Y-chromosome-bearing sperm, according to their DNA content, has been achieved in 23 mammalian species, whereas normal-appearing offspring have been produced from the sexed sperm of at least seven species (43).

Gonadotrope Isolation

The adenohypopyseal cells of the pituitary gland that produce the gonadotropins—luteinizing hormone (LH) and follicle stimulating hormone—are known as gonadotropes. Release of LH from isolated gonadotropes has been monitored in cells prepared by fluorescence-activated cell sorting and laser treatment (79). First, gonadotropes were labeled with anti-LH antibody and flow sorted to increase their concentration from 5%–10% to 70%–85% of whole pituitary extracts. Following sorting, the cells then were cultured for 24–48 hr before being relabeled, so that any cells not positive for LH could be eliminated with an argon laser to produce pure cultures. The purified gonadotropes were used to show that intracellular free calcium concentrations, $[Ca^{2+}]$, as indicated by imaging fluorescence microscopy, responded to exogenous gonadotropin-releasing hormone stimulation with increased release of LH (79).

Future Directions

Flow cytometry is increasingly being used to provide information on the reproductive systems of animals. Sperm viability assays are used routinely by some cattle artificial insemination organizations to monitor the quality of cryopreserved spermatozoa (11). The use of flow cytometry in selecting gametes for preselection of sex likely will be expanded to include selection for the presence of desirable genes as well as the absence of detrimental genes. Another important application will likely be the selection of specific cells for cloning.

References

1. Adinolfi, G., Carozza M., De Nuptiis, T. 1974. La nostra esperienza sulle gravidanze ad alto rischio. *Minerva Ginecol.* 26:425–426.
2. Ballachey, B.E., Hohenboken, W.D., Evenson, D.P. 1987. Heterogeneity of sperm nuclear chromatin structure and its relationship to fertility of bulls. *Biol. Reprod.* 36:915–925.
3. Ballachey, B.E., Saacke, R.G., Evenson, D.P. 1988. The sperm chromatin structure assay: relationship with alternative test of sperm quality and heterospermic performance of bulls. *J. Androl.* 109–115.
4. Benaron, D.A., Gray, J.W., Gledhill, B.L., Lake, S., Wyrobek, A.J., Young, I.T. 1982. Quantification of mammalian sperm morphology by slit-scan flow cytometry. *Cytometry* 2:344–349.
5. Berger, T., Miller, M.G., Horner, C.M. 2000. In vitro fertilization after in vivo treatment of rats with three reproductive toxicants. *Reprod. Toxicol.* 14:45–53.

6. Bretland, A.J., Reid, S.V., Chapple, C.R., Eaton, C.L. 2001. Role of endogenous trans forming growth factor beta (TGF beta) 1 in prostatic stromal cells. *Prostate* 48:297–304.

7. Casey, P.J., Hillman, R.B., Robertson, K.R., Yudin, A.I., Lui, I.K., Drobnis, E. 1993. Validation of an acrosomal stain for equine sperm that differentiates between living and dead sperm. *J. Androl.* 14:289–297.

8. Cavacini, L.A., Duval, M., Eder, J.P., Posner, M.R. 2002. Evidence of determinant spreading in the antibody response to prostate cell surface antigens in patients immunized with prostate-specific antigen. *Clin. Cancer Res.* 8:368–373.

9. Cheng, F.P., Fazeli, A., Voorhout, W.F., Marks, A., Bevers, M.M., Colenbrander, B. 1996. Use of peanut agglutinin to assess the acrosomal status and the zona pellucida-induced acrosome reaction in stallion spermatozoa. *J. Androl.* 17:674–682.

10. Cheng, F.P., Gadella, B.M., Voorhout, W.F., Marks, A., Fazeli, W.F., Bevers, M.M., Colenbrander, B. 1998. Progesterone-induced acrosome reaction in stallion spermatozoa is mediated by a plasma membrane progesterone receptor. *Biol. Reprod.* 59:733–774.

11. Christensen, P. 2002. Danish semen analysis: fertility vs. quality tests. Pages 96–101 in *Proceedings of the Nineteenth Technical Conference on Artificial Insemination and Reproduction.* National Association of Animal Breeders, Columbia, MO.

12. Corver, W.E., Koopman, L.A., Mulder, A., Cornelisse, C.J., Fleuren, G.J. 2000. Distinction between HLA class I-positive and -negative cervical tumor subpopulations by multiparameter DNA flow cytometry. *Cytometry* 41:73–80.

13. D'Anna, R., Le Buanec, H., Alessandri, G., Caruso, A., Burny, A., Gallo, R., Zagury, J.F., Zagury, D., D'Alessio, P. 2001. Selective activation of cervical microvascular endothelial cells by human papillomavirus 16-e7 oncoprotein. *J. Natl. Cancer Inst.* 93:1843–1851.

14. Darzynkiewicz, Z., Traganos, F., Sharples, T., Melamed, M. 1975. Thermal denaturization of DNA in Situ as studied by acridine orange staining and automated cytofluorometry. *Exp. Cell Res.* 90:411–428.

15. Davis, R.O., Gravance, C.G., Thal, D.M., Miller, M.G. 1994. Automated analysis of toxicant-induced changes in rat sperm head morphometry. *Reprod. Toxicol.* 8:521–529.

16. Dean, P.N., Pinkel, D., Mendelsohn, M.L. 1978. Hydrodynamic orientation of sperm heads for flow cytometry. *Biophys. J.* 23:7–13.

17. den Daas, N. 1992. Laboratory assessment of semen characteristics. *Anim. Reprod. Sci.* 28:87–94.

18. Donoghue, A.M., Garner, D.L., Donoghue, D.J., Johnson, L.A. 1996. Viability assessment of turkey sperm using fluorescent staining and flow cytometry. *Poultry Sci.* 74:1191–1200.

19. Donoghue, A.M., Garner, D.L., Donoghue, D.J., Johnson, L.A. 1996. Flow cytometric assessment of the membrane integrity of fresh and stored turkey sperm using a combination of hypo-osmotic stress, fluorescent staining and flow cytometry. *Theriogenology* 46:153–163.

20. Ericsson, S.A., Garner, D.L., Johnson, L.A., Redelman, D., Ahmad, K. 1990. Flow cytometric evaluation of cryopreserved bovine spermatozoa processed using a new antibiotic combination. *Theriogenology* 33:1211–1220.

21. Ericsson, S.A., Garner, D.L., Thomas, C.A., Downing, T.W., Marshall, C.E. 1993. Interrelationships among fluorometric analyses of spermatozoal function, classical seminal quality parameters and the fertility of cryopreserved bovine spermatozoa. *Theriogenology* 39:1009–1024.

22. Eustache, F., Jouannet, P., Auger, J. 2001. Evaluation of flow cytometric methods to measure human sperm concentration. *J. Androl.* 22:558–567.

23. Evenson, D.P. 1989. Flow cytometric analysis of toxic chemical induced alteration in testicular cell kinetics and sperm chromatin structure. In: *New Trends in Genetic Risk*, Jolles, G., and Cordier, A. (eds.). Academic Press, New York, pp. 343–365.

24. Evenson, D.P., Baer, R.K., Jost, L.K. 1989. Long-term effects of triethylenemelamine exposure on mouse testis cells and sperm chromatin structure assayed by flow cytometry. *Environ. Mol. Mutagen* 14:79–89.

25. Evenson, D.P., Darzynkiewicz, Z., Malamed, M.R. 1982. Simultaneous measurement by flow cytometry of sperm cell viability and mitochondrial membrane potential related to sperm motility. *J. Histochem. Cytochem.* 30:279–280.

26. Evenson, D.P., Darzynkiewicz, Z., Melamed, M.R. 1980. Regulation of mammalian sperm chromatin heterogeneity to fertility. *Science* 210:1131–1133.

27. Evenson, D.P., Higgins, P.H., Grueneberg, D., Ballachey, B.E 1985. Flow cytometric analysis of mouse spermatogenic function following exposure to ethylnitrosourea. *Cytometry* 6:238–253.

28. Evenson, D.P., Janca, F.C., Jost, L.K. 1987. Effects of the fungicide methyl-benzimidazol-2-yl carbamate (MBC) on mouse germ cells as determined by flow cytometry. *J. Toxicol. Environ. Health* 20:387–399.

29. Evenson, D.P., Parks, J.E., Kaproth, M.T., Jost, L.K. 1993. Rapid determination of sperm cell concentration in bovine semen by flow cytometry. *J. Dairy Sci.* 76:86–94.

30. Evenson, D.P., Sailer, B.L., Jost, L.K. 1995. Relationship between stallion sperm deoxyribonucleic acid (DNA) susceptibility to denaturation in situ and presence of DNA strand breaks: Implications for fertility and embryo viability. *Biol. Reprod. Mono.*1:655–659.

31. Farlin, M.E., Jasko, D.J., Graham, J.K., Squires, E.L. 1992. Assessment of *Pisum sativum* agglutinin in identifying acrosomal damage in stallion spermatozoa. *Mol. Reprod. Dev.* 32:23–27.

32. Flesch, F.M., Colenbrander, B., van Golde, L.M.G., Gadella, B.M. 2000. Capacitation induced molecular alterations in plasma membrane of boar spermatozoa in relation to zona pellucida affinity. In: L.A. Johnson and H.D. Guthrie. Boar Semen Preservation IV. Allen Press, Lawrence, KS, pp. 21–31.

33. Fulwyler, M.J. 1977. Hydrodynamic orientation of cells. *J. Histochem. Cytochem.* 25:781–783

34. Garner, D.L. 2001. Sex-sorting mammalian sperm: concept to application in animals. *J. Androl.* 22:519–526.

35. Garner, D.L., Dobrinsky, J.R., Welch, G.R., Johnson, L.A. 1996. Porcine sperm viability, oocyte fertilization and embryo development after staining sperm with SYBR-14. *Theriogenology* 45:1103–1113.

36. Garner, D.L., Gledhill, B.L., Pinkel, D., Lake, S., Stephenson, D., van dilla, M.A., Johnson, L.A. 1983. Quantification of the X- and Y-chromosome-bearing spermatozoa of domestic animals by flow cytometry. *Biol. Reprod.* 28:312–321.

37. Garner, D.L., Hurtado, M. 2000. Assessment of mitochondrial function in cryopreserved stallion sperm. 14th International Congress on Animal Reprod., Stockholm, Sweden. Abstracts 2:309

38. Garner, D.L., Johnson, L.A. 1995. Viability assessment of mammalian sperm using SYBR-14 and propidium iodide. *Biol. Reprod.* 53:276–284.

39. Garner, D.L., Johnson, L.A., Allen, C.H. 1988. Fluorometric evaluation of cryopreserved bovine spermatozoa extended in egg yolk and milk. *Theriogenology* 30:369–378.

40. Garner, D.L., Johnson, L.A., Yue, S.T., Roth, B.L., Haugland, R.P. 1994. Dual DNA staining assessment of bovine sperm viability using SYBR-14 and propidium iodide. *J. Androl.* 15:620–629.

41. Garner, D.L., Pinkel, D., Johnson, L.A., Pace, M.M. 1986. Assessment of spermatozoal function using dual fluorescent staining and flow cytometric analyses. *Biol. Reprod.* 34:127–138.

42. Garner, D., Schenk, J., Seidel, G. Jr. 2001. Chromatin stability in sex-sorted sperm. Pages 3–7 in *Andrology in the 21st Century, Proceedings of the VIIth International Congress of Andrology Montreal, Canada*. Medimond Medical Publications, Englewood, NJ.

43. Garner, D.L., Seidel, G.E., Jr. 2002. Past, present, and future perspectives in sexing sperm. CSAS-Symposium-SCSA, Québec 2002. *Amino Acids: Meat, Milk and More/Improving Animal Production with Reproductive Physiology*. Organization Committee for Congress CSAS 2002, Bibilothèque Nationals du Canada, 2002, pp. 67–78.

44. Garner, D.L., Suh, T.K. 2002. Effect of Hoechst 33342 staining and laser illumination on viability of sex-sorted bovine sperm. *Theriogenology* 57:746.

45. Garner, D.L., Thomas, C.A. 1999. Quantification of relative mitochondrial membrane potentials in bovine spermatozoa using the fluorescent probe, JC-1. *Mol. Reprod. Dev.* 53:222–229.

46. Garner, D.L., Thomas, C.A., Gravance, C.G. 1999. The effect of glycerol on the viability,

mitochondrial function and acrosomal integrity of bovine spermatozoa. *Reprod. Dom. Anim.* 34:399–404.

47. Garner, D.L., Thomas, C.A., Gravance, C.G., Marshall, C.E., DeJarnette, J.M., Allen, C.H. 2001. Seminal plasma addition attenuates the dilution effect in bovine sperm. *Theriogenology* 56:31–40.
48. Garner, D.L., Thomas, C.A., Joerg, H.W., DeJarnette, J.M., Marshall, C.E. 1997. Fluorometric assessments of mitochondrial function and viability in cryopreserved bovine sperm. *Biol. Reprod.* 57:1401–1406.
49. Gledhill, B.L., Lake, S., Steinmetz, L.L., Gray, J.W., Crawford, J.W., Dean, P.N., and Van Dilla, M.A. 1976. Flow microfluorometric analysis of sperm DNA content: effect of cell shape on the fluorescence distribution. *J. Cell Physiol.* 87:367–376.
50. Graham, J.K. 1996. Analysis of stallion semen and it relation to fertility. Pages 119–130 in E.L. Squires (ed.), *Diagnostic Techniques and Assisted Reproductive Technology.* W.B. Saunders, Philadelphia.
51. Goh, M., Chen, F., Paulsen, M.T., Yeager, A.M., Dyer, E.S., Ljungman, M. 2001. Phenylbutyrate alternates the expression of Bcl-X(L), DNA-PK, caveolin-1 and VEGF in prostate cancer cells. *Neoplasia* 3:331–338.
52. Gravance, C.G., Garner, D.L., Miller, M.G., Berger, T. 2001. Fluorescent probes and flow cytometry to assess rat sperm integrity and mitochondrial function. *Reprod. Toxicol.* 15:5–10.
53. Grundhoefer, D., Patterson, B.K. 2001. Determination of liquid-based cervical cytology specimen adequacy using cellular light scatter and flow cytometry. Cytometry 46:340–344.
54. Gupta, S., Afaq, F., Mukhtar, H. 2001. Selective growth-inhibitory, cell cycle deregulatory, and apoptotic response of apigenin in normal versus human prostate carcinoma cells. *Biochem. Biophys. Res. Commun.* 287:914–920.
55. Guthrie, H.D., Cooper, B.S., Welch, G.R., Zakaria, A.D., Johnson, L.A. 1995. Atresia in follicles grown after ovulation in the pig: measurement of increased apoptosis in granulosa cells and reduced follicular fluid estradiol-17β. *Biol. Reprod.* 52:920–927.
56. Guthrie, H.D., Grimes, R.W., Cooper, B.S., Hammond, J.M. 1995. Follicular atresia in pigs: measurement and physiology. *J. Animal Sci.* 73:2834–2844.
57. Guthrie, H.D., Johnson, L.A., Garrett, W.M., Welch, G.M., Dobrinsky, J.R. 2002. Flow cytometric sperm sorting: effects of varying laser power on embryo development in swine. *Mol. Reprod. Dev.* 61:87–92.
58. Harrison, R.A.P., Gadella, B.M. 1995. Membrane changes during capacitation with special reference to lipid architecture. Pages 45–65 in P. Fénichel and J. Parinaud (eds.), *The Human Sperm Acrosome Reaction.* Coloque INSERM/John Libbey Eurotext Ltd., Montrouge, France.
59. Haugland, R.P. 1996. *Handbook of Fluorescent Probes*, 6th ed., 8.1 Nuclei Acid Stains. Molecular Probes, Eugene, OR, p.150
60. Hellstrom, W.J.G., Deitch, A.D., deVere, White, R.W. 1989. Evaluation of vasovasostomy candidates by deoxyribonucleic acid flow cytometry of testicular aspirates. *Fertil. Steril.* 51:54336–54548.
61. Henely, N., Baron, C., Roberts, K.D. 1994. Flow cytometric evaluation of the acrosome reaction of human spermatozoa: a new method using photoactivated supravital stain. Int. J. Androl. 17:78–84.
62. Higuchi, K.H., Nakano, T., Tsuboi, A., Suzuki, Y., Ohno, T., Oka, K. 2001. Flow cytometric and Ki-67 immunohistochemical analysis of cell cycle distribution of cervical cancer during radiation therapy. *Anticancer Res.* 21:2511–2518.
63. Jansen, M.W., Korver-Hakkennes, K., van Leenen, D, Grandenburg, H., Wildschut, H.I. Wladimiroff, J.W., Ploemacher, R.E. 2001. How useful is the in vitro expansion of fetal CD34+ progenitor cells from maternal blood samples for diagnostic purposes. *Prenat. Diagn.* 20:725–731.
64. Johnson, K.P., Rowe, G.C., Jackson, B.A., D'Agustino, J.L., Campbell, P.E., Guillory, B.O., Williams, M.V., Matthews, Q.L., McKay, J., Charles, G.M., Verret, C.R., Deleon, M., Johnson, D.E., Cooke, D.B. 2001. Novel antineoplastic isochalcones inhibit the expression of cyclooxygenase 1,2 and EGF in human prostate cancer cell line LNCaP. *Cell Mol. Biol.* 47:1039–1045.

65. Johnson, L.A. 1992. Gender preselection in domestic animals using flow cytometrically sorted sperm. *J. Animal Sci.* 70(Suppl 2):8–18.
66. Johnson, L.A. 2000. Sexing mammalian sperm for production of offspring: the state of the art. *Anim. Reprod. Sci.* 60–61:93–107.
67. Johnson, L.A., Flook, J.P., Hawk, H.W. 1989. Sex preselection in rabbits: live births from X and Y sperm separated by DNA and cell sorting. *Biol. Reprod.* 41:199–203.
68. Karabinus, D., Volgler, C.J., Saacke, R.G., Evenson, D.P. 1997. Chromatin changes in bovine sperm after scrotal insulation of Holstein bulls. *J. Androl.* 18:549–555.
69. Kiefer, J.A., Farach-Carson, M.C. 2001. Type I collagen-mediated proliferation of PC3 prostate carcinoma cell line: implications for enhanced growth in the bone microenvironment. *Matrix Biol.* 20:429–437.
70. Kobayashi, T., Nakata, T., Kuzumaki, T. 2002. Effect of flavonoids on cell cycle progression in prostate cancer cells. *Cancer Lett.* 176:17–23.
71. Kramer, R.Y., Garner, D.L., Ericsson, S.A., Redelman, D., Downing, T.W. 1991. The effect of cottonseed components on testicular development in pubescent rams. *Vet. Hum. Toxicol.* 33:11–19.
72. Krishan, A., Wen, J., Thomas, R.A., Sridhar, K.S., Smith, W.I. Jr. 2001. NASA/American Cancer Society High Resolution flow cytometry project—III. Multiparameter analysis of DNA content and electronic nuclear volume in human solid tumors. *Cytometry* 43:16–22.
73. Kumar, A.P., Garcia, G.E., Slaga, T.J. 2001. 2-methoxyestradiol blocks cell-cycle progression at G(2)/M phase and inhibits growth of human prostate cancer cells. *Mol. Carcinogen.* 31:111–124.
74. Lee, Y.G., Korenchuk, S., Lehr, J., Whitney, S., Vessela, R., Pienta, K.J. 2001. Establishment and characterization of a new human prostatic cancer cell line: DuCaP. *In Vivo* 15:157–162.
75. Libbus, G.L, Perreault, S.D., Johnson, L.A., Pinkel, D. 1987. Incidence of chromosome aberrations in mammalian sperm stained with Hoechst 33342 and UV-laser irradiated during flow sorting. *Mutat. Res.* 182:265–274.
76. Ling, Y.H., Donato, N.J., Perez-Soler, R. 2001. Sensitivity to topoisomerase I inhibitors and cisplatin is associated with epidermal growth factor receptor expression in human cervical squamous carcinoma ME 180 sublines. *Cancer Chemother. Pharmacol.* 47:473–480.
77. Lu, S., Epner, D.E. 2000. Molecular mechanisms of cell cycle block by methionine restriction in human prostate cancer cells. *Nutr. Cancer* 38:123–130.
78. Mahmoud, A.I., Parrish, J.J. 1996. Oviduct fluid and heparin induce similar surface changes in bovine sperm during capacitation: a flow cytometric study using lectins. *Mol. Reprod. Dev.* 43:554–560.
79. Masumoto, N., Tasaka, K., Kasahara, K., Miyake, A., Tanizawa, O. 1991. Purification of gonadotropes and intracellular calcium oscillation. Effects of gonadotropin-releasing hormone and interleukin 6. *J. Biol. Chem.* 266:6485–6488.
80. Maxwell, W.M.C., Long, C.R., Johnson, L.A., Dobrinsky, J.R., Welch, G.R. 1998. The relationship between membrane status and fertility of boar spermatozoa after flow cytometric sorting in the presence or absence of seminal plasma. *Reprod. Fertil. Dev.* 10:433–440.
81. McFarland, M., Holcombe, D.W., Redelman, D., Garner, D.L., Allen, J.R., Surian, M.E, King, D.J. 2000. Quantification of subclinical mastitis using flow cytometry in sheep. In: *Proceedings of the Western Section American Society of Animal Science*, Davis, CA, 15:380–384. American Society of Animal Science, Savoy, IL.
82. Meistrich, M.L. 1996. Possible genetic risks of sperm sorting. *Cytometry* Suppl. 8, Abst. S113, p. 134.
83. Meistrich, M.L., Lake, S., Steinmetz, L.L., Gledhill, B.L. 1978. Flow cytometry of DNA in mouse sperm and testis nuclei. *Mutat. Res.* 49:383–396.
84. Meyers, S.A., Overstreet, J.W., Liu, I.K.M., Drobnis, E. 1995. Capacitation in vitro of stallion spermatozoa: comparison of progesterone-induced acrosome reactions in fertile and subfertile males. *J. Androl.* 16:47–54.
85. Mixner, J.P., Saroff, J. 1954. Interference by glycerol with differential staining of bull spermatozoa. *J. Dairy Sci.* 37:652.

86. Morrell, J.M., Dresser, D.W. 1989. Offspring from inseminations with mammalian sperm stained with Hoechst 33342, either with or without flow cytometry. *Mutat. Res.* 224:177–183.
87. Otto, F.J., Hacker, U., Zante, J., Schumann, J., Göhde, W., Miestrich, M.L. 1979. Flow cytometry of human spermatozoa. *Histochem.* 61:249–254.
88. Ozen, M., Giri, D., Ropiquet, F., Mansakhani, A. Ittmann, M. 2001. Role of fibroblast growth factor receptor signaling in prostate cancer cell survival. *J. Natl. Cancer Inst.* 93:1783–1790.
89. Palencia, D.D., Garner, D.L., Hudig, D., Holcombe, D.W., Burner, C.A., Redelman, D., Fernandez, G.C.J., Abuelyaman, A.S., Kam, C.M., Powers, J.C. 1996. Determination of activable proacrosin using an irreversible isocoumarin serine protease inhibitor. *Biol. Reprod.* 55:536–542.
90. Papaioannou, K.Z., Murphy, R.P., Monks, R.S., Hynes, N., Ryan, M.P., Boland, M.P., Roche, J.F. 1997. Assessment of viability and mitochondrial function of equine spermatozoa using double staining and flow cytometry. *Theriogenology* 48:299–312.
91. Peltier, M.R., Liu, W.J., Hansen, P.J. 2000. Regulation of lymphocyte proliferation by uterine serpin: interleukin-2 mRNA production, CD25 expression and responsiveness to interleukin-2. *Proc. Soc. Exp. Biol. Med.* 223:75–81.
92. Peña, A.I., Quintela, L.A., Herradón, P.G. 1998. Viability assessment of dog spermatozoa using flow cytometry. *Theriogenology* 50:1211–1220.
93. Pinkel, D., Lake, S., Gledhill, B.L., Van Dilla, M.A., Stephenson D., Watchmaker, G. 1982. High resolution DNA content measurements of mammalian sperm. *Cytometry* 3:1–9.
94. Pinkel, D., Garner, D.L., Gledhill, B.L., Lake, S., Stephenson, D., Johnson, L.A. 1985. Flow cytometric determination of the proportions of X- and Y-chromosome-bearing sperm in samples of purportedly separated bull sperm. *J. Animal Sci.* 60:1303–1307.
95. Prakash, M., Kapembwa, M.S., Gotoh, F., Patterson, S. 2001. Higher levels of activation markers and chemokine receptors on T lymphocytes in the cervix than peripheral blood of normal healthy women. J. Reprod. Immunol. 52:101–111.
96. Rachmani, M., Deutsch, M., Ron, I., Tirosh, R., Weinreb, A., Chaitchik, S., Lalchuk, S. 1996. Efficacy of the cellscan in breast cancer detection in a hospital environment. *Cytometry*, Suppl. 8, Abst. AC 55, p. 109.
97. Reader, S., Menard, S., Filion, B., Filion, M., Phillips, N.C. 2001. Pro-apoptotic and immunomodulatory activity of a mycobacterial cell wall DNA complex LNCaP prostate cancer cells. *Prostate* 49:155–165.
98. Redelman, D., Butler, S., Robinson, J.D., Garner, D.L. 1988. Identification of inflammatory cells in bovine milk by flow cytometry. *Cytometry* 9:463–468.
99. Redkar, A., Krishan, A. 1999. Flow cytometric analysis of estrogen, progesterone receptor expression with DNA content of formalin fixed, paraffin embedded human breast tumors. *Cytometry* 38:61–69.
100. Reers, M.K., Smith, T.W., Chen, L.B. 1991. J-aggregate formation of a carbocyanine as a quantitative fluorescent indicator of membrane potential. *Biochemistry* 30:4480–4486.
101. Rens, W., Welch, G.R., Houck, D.W., van Oven, C.H., Johnson, L.A. 1996. Slit-scan flow cytometry for consistent high resolution DNA analysis of X- and Y-chromosome bearing sperm. *Cytometry* 25:191–199.
102. Rouas-Freiss, N., Gonçalves, R.M., Menier, C., Dausset, J., Carosella, E.D. 1997. Direct evidence to support the role of HLA-G in protecting the fetus from maternal uterine natural killer cytolysis. *Proc. Natl. Acad. Sci. USA* 94:11520–11525.
103. Seidel, G.E. Jr., Garner, D.L. 2002. Sexing mammalian sperm. *Reproduction* 124:733–743.
104. Seidel, G.E. Jr., Schenk, JL., Herickhoff, L.A, Doyle, S.P., Brink, Z., Green, R.D. 1999. Insemination of heifers with sexed sperm. *Theriogenology* 52:1407–1420.
105. Sheridan, M.T., West, C.M. 2001. Ability to undergo apoptosis does not correlate with the intrinsic radiosensitivity (SF2) of human cervix tumor cell lines. *Int. J. Radiat. Oncol. Biol. Phys.* 50:503–509.
106. Slott, V.L., Suarez, J.D., Poss, P.M., Linder, R.R., Strader, L.F., Perreault, S.D. 1993. Optimization of the Hamilton-Thorn computerized sperm motility analysis system for use with rat spermatozoa in toxicological studies. *Fund. Appl. Toxicol.* 21:298–307.

107. Suter, L., Koch, E., Bechter, R., Bobadilla, M. 1997. Three-parameter flow cytometric analysis of rat spermatogenesis. *Cytometry* 27:161–168.
108. Suzuki, F., Yanagimachi, R. 1989. Changes in the distribution of intramembranous particles and filipin-reactive membrane sterol during in vitro capacitation of golden hamster spermatozoa. *Gamete Res.* 23:335–347.
109. Takacs, T. Szollosi, J., Balazs, M., Gaspac, R., Matyus, C., Szabo, G., Tron, L., Resli, I., Damjanovich, S. 1987. Flow cytometric determination of sperm cell number in diluted bull semen by DNA staining method. *Acta Biochem. Biophys. Hung.* 22:45–57.
110. Tekur, S., Ho, S.M. 2002. Ribozyme-mediated down regulation of human metallothionein IIa induces apoptosis in human prostate and ovarian cancer cell lines. *Mol. Carcinog.* 33:44–55.
111. Terasawa, H., Tsang, K.Y., Gulley, J., Arlen, P., Schlom, J. 2002. Identification and characterization of a human agonist cytotoxic T-lymphocyte epitope of human prostate-specific antigen. *Clin. Cancer Res.* 8:41–53.
112. Thibodeaux, J.K., Myers, M.W., Goodeaux, L.L., Menezo, Y., Roussel, J.D., Broussard, J.R., Godke, R.A. 1992. Evaluating an in vitro culture system of bovine uterine and oviduct epithelial cells for subsequent embryo co-culture. *Reprod. Fertil. Dev.* 4:573–583.
113. Thomas, C.A., Garner, D.L., DeJarnette, J.M., Marshall, C.E. 1998. Effect of cryopreservation on bovine sperm organelle function and viability as determined by flow cytometry. *Biol. Reprod.* 58:786–793.
114. Tiemann, U., Viergutz, T., Jonos, L. Wollenhaupt, K., Pohland, R., Kanitz, W. 2001. Fluorometric detection of platelet activating factor receptor in cultured oviductal epithelial and stromal cells and endometrial stromal cells from bovine at different stags of the estrous cycle and early pregnancy. *Domest. Anim. Endocrinol.* 20:149–164.
115. Toppari, J., Eerola, E., Parvinen, M. 1985. Flow cytometric DNA analysis of defined stages of rat seminiferous epithelial cycle during in vitro differentiation. *J. Androl.* 6:325–333.
116. van Dilla, M.A., Gledhill, B.L., Lake, S., Dean, P.N., Gray, J.W., Kachel, V., Barlogie, B., Göhde, W. 1977. Measurement of mammalian sperm deoxyribonucleic acid by flow cytometry. *Cytometry* 25:763–773.
117. van dilla, M.A., Steinmetz, L.L., Davis, D.T., Calvert, R., Gray, J.W. 1974. High speed cell analysis and sorting systems: biological applications and new approaches. *IEEE Trans. Nucl. Sci.* NS-21:714–720.
118. Varner, D.D., Ward, C.R., Storey, B.T., Kenney, R.M. 1987. Induction and characterization of the acrosome reaction in equine sperm. *Am. J. Vet. Res.* 48:1383–1389.
119. Wachtel, S.S., Shulman, L.P., Sammons, D. 2001. Fetal cells in maternal blood. *Clin. Genet.* 59:74–79.
120. Wartenberg, M., Fisher, K., Heschler, J., Sauer, H. 2002. Modulation of intrinsic P-glycoprotein expression in multicellular prostate tumor spheroids by cell cycle inhibitors. *Biochim. Biophys. Acta* 1589:49–62.
121. World Health Organization. 1992. *Laboratory Manual for the Examination of Human Semen and Semen-Cervical Mucus Interaction.* 4th ed. Cambridge University Press, New York.
122. Yamamoto, T., Mori, S., Yoneyama, M., Imanishi, M., Takeuchi, M. 1998. Evaluation of rat sperm by flow cytometry: simultaneous analysis of sperm count and sperm viability. *J. Toxicol. Sci.* 23:373–378.
123. Zorzi, W., Thellin, O., Coumans, B., Melot, F., Hennen, G., Lakaye, B., Igout, A., Heinen, E. 1998. Demonstration of the expression of CD95 ligand transcript and protein in human placenta. *Placenta* 19:269–277.

14

Genetic Analysis Using Microsphere Arrays

JOHN P. NOLAN, ALINA DESPHANDE, AND FENG ZHOU

Introduction

The use of DNA microarrays to analyze simultaneously many genetic features from a sample is providing a new perspective on the effect of the genome on normal biological processes and disease states. The translation of this new perspective into new diagnostics and treatments will require development of microarray technology to enable cost-effective, high-throughput analysis of many samples. Microsphere arrays—sets of optically encoded microparticles—measured by flow cytometry are an experimentally flexible alternative to the flat-surface microarrays that are being used in a variety of biomedical analyses. In this chapter, we review the major areas where microsphere arrays are being used for genetic analysis and discuss the key experimental considerations that enable these and future applications.

The human genome project and other genome projects have opened great new possibilities for biomedical research. Along with these new possibilities have come new technical challenges involving the conversion of raw DNA sequence data into useful information for basic research, diagnostic, and therapeutic applications. Applications include the discovery of new associations between diseases and specific genes, the evaluation of genetic predisposition to disease, and the prediction of the response of individual patients to drugs. In the areas of public health, addressing issues such as infectious disease, food safety, and bioterrorism can all be enhanced through molecular analysis. Perhaps the most daunting challenge involves the specific analysis of hundreds or thousands of genetic features in hundreds or thousands of individual samples.

To address these challenges, a variety of analytical methods and platforms are being developed, ranging from laboratory methods such as mass spectrometry to portable microfluidic devices. Each of these assay platforms has characteristic features that determine sensitivity, throughput, and flexibility. Flow cytometry is among the most versatile of analytical platforms, providing sensitive and quantitative multiparameter fluorescence measurements of cells and microparticles with high analysis rates (25–27). These features underlie the development of several current and emerging genetic analysis tools.

Flow Cytometry and Cellular Genetic Analysis

One of the earliest flow cytometric measurements was a genetic analysis of a sort—the measurement of relative DNA content in individual cells using fluorescent DNA binding dyes (39). Though it gives little to no resolution at the sequence level, this approach has been critical to our understanding of the large-scale mechanics of the genome. Metabolic incorporation of nucleotide analogues gives information about the rate of DNA synthesis (14), and enzymatic labeling of the genome reports strand breaks (12), but again at a very coarse level. The analysis of the finer-scale structure of the genome to the gene and sequence level requires more specificity.

This specificity can be provided by synthetic oligonucleotide probes. When designed and labeled appropriately, oligonucleotides can be used to detect short stretches of sequence in fixed or permeabilized cells via in situ hybridization, whereas for low-copy sequence targets, polymerase chain reaction (PCR) can provide the necessary amplification needed for detection. These approaches have been used to identify classes of bacteria in complex mixtures (1, 49) and to detect viral DNA (4, 9) and specific mRNA (7, 30, 36, 38, 43, 47) in individual cells. As will be described, cell-based methods for genetic analysis and screening have great potential in medical diagnostics and basic research. However, there is another growing class of genetic analysis applications that are not cell based but, rather, employ microspheres as solid supports for the analysis of genes, RNA, and single nucleotides.

Flow Cytometry and Microsphere Array–Based Genetic Analysis

The notion of using microsphere as solid supports for molecular analysis has been in the flow cytometry literature since the 1970s (15) and has been developed through the 1980s and 1990s especially for use with immunoassays (24, 41). In the mid-1980s, Saunders et al. used DNA-bearing microspheres to measure drug binding (31), and by the early 1990s, microspheres had been used to detect fluorescent PCR products (42, 44). With the advent of optically encoded microsphere arrays (11, 21) to support highly multiplexed analysis, and the progress of various genome sequencing projects, new bead-based assay chemistries have been developed to support a variety of genomic applications including gene expression analysis, single-nucleotide polymorphism (SNP) genotyping, and detection and analysis of pathogenic and infectious agents. The highly parallel analysis of many genetic features simultaneously using microarray technology is likely to be a major tool for both basic and clinical research in the coming years, and microsphere arrays and flow cytometry will likely be a preferred format for many microarray applications.

Overview

Microarrays: From Flat to Multidimensional

The highly parallel analysis of many genetic features using microarray approaches has emerged as a central tool in a number of emerging applications that exploit genomic sequence information. Gene expression analysis on cDNA or oligonucleotide microarrays represents a first cut at understanding the integrated network of pathways and regulatory elements within the cell, while large scale analysis of genetic variation, especially of SNPs, holds the potential for better understanding disease and susceptibility and to improve diagnostics and treatment. Over the past decade, two main microarray formats have emerged: the high-density, slide-based arrays created by mechanical spotting of material (10) and the even higher density Affymetrix-type arrays created using photolithography (22). As these technologies continue to evolve, a third type of microarray is coming into use—one composed of optically encoded microspheres that are analyzed by flow cytometry.

Microsphere arrays comprise preparations of microspheres exhibiting discrete intensities of one or more optical properties, such as fluorescence and light scatter (figure 14.1). Originally proposed and demonstrated using a few different-sized beads (15, 24), the current generation of microspheres uses one or two fluorescent probes exhibiting as many as 10 discrete intensities (21), creating arrays ranging from a dozen to a hundred array elements. In principle, the use of more dyes increases the multiplexing exponentially, and a theoretical array composed of beads bearing 10 discrete intensities for each of six different optical parameters would contain one million array elements. In practice, many important questions can be effectively addressed with ar-

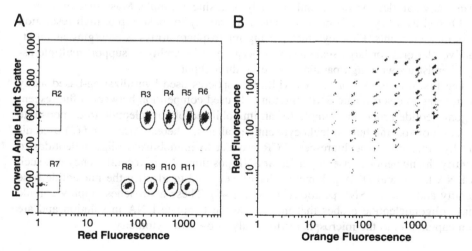

Figure 14.1. Microsphere arrays analyzed by flow cytometry. Flow cytometry histograms of microsphere arrays employing (A) five discrete intensities of one fluorophore plus two discrete light scatter intensities from two different sized beads and (B) eight discrete intensities of two colors of fluorescence for encoding.

rays of one hundred, or even 10, array elements. Originally demonstrated for the analysis of immunoassays, microsphere arrays can potentially exploit all of the bead-based assays used for molecular assembly described elsewhere in this volume (see chapter 9), with nucleic acid hybridization being the critical molecular interaction for genomic analysis. In this chapter, we highlight some of the emerging applications areas in which microsphere arrays are being used, as well as some of the key experimental concepts that underlie these applications.

Genetic Variation and SNPs

The Human Genome Project presents great possibilities for better understanding of the genetics of human health and disease, but it also presents great analytical challenges. In particular, to apply all of our knowledge of the genome to important biomedical problems, the scale of analyses must be very large. A case in point is the analysis of individual genetic variation (32). The most common form of genetic variation is the SNP, a nucleotide position in genomic DNA in which the nucleotide base varies within a population (5). In humans, it is estimated that between any two genomes, 1 in 1000 bases will be a SNP. Many of these SNPs will have medical applications ranging from identification of potential drug targets to diagnosis of disease susceptibilities and targeting of therapies. The analysis of millions of SNPs in millions of patients presents a daunting challenge, with new types of considerations, and requiring the development of new approaches.

Most methods to score SNPs can be broken down into two parts: the assay chemistry and the analysis platform. In general, several assay chemistries can be adapted to any particular platform. The traditional analysis platform for genetic research is gel electrophoresis, and most of the human genome was sequenced using this approach. In recent years, flat DNA microarrays have attracted attention for their ability to perform many parallel analyses simultaneously on a single sample. Mass spectrometry is established as a key platform as a result of its ability to make rapid, high-resolution mass measurements. Flow cytometry using microsphere arrays is emerging as an attractive platform for large-scale analysis because of its ability to support multiple assay chemistries with high parallel and serial throughput.

The first efforts to score SNPs with microspheres used hybridization-based assay chemistries to discriminate SNPs. Fulton (11) used competition between a fluorescent oligonucleotide probe and sample for an immobilized oligonucleotide to demonstrate detection of specific human leukocyte antigen alleles, whereas Armstrong (2) used direct hybridization of a fluorescent PCR product to immobilized oligonucleotides to identify the nucleotide base at particular SNP positions. More recently, enzymes such as DNA ligase and DNA polymerase have been used to identify the nucleotide base identity and specific SNP positions (6, 8, 17, 46). These approaches employ soluble phase oligonucleotide probes that interrogate the template DNA in solution and are then captured onto the microspheres for analysis by flow cytometry.

Gene Expression Analysis

Another area of genomic analysis that has attracted a great deal of effort is the measurement of gene expression through the detection of mRNA. Such efforts are aimed at understanding the behavior of regulatory networks in the cell, identifying potential drug targets, and developing diagnostic tools. The simultaneous measurement of changes in the relative abundance of many specific mRNAs has come into wide use recently, typically using flat microarrays. In this case (10), the microarray surface is patterned with hundreds to thousands of DNA probes. Sample mRNA serves as the template, and a variety of assay chemistries are used to generate labeled reporter molecules that report on individual message levels when hybridized to the microarray surface. The current generation of DNA microarray technology represents a powerful tool, but improvements in throughput, quantification, and ease of use are still needed to realize the potential of gene expression analysis. Flow cytometry–based analyses are addressing several of these issues.

For example, in a typical analysis of gene expression, the relative abundance of hundreds or thousands of different mRNA transcripts is compared in a limited number of samples; for example, diseased tissue versus healthy tissue or cells treated with some drug or compound versus control cells. From such a comparison, dozens or hundreds of potentially interesting genes may be identified. To prioritize which of these drugs are key regulators or high-value drug targets, considerably more work is required. Biological conclusions typically will require analysis of complete time courses and dose responses, on duplicate or triplicate samples, with a full set of control experiments. This will require hundreds of analyses, taxing the flexibility and throughput of flat microarrays. For this type of analysis, microsphere arrays offer an attractive alternative to flat microarrays. By immobilizing oligonucleotide probes on optically encoded microspheres and using these probes to capture labeled cDNA from samples, Yang et al. (45) have configured a gene expression microsphere array that offers comparable sensitivity to the flat array methods, with several advantages in speed and flexibility. These advantages will become important as large-scale gene expression analysis becomes a routine tool in biomedical research.

Pathogen Detection and Identification

Another area in which microsphere-based flow cytometry is emerging as an important new tool is in the area of bacterial detection and forensics. The analytical needs can vary depending on the application. For example, field or clinic use may require the ability to identify a specific species of organism rapidly, whereas identification of a particular strain or drug sensitivity might be performed in a clinical laboratory setting. The versatility of the microsphere array platform makes it suitable for a range of applications and analyses. Although antibody-based methods are very widely used for detection of bacterial and viral pathogens, nucleic acid–based assays are increasing in popularity because of the sensitivity and specificity they offer.

The objective of a nucleic acid–based pathogen assay is to detect specific genetic features that serve as signatures for a particular organism or phenotype. A variety of approaches have been developed for genetic analysis, and they vary according to the nature and number of genetic features analyzed. These methods also vary in their ease of use, flexibility, scalability, and instrumentation requirements, and these factors will affect their suitability for various applications.

Conventional nucleic acid–based analysis aims to detect specific genetic features that serve as signatures for a particular organism. Methods involving the analysis of DNA fragment patterns have been used to great advantage in the characterization and identification of human pathogens. Restriction-enzyme-based methods such as restriction fragment–length polymorphism and amplified restriction fragment–length polymorphisms (18) reveal genetic variation without direct sequencing. Often, these tools are used in conjunction with fingerprint databases to match unknowns with previously characterized samples. As described in chapter 7, high-sensitivity flow cytometry offers significant advantages in speed and sample size compared with gel electrophoresis–based methods for DNA fragment size analysis. Fragment length analysis can be very valuable for the characterization of previously uncharacterized organisms, as these methods do not generally require prior knowledge of the DNA sequence. However, these methods require a fairly high level of technical expertise, and the "fingerprints" that are developed are platform specific and are not readily transferable to other detection platforms. For this reason, these approaches are most valuable when characterizing samples for which sequence data is not available.

When DNA sequence is available for a target organism, a variety of methods that use specific oligonucleotide probes can be employed. PCR is very widely used, providing assay sensitivity resulting from exponential amplification and specificity through the use of specific primer pairs. Conventionally, PCR products are analyzed by electrophoresis and identified by DNA fragment length. This type of approach is used to analyze variable nucleotide tandem repeats (20), genetic sequence characters that can be used to infer phylogenetic relationships between isolates. Often, a PCR product is interrogated using a hybridization probe, as in 5-nuclease or TaqMan assays (33). The addition of a hybridization probe to the two PCR primers creates a DNA "signature" with even more specificity.

Several methods have been developed to adapt hybridization-based assays to microsphere arrays. One approach uses an immobilized hybridization probe to capture a labeled PCR product onto microspheres. In this approach (figure 14.2), PCR is performed with a fluorescently labeled primer to generate a double-stranded DNA target, which is then digested with an exonuclease to generate single-stranded DNA, followed by hybridization to the immobilized primer. This approach has been used to detect 16S rDNA sequences in PCR products from bacteria in groundwater (34, 35). Another approach uses polymerase-mediated probe extension in solution and capture onto universal arrays in a manner similar to the SNP genotyping methods discussed above. In this method, following PCR, capture-tagged hybridization probes are added that anneal to the PCR product. DNA polymerase then extends the hybridized probes with labeled nucleotide analogues (28). This enzymatic extension of hybridized probes can be performed with a thermocycler, allowing amplification through multiple rounds of annealing and extension. The labeled hybridization probes are then captured onto microspheres for analysis by flow cytometry.

Although the hybridization assays described above are useful for detecting target-specific PCR amplicons, as a result of the increase in the amount of microbial DNA sequence in databases it is reasonable to consider a new type of DNA signature, not based on indirect measures of genetic sequence variation (fragment length–based approaches) or on the presence or absence of whole PCR amplicons but using the DNA sequence itself as the signature. The increasing understanding of the nature of genetic variation at the single nucleotide level, and the proliferation of methods to assess SNPs, makes it possible to envision DNA signatures that consist of collections of SNPs that would be completely portable between labs and among assay platforms. Being platform independent, these sequence-based signatures would be assessable by any num-

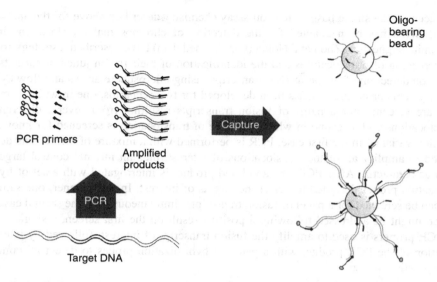

Figure 14.2. Direct hybridization of a labeled target molecule schematic diagram of nucleic acid target detection by direct hybridization to immobilized oligonucleotides. In a typical application, a specific target sequence is amplified by PCR using a labeled primer, which is incorporated into the PCR product. The labeled PCR product is then captured by specific hybridization to the immobilized oligonucleotide.

ber of current and yet-to-be-developed sequencing methods, avoiding signature obsolescence. The hybridization-based binding assays have been adapted to microspheres for rapid analysis by flow cytometry (2). More recently, allele-specific chain extension has been adapted to detection platforms that are more compatible with large-scale analysis, including microsphere arrays (46). Multiplex minisequencing has been successfully configured for microsphere arrays and flow cytometry (6, 8).

Diagnostic Assays: Chromosomal Translocations in Leukemia

Over the last decade, the role of chromosomal abnormalities, particularly chromosomal translocations, in leukemia has become increasingly apparent, as reflected in the World Health Organization's recent efforts to classify leukemia according to translocations and other genetic abnormalities (19). Not only will this approach significantly enhance diagnosis of disease, but information about the genetic lesions that underlie leukemia can have prognostic value and direct treatment strategies (23).

Traditionally, chromosomal abnormalities are detected by microscopy-based cytogenetic approaches such as fluorescent in situ hybridization, a slow and labor-intensive process that can take several days. More recently, PCR-based methods, and especially reverse transcriptase–PCR, to detect the mRNA fusion products of chromosomal translocations have been introduced (29, 40). These methods are faster and more sensitive than cytogenetic methods, but they employ gel electrophoresis, which is difficult to standardize and automate.

Recently, the single base–extension assay chemistry described above for the analysis of SNPs has been adapted for the detection of chromosomal translocations in leukemia diagnostics. Zhou and Nolan (unpublished data) have described a strategy for the screening of translocations and the identification of their fusion junction variants, based on detection of fusion mRNA transcripts using microsphere arrays and flow cytometry. This basic strategy has been developed for two scenarios, one in which samples are screened for a panel of fusion transcripts resulting from several different translocations and a second in which one type of translocation is screened for known junction variants. In the first case, PCR is performed with a mixture of primer sets designed to amplify all of the translocations of interest, plus an internal control target such as beta-actin. After PCR, the amplified product is interrogated with a set of hybridization probes designed to detect the targets of interest. In this manner, one sample can be screened for a panel of fusion transcripts simultaneously. In the second case, which might be addressed following a positive result on the first screen, a single set of PCR primers is used to amplify the fusion transcript of interest, followed by interrogation of the PCR product with a panel of hybridization probes to detect all common junction variants.

Summary

The applications described above demonstrate the versatility of microsphere arrays for a variety of genetic analyses in environments ranging from the basic research lab to the clinic. Yet these represent only a small fraction of the potential uses of this platform. In the coming years, as the information contained in genomic DNA sequence is converted to knowledge about the biological mechanisms and the causes of disease, the efficiency and flexibility of microparticle array-based analysis is likely to find many more uses. In the next section, we examine some of the essential features of microsphere arrays relevant to genetic analysis.

Implementation

The advantages of highly parallel assays using microspheres as array elements, combined with the high analysis rate and homogeneous measurement capability of flow cytometry, make microsphere arrays extremely attractive for large-scale genetic analysis. As described above, microsphere arrays can serve as an analysis platform for a wide range of assay chemistries and applications. In this section, we use the applications described above to illustrate some of the key technical features of microsphere array–based genomic analysis.

Hybridization: Sequence Detection and Gene Expression Analysis

Nucleic acid hybridization is a fundamental feature of all genetic analysis. Nearly every assay chemistry used for genetic analysis (PCR, sequencing, Southern and Northern blots, microarray analysis, etc.) involves the interaction of synthetic oligonucleotide probes with nucleic acid targets. The implementation of hybridization in solid-phase assays, whether on flat surfaces or microspheres, can have a great effect on the performance of an assay.

To illustrate, we will use a very common and general type of application: the detection of a specific stretch of nucleic acid sequence. The sequence might be genomic DNA, mRNA or cDNA, or PCR-amplified fragments of DNA. This type of measurement underlies many analyses including gene expression analysis, detection of some types of nucleic acid signatures for pathogens, and many others. The most common implementation of this type of measurement involves gel electrophoresis to separate target nucleic acid by size, followed by hybridization of labeled probe to detect a specific string of DNA (Southern blot) or RNA (Northern blot) sequence. Although effective, this approach is slow, labor intensive, and not readily scaled up for the analysis of many sequences. To support large-scale analysis, solid-phase assays have been developed to facilitate automation and, through the use of microarrays, highly parallel or multiplexed analysis of many sequences simultaneously.

Hybridization on Surfaces

The simplest approach to detecting a DNA sequence using microspheres is to label it and capture it onto the microsphere surface (figure 14.2). The target can be labeled by using labeled PCR primer during amplification or with a labeled hybridization probe. Capture of the labeled target onto an oligonucleotide immobilized on the microsphere surface can then be measured by flow cytometry. However, this simplest approach has some practical complications.

For assay chemistries adapted to solid supports, whether flat surface or microparticles, several considerations come into play. First, the concentration dependence and kinetic aspects of the reagents and processes involved must be recognized. Almost every assay, whether nucleic acid based or involving antibodies or any other biomolecule, involves a diffusion-limited binding step. For genomic analysis, this typically involves hybridization of an oligonucleotide primer or probe to a template molecule. The rate of this reaction will depend on the diffusion rates of the two components as well as on their respective concentrations.

By immobilizing one of the components on a surface, its diffusion rate is greatly reduced. In the case of a flat surface (i.e., a DNA chip), the diffusion rate of the immobilized probe goes to zero. The situation improves when an oligonucleotide is bound to a microsphere, but the diffusion rate is still much slower than free DNA in solution. Thus, the kinetics of any hybridization reaction will be slower on a surface than in solution. Adding to the diffusion problem is the issue of concentration. When oligonucleotides are immobilized at reasonable densities (10^5–10^6 per bead) on microspheres at concentrations typically used for flow-based assays (10^5–10^7 beads per milliliter), the effective solution concentration will be 10 nM or lower. Similar calculations can be performed for flat surfaces (microarrays or microwells). Compare these concentrations with those of primers and probes typically used in solution assays (tens of nanomoles to micromolars), and one can see that the rate of the surface reaction would be much less than that of reactions occurring free in solution.

Slow reaction rates can generally be dealt with by increasing incubation times, but this is not always the case. Take, for example, the problem of annealing a hybridization probe to a PCR-amplified target molecule—a general step in almost every genomic analysis. The PCR product is double stranded, and thus for every molecule of target, there is a complementary strand competing with a probe for hybridization. In general, the hybridization probe is smaller and supplied at a high enough concentration that after heating the sample to denature the double-stranded target, the hybridization probe will out-compete the complementary strand for binding on renaturation. However, if the hybridization probe is immobilized and present at low concentration, capture of the long, double-stranded target to the surface will be inefficient compared to capture of short, single-stranded oligonucleotides.

Solution Reactions versus Surface Reactions

For these reasons, many genomic analyses that involve a surface-based measurement are configured so that assay chemistry is performed in solution, where all the kinetic and concentration advantages of freely diffusing molecules can be exploited and then captured onto a surface for measurement. The simplest version of this approach is the competitive hybridization assay (figure 14.3), in which target in solution and an immobilized target sequence compete for the labeled probe. In this case, an increase in target concentration results in a decrease in microsphere fluorescence.

A more sophisticated approach uses enzyme-based assay chemistries performed in solution, followed by capture of modified probes onto the microsphere. A key to this approach is the development of sets of compatible oligonucleotide tags used to address the microarray. These oligonucleotide tags, often referred to as address tags or "Zip codes," have been designed by several methods (3, 6, 17), but they all share the same

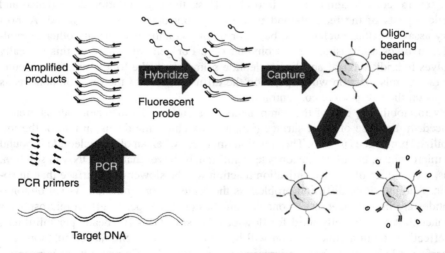

Figure 14.3. Competitive hybridization assay. An alternative to direct capture of a PCR product, the competitive hybridization assay format detects a specific nucleic acid sequence as the loss of fluorescence from a microsphere. A microsphere bearing an oligonucleotide with the sequence of the target and a labeled probe, complementary to the target sequence, is designed such that in the absence of target, the probe binds to the microsphere to generate a signal. If the target is present, the target sequence competes with the immobilized oligonucleotide for the labeled probe, and the signal from the microsphere is reduced.

objective of capturing only the primer of interest, and no other tagged primer. For use with microsphere arrays, address tags encode the microspheres, and primers have the reverse complement capture tag at their 5' ends. This enables the addressed-tagged microspheres to be used as a universal reagent for any number of different assays in which new sequences are targeted by designing new primers encoded with the appropriate capture tags. With this general background regarding microsphere-based analysis, we will discuss some specific assay chemistries for nucleic acid sequence detection and analysis that have been adapted to the microsphere array format.

Enzymatic Modification: SNP Genotyping and Sequence Detection

Beyond simple hybridization reactions, many genetic analysis applications employ enzymatic modification of synthetic oligonucleotides as a means to interrogate a target DNA. Nucleic acid polymerases and ligases perform their functions with sequence specificity, and this specificity, combined with the specificity provided by the hybridization of an oligonucleotide, can be used to detect sequences with the ability to resolve a single nucleotide-base difference. These approaches have been exploited in the development of several methods for SNP genotyping, an application that serves to illustrate several important features of nucleic acid detection.

Assays for SNP Genotyping

The simplest assay chemistry for scoring SNPs is hybridization based and is similar to the approach for detection described above. The melting temperature for hybridization of an oligonucleotide of modest length (15–20 mer) can differ by several degrees between a fully complementary template and a template that has a 1-base difference (i.e., a SNP). By carefully designing oligonucleotide probes and choosing hybridization conditions that allow the probe to bind to the fully complementary template, but not a template containing a SNP, it is possible to use hybridization to distinguish a single-base mismatch. This approach has been configured for microspheres in a competitive hybridization format (11) and in a direct hybridization format (2).

In practice, the hybridization approach generally requires careful optimization of hybridization conditions, because probe melting temperatures are very sensitive to sequence context as well as to the nature and position of the mismatched base. This difficulty in choosing optimum hybridization conditions (buffer and temperature) can be eliminated by continuously monitoring probe hybridization as the temperature is increased, allowing the melting temperature for matched and mismatched probes to be determined directly in a process known as dynamic allele-specific hybridization (16). The recent development of dynamic temperature-control capabilities for flow cytometry (13) makes this approach possible for microsphere-based hybridization assays, but the issues associated with hybridization of a large, double-stranded target to a surface-immobilized oligonucleotide remain.

An alternative, and generally more robust, set of assay chemistries for SNP scoring uses enzymes to reveal DNA sequence information. Single-base extension uses an oligonucleotide probe designed to anneal immediately adjacent to the SNP site on the template DNA (figure 14.4). In the presence of labeled dideoxynucleoside triphos-

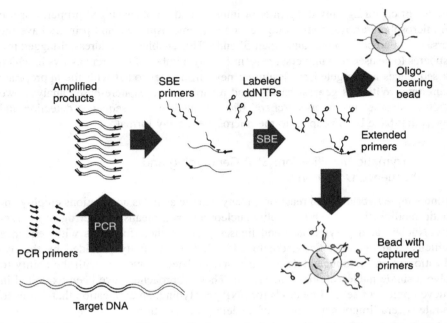

Figure 14.4. Enzymatic modification of primers: single-base extension. For solution-based sequence detection or analysis (single-nucleotide polymorphism) assays, enzymatic modification of oligonucleotide primers provides a robust assay chemistry. Capture-tagged primers in solution hybridize to the target sequence immediately adjacent to the site of interest, and DNA polymerase extends the primer by one labeled dideoxynucleotide analogue. Thermal cycling in this process allows each target molecule to be interrogated by many primers. After primer extension, the primers are captured onto the address-tagged microspheres and the labels measured by flow cytometry. Similar approaches have been developed for allele-specific chain extension and oligonucleotide ligation-based assays.

phates, DNA polymerase will extend the annealed primer by 1 base. The identity of the incorporated base reveals the base at the SNP site on the template. This is essentially the same dye terminator chemistry widely used in conventional sequencing applied to a single base, hence its nickname "minisequencing." There are several examples of the single-base extension approach being successfully adapted to microsphere arrays (6, 8, 45). In these applications, multiplexed single-base extension was performed in solution on several SNPs simultaneously, followed by capture of the labeled primers onto microspheres bearing unique address tags for analysis by flow cytometry.

A related assay chemistry that also employs DNA polymerase is allele-specific chain extension. In this approach, primers are designed such that the 3′ nucleotide pairs with the SNP site. If the base at the SNP site on the template DNA is complementary to the 3′ base of the allele-specific primer, then DNA polymerase will efficiently extend the primer. If the SNP base is not complementary, polymerase extension of the primer is inefficient, and under the appropriate conditions, no extension will occur. This approach has also been adapted to encoded microsphere arrays (46), and although it requires a different primer for each allele of a particular SNP, it does not require the removal of excess primers and unincorporated deoxynucleotide triphosphates from the PCR amplified template before analysis.

Another enzymatic assay chemistry that can be employed for SNP scoring uses DNA ligase to discriminate between matched and mismatched probe bound to a template DNA molecule, which serves as the basis for oligonucleotide ligation assays in which the ligation of two adjacent probes is detected (17). The ligation assay requires more primers to detect a SNP, which can increase the cost, but this assay format has the potential to detect small insertions and deletions as well as SNPs.

Recently, the 5′ endonuclease assay known as Invader has also been adapted to microspheres (37) and is compatible with flow cytometry. This assay takes advantage of a structure specific endonuclease, flap endonuclease-1, which recognizes and cleaves a specific DNA structure that can be formed when oligonucleotide probes are assembled on a DNA template. With careful design of the oligonucleotide probes, this cleavage can be made allele specific and thus be used to genotype SNPs. Probe design for this approach is critical, but it can enable the isothermal amplification of the signal and therefore provide sensitivity that should allow genotyping to be performed directly on a genomic DNA template without the need for PCR. It should be noted however, that PCR contributes more to an assay than sensitivity. By amplifying a specific segment of a genome, PCR greatly reduces the complexity of the template and significantly enhances the specificity of all of the assays that employ it.

Conclusions and Prospects

The development of microsphere array–based applications for flow cytometry represents a major expansion in the capabilities of this analytical platform that has the potential to affect many areas of biomedical research. In the area of genomic analysis and genetic testing specifically, it is clear that the general features of the flow cytometry–based microsphere approaches have several advantages over other platforms for genomic analysis that are worth restating. First, because flow cytometry provides intrinsic resolution between free and particle-bound fluorophore, samples can be analyzed without any separation or wash steps. Second, flow cytometry is a very sensitive method of fluorescence detection. Most commercial instruments can easily measure a few thousand fluorescent molecules per particle. In this assay, this sensitivity enables the analysis of DNA template at subnanomolar concentrations. Third, we gain efficiency by performing hybridization and primer extension in solution. Hybridization on a surface is much slower than hybridization in solution (48). By performing hybridization and extension in solution, followed by capture on microspheres for analysis, we further increase the assay sensitivity and speed. Finally, because flow cytometry is a multiparameter detection platform, it is possible to measure several features of a particle simultaneously. For example, it is possible to label each of the four dideoxynucleoside triphosphates with a different fluorophore, as is the case for dye-terminator sequencing, and to detect them simultaneously in a single reaction. These features should make flow cytometry of microsphere arrays a key measurement platform for genomic analysis in the coming years.

These microsphere-based approaches complement flow cytometric cellular analysis. Although cell-based assays have been developed to detect specific nucleic acids inside of single cells, which is very useful for assessing cell-to-cell variability in molecular features, these methods cannot assess multiple molecular targets, nor do they provide resolution that would enable the identification of single-base changes. Microsphere array–based methods can have single-base resolution and can assess dozens of targets simultaneously, but in general, they are most useful for reporting population averages from large numbers of cells. In principle, fluorescence-activated sorting of individual cells can be followed by single-cell PCR and application of microsphere array–based assays, but the development of improved in situ assay chemistries is the most likely route to high-throughput genetic and genomic analysis of individual cells.

The types of analysis described above are expected to have a broad affect from the research lab to the clinic. In the research lab, experimental flexibility is the key to decoding the genetic features that influence health and disease. Such efforts will involve the analysis of tens of thousands of mRNA transcripts or SNPs in thousands of samples. Once clinically important molecular markers are identified, disease-specific panels will be configured for standardized, routine testing in clinical laboratories. Tests are envisioned for disease susceptibility, disease diagnostics, and drug response. In other areas, similar tools will be used for the identification of pathogenic organisms for food safety and public health. Although there are many potential detection platforms available for such analyses, for applications in which large numbers of genetic features need to be identified in large numbers of samples, flow cytometry of microsphere arrays is a very attractive solution. Beyond genetics and genomics, the microsphere array–based platform will be extremely useful for the analysis of proteins and other molecular entities as well. Combined with the ability to make sensitive and quantitative measurements of molecular entities in living cells, microsphere array analysis makes flow cytometry a very versatile platform for biotechnology research.

References

1. Amann, R.I., Binder, B.J., Olson, R.J., Chisolm, S.W., Devereux, R., Stahl, D.A. 1990. *Appl. Environ. Microbiol.* 56:1919–1925.
2. Armstrong, B., Stewart, M., Mazumder, A. 2000. Suspension arrays for high throughput, multiplexed single nucleotide polymorphism genotyping. *Cytometry* 40:102–108.
3. Barany, F. 1991. Genetic disease detection and DNA amplification using cloned thermostable ligase. *Proc. Natl. Acad. Sci. USA* 88:189–193.
4. Borzi, R.M., Piacentini, A., Monaco, M.C.G., Lisignoli, G., Degrassi, A., Cattini, L., Santi, S., Facchini, A. 1996. A fluorescent in situ hybridization method in flow cytometry to detect HIV-1 specific RNA. *J. Immunol. Methods* 193:167–176.
5. Brookes, A.J. 1999. The essence of SNPs. *Gene* 234:177–186,
6. Cai, H., White, P.S., Torney, D.C., Deshpande, A., Wang, Z., Keller, R.A., Marrone, B.L., Nolan, J.P. 2000. Flow cytometry based minisequencing: a new platform for high throughput single nucleotide polymorphism analysis. *Genomics* 66:135–143.
7. Chen, F., Binder, B., Hodson, R.E. 2000. Flow cytometric detection of specific gene expression in prokaryotic cells using in situ RT-PCR. *FEMS Microbiol. Lett.* 184: 291–295.
8. Chen, J., Iannone, M.A., Li, M.S., Taylor, J.D., Rivers, P., Nelsen, A.J., Slentz-Kesler, K.A., Roses, A., Weiner, M.P. 2000. A microsphere-based assay for multiplexed single nucleotide polymorphism analysis using single base chain extension. *Genome Res.* 10:549–557.
9. Crouch, J., Leitenberg, D., Smith, B.R., Howe, J.G. 1997 Eptstein-Barr virus suspension cell assay using in situ hybridization and flow cytometry. *Cytometry* 29:50–57.

10. Eisen, M.B., Brown, P.O. 1999. DNA arrays for analysis of gene expression. *Methods Enzymol.* 303:179–205.
11. Fulton, R.J., McDade, R.L., Smith, P.L., Kienker, L.J., Kettman, J.R. 1997. Advanced multiplexed analysis with the FlowMetrix system. *Clin. Chem.* 43:1749–1756.
12. Gorczyca, W., Gong, J., Darzenkiewicz, Z. 1993. Detection of DNA strand breaks in individual apoptotic cells by the in situ terminal deoxynucleotidyl transferase and nick translation assays. *Cancer Res.* 53:1945–1951.
13. Graves, S.W., Habbersett, R.C., Nolan, J.P. 2001. A dynamic inline sample thermoregulation unit for flow cytometry. *Cytometry* 43:23–30.
14. Gray, J.W., Mayall, B.H. 1985. Monoclonal antibodies against bromodeoxyuridine. New York, Alan R. Liss.
15. Horan, P.K., Wheeless, L.L. 1977. Quantitative single cell analysis and sorting. *Science* 198:149–157.
16. Howell, W.M., Jobs, M., Gyllensten, U., Brookes, A.J. 1999. Dynamic allele-specific hybridization: a new method for scoring single nucleotide polymorphisms. *Nat. Biotechnol.* 17:87–88.
17. Iannone, M.A., Taylor, J.D., Chen, J., Li, M.S., Rivers, P., Slentz-Kesler, K.A., Weiner, M.P. 2000. Multiplexed single nucleotide polymorphism genotyping by oligonucleotide ligation and flow cytometry. *Cytometry* 39:131–140.
18. Jackson, P.J., Hill, K.K., Laker, M.T., Ticknor, L.O., Keim, P. 1999. Genetic comparison of Bacillus anthracis and its close relatives using amplified fragment length polymorphisms and polymerase chain reaction analysis. *J. Appl. Microbiol.* 87:263–269.
19. Jaffe, E.S., Harris, N.L., Diebold, J., Muller-Hermelink, H.K. 1999. World Health Organization classification of neoplastic diseases of the hematopoietic and lymphoid tissues: a progress report. *Am. J Clin. Pathol.* 111, suppl. 1:S8–S12.
20. Keim, P., Klevytska, A.M., Price, L.B., Schupp, J.M., Zinser, G., Smith, K.L., Hugh-Jones, M.E., Okinaka, R., Hill, K.K., Jackson, P.J. 1999. Molecular diversity in *Bacillus anthracis*. *J. Appl. Microbiol.* 87:215–217.
21. Kettman, J.R., Davies, T., Chandler, D., Oliver, K.G., Fulton, R.J. 1998. Classification and properties of 64 multiplexed microsphere sets. *Cytometry* 33:234–243.
22. Lipshutz, R.J., Fodor, S.P.A., Gingeras, T.R., Lockhart, D.J. 1999. High density synthetic oligonucleotide arrays. *Nat. Genet.* 21: 20–24.
23. Look, A.T. 1997. Oncogenic transcription factors in the human acute leukemias. *Science* 278:1059–1064.
24. McHugh, T.M. 1994. Flow microsphere immunoassay for the quantitative and simultaneous detection of multiple soluble analytes. *Methods Cell Biol.* 42, Pt B:575–595.
25. Nolan, J.P., Sklar, L.A. 2002. Suspension array technology: evolution of the flat array paradigm. *Trends Biotechnol.* 20:9–12.
26. Nolan, J.P., Chambers, J.D., Sklar, L.A. 1998. Cytometric approaches to the study of receptors. In, *Cytometric Cellular Analysis, Vol. 1. Phagocyte Function: A Guide for Research and Clinical Evaluation* (Eds. J.P. Robinson and G. Babcock), pp. 19–46. John Wiley and Sons, New York.
27. Nolan, J.P., Mandy, F.F. 2002. Suspension array technology: new tools for gene and protein analysis. *Cell. Mol. Biol.* 47:1241–1256.
28. Nolan, J.P., Gallegos, L., Nolan, R.L., Graves, S.G., Cai, H., White, P.S. 2002. Molecular microbiology: detection and identification of bacterial and viral pathogens. *Cytometry* (S11):27–28.
29. Pallisgaard, N., Hokland, P., Riishoj, D.C., Pedersen, B., Jorgensen, P. 1998. Multiplex reverse transcription-polymerase chain reaction for simultaneous screening of 29 translocations and chromosomal aberration in acute leukemia. *Blood* 92:574–588.
30. Patterson, B.K., Till, M., Otto, P., Goolsby, C., Furtado, M.R., McBride, L.J., Wolinsky, S.M. 1993. Detection of HIV-1 DNA and messenger RNA by PCR-driven in situ hybridization and flow cytometry. *Science* 260:976–981.
31. Saunders, G.C., Martin, J.C., Jett, J.H., Perkins, A. 1990. Flow cytometric competitive binding assay for determination of actinomycin-D concentrations. *Cytometry* 11:311–313.

32. Schaffer, A.J., Hawkins, J.R. 1998. DNA variation and the future of human genetics. *Nat. Biotechnol.* 16:33–39.
33. Sen, K. 2000. Rapid identification of *Yersinia enterocolitica* in blood by the 5' nuclease PCR assay. *J. Clin. Microbiol.* 38:1953–1958.
34. Spiro, A., Lowe, M., Brown, D. 2000. A bead-based method for multiplexed identification and quantitation of DNA sequences using flow cytometry. *Appl. Environ. Microbiol.* 66:4258–4265.
35. Spiro, A., Lowe, M. 2002. Quantitation of DNA sequences in environmental PCR products by a multiplexed bead-based method. *Appl. Environ. Microbiol.* 68:1010–1013.
36. Stemme, V., Rymo, L., Risberg, B., Stemme, S. 2001. Quantitative analysis of specific mRNA species in minute cell samples by RT-PCR ad flow cytometry. *J. Immunol. Methods* 249:223–233.
37. Stevens. P.W., Hall. J.G., Lyamichev, V., Neri, B.P., Lu, M.C., Wang, L.M., Smith, L.M., Kelso, D.M. 2001. Analysis of single nucleotide polymorphisms with solid phase invasive cleavage reactions. *Nucleic Acids Res.* 29:U13–U20.
38. Timm, E.A., Stewart, C.C. 1992. Fluorescent in situ hybridization en suspension (FISHES) using digoxigenin-labeled probes and flow cytometry. *Biotechniques* 12:362–365.
39. van Dilla, M.A., Trujillo, T.T., Mullaney, P.F., Coulter, J.F. 1969. Cell microfluorimentry: a method for rapid fluorescence measurement. *Science* 163:1213–1214.
40. van Dongen, J.J.M., Macintyre, E.A., Gabert, J.A., Delabesse, E., Rossi, V., Saglio, G., Gottardi, E., Rambaldi, A., Dotti, G., Griesinger, F., et al. 1999. Standardized RT-PCR analysis of fusion gene transcripts from chromosome aberrations in acute leukemia for detection of minimal residual disease: Report of the BIOMED-1 Concerted Action: investigation of minimal residual disease in acute leukemia. *Leukemia* 13:1901–1928.
41. Vignali, D.A. 2000. Multiplexed particle-based flow cytometric assays. *J. Immunol. Methods* 243, 243–255.
42. Vlieger, A.M., Medenblik, A.M., van Gijlswijk, R.P., Tanke, H.J., van Der, P.M., Gratama, J.W., Raap, A.K. 1992. Quantitation of polymerase chain reaction products by hybridization-based assays with fluorescent, colorimetric, or chemiluminescent detection. *Anal. Biochem.* 205:1–7.
43. Wieckiewicz, J., Krzeszowiak, A., Ruggiero, Pituch-Noworolska, A., Zembaia, M. 1998. Detection of cytokine gene expression in human monocytes and lymphocytes by fluorescence in situ hybridization in cell suspension in flow cytometry. *Int. J. Mol. Med.* 1:995–999.
44. Yang, G., Garwahl, S., Olson, J.C., Vyas, G.N. 1994. Flow cytometric immunodetection of human immunodeficiency virus type 1 proviral DNA by heminested PCR and digoxigenin labeled probes. *Clin. Diagnostic Lab. Immunol.* 1, 26–31.
45. Yang, L., Tran, D.K., Wang, X. 2001. BADGE, BeadsArray for the detection of gene expression, a high throughput diagnostic assay. *Genome Res.* 11:1888–1898.
46. Ye, F., Li, M.S., Taylor, D., Nguyen, Q., Coulton, H., Casey, W.M., Wagner, M., Weiner, M.P., Chen, J. 2001. Fluorescent microsphere-based readout technology for multiplexed single nucleotide polymorphism analysis and bacterial identification. *Hum. Mutat.* 17:305–316.
47. Yu, H., Ernst, L., Wagner, M., Waggoner, A. 1992. Sensitive detection of RNAs in single cells. *Nucleic Acids Res.* 20:83–88.
48. Zammatteo, N., Alexandre, I., Ernest, I., Le, L., Brancart, F., Remacle, J. 1998. Comparison between microwell and bead supports for the detection of human cytomegalovirus amplicons by sandwich hybridization. *Anal. Biochem.* 253:180–189.
49. Zarda, B., Amann, R., Wallner, G., Schleifer, K.H. 1991. Identification of single bacterial cells using digoxigenin-labeled, ribosomal RNA targeted oligonucleotides. *J. Gen. Microbiol.* 137:2823–2830.

15

Uses of Flow Cytometry in Preclinical Safety Pharmacology and Toxicology

SCOTT W. BURCHIEL AND JAMES L. WEAVER

Introduction

During the last decade, commercial bench-top and multilaser flow cytometers have become widely available in academic, industrial, and government research laboratories. These flow cytometers provide a broad range of applications for multilaser and multiparameter analyses with user-friendly software interfaces. The purpose of this chapter is to review currently available and specialized applications of flow cytometry in preclinical pharmacology and toxicology.

It is now common for most research and development, safety pharmacology, and safety assessment/toxicology groups to have direct access to flow cytometry instrumentation and technology. Many pharmaceutical companies have developed preclinical research groups that combine certain aspects of pharmacology and toxicology into early stages of drug discovery and development. The reason for this is that many new chemical entities (these are usually xenobiotics, i.e., nonindigenous chemicals) with interesting pharmacologic properties have proved to have unsuspected toxicities, but these effects are often undetected until late in development. Because drug toxicity is often not associated with the pharmacology of chemical agents, it is sometimes possible to select novel chemicals with excellent pharmacologic efficacy with minimal or reduced toxicity or potential for drug interactions. As will be discussed later in this chapter, flow cytometry has many desirable features that allow for high-throughput analysis of xenobiotics, as well as a definition of rare target cells and the measurement of biochemical endpoints. Thus, flow cytometry is finding increasing numbers of applications in preclinical pharmacology and toxicology and is being incorporated into all phases of drug discovery: development, preclinical, and clinical safety evaluation.

Because of the longstanding use of flow cytometry in the evaluation of peripheral blood and hematopoietic cells in humans and other species, immunotoxicology was one of the first areas of toxicology to use flow cytometry and to develop routine testing procedures (1–3). Therefore, this chapter will largely focus on numerous applications of flow cytometry in immunotoxicity evaluation. Another factor that is driving the use of flow cytometry in immunotoxicology is that new regulations have been implemented in Europe and are being proposed in the United States and Japan to incorporate flow cytometry surface marker evaluation of circulating leukocytes in routine preclinical toxicology testing of pharmaceuticals in animals (4). Several preclinical toxicology regulatory documents specifically state measurement of lymphocyte subpopulations using flow cytometry. A document from European Agency for Evaluation of Medicinal Products, the Note for Guidance on Repeated Dose Toxicity (CPMP/SWP/1042/99) (http://www.emea.eu.int/pdfs/human/swp/104299en.pdf) specifically requests "distribution of lymphocyte subsets and NK [natural killer] activity." The draft guidance from the U.S. Food and Drug Administration (http:///www.fda.gov/cder/guidance/4945fnl.pdf) suggests that phenotypic analysis of leukocyte surface markers is useful in selected circumstances. The U.S. Environmental Protection Agency also requires immunotoxicology testing for some agents (http://www.epa.gov/OPPTS_Harmonized/870_Health_Effects_Test_Guidelines/Series/870-7800). These documents all state that flow cytometry will be used in the immunotoxicologic evaluation of investigational drugs. Immunotoxicology has been proposed as a subject for the International Conference on Harmonization of Technical Requirements for Registration of Pharmaceuticals for Human Use. The details of the regulatory requirements are still in a state of flux, but it is apparent that flow cytometry immunophenotypic analysis may play a significant role in regulatory decisions in the future.

Flow cytometry has proven useful in the evaluation of the effects of certain environmental agents in immunotoxicity testing (1); the sensitivity and specificity of standard immunophenotypic biomarkers in detecting preclinical toxicity of pharmaceuticals have not been established to the same degree. There are well-known examples of agents that alter immune function without producing changes in the level of expression of surface markers, as well as examples of agents that produce profound immunotoxicity without concomitant alteration of the distribution of cells among lymphoid subsets (5, 6). The consensus of a conference on this subject (Application of Flow Cytometry to Immunotoxicity Testing) was that routine surface marker analysis may be of limited usefulness to detect drug toxicity of new chemical entities that are not cytotoxic or that do not alter lymphoid cell differentiation and proliferation [(7) and located at http://hesi.ilsi.org/file/h1_866000flow.pdf]. Thus, a new generation of flow cytometry assays that specifically assess immune function may be useful in evaluating the immunotoxicity of drugs and chemicals. We refer to this new technology as functional flow cytometry. Flow cytometry has often been used to identify biochemical mechanisms of drug action and toxicity in defined subsets of lymphoid and nonlymphoid cells (3, 5, 6). Thus, some of these assays may be modified in useful ways to give information about immune or other cell and organ functions.

In this chapter, we review the use of flow cytometry in safety pharmacology and mechanism assessment and suggest areas in which flow cytometry may be useful in preclinical and clinical toxicology. Although many of the current applications specifically address issues of immunotoxicity, there are many potential cell targets that can

Table 15.1. Common markers for immunophenotyping peripheral blood, spleen cells, or lymph nodes

Cell Population	CD or Other Name
Common leukocyte marker	CD45
Lymphocytes	
Pan-T cell marker	CD3, CD2
	Thy 1.1/1.2, TcR-beta (Mouse)
Helper T cell	CD4
Cytotoxic T cell	CD8
Pan-B cell marker	CD45R/B220 (mouse)
	CD45RA (Human, rat)
	CD19, CD20 (not rodents)
	mIg (μ, δ, γ, α, ϵ heavy chain, or κ, λ light chain)
Monocytes	CD14
Natural killer cells	CD56, CD161
Granulocytes	CD11b & Gr-1 in mice, CD11b & CD45 in rats; or light scatter

be evaluated in flow cytometry assays. The only requirement is that cells must be analyzed as single-cell suspensions. With the advent of new single-cell suspension techniques employing collagenase and other tissue-dissociative enzymes, many different cells types obtained from animal or human tissues can be analyzed by flow cytometry, including epithelial and endothelial cells. We will mention a few new applications of flow cytometry in the evaluation of toxicity to nonlymphoid cells.

This chapter concludes with a discussion of new applications for flow cytometry in the general areas of pharmacology and toxicology. Flow cytometry is well suited for high-throughput screening applications, as well as for purification of unique or rare cell populations through cell sorting. Combined with new quantitative polymerase chain reaction (PCR), genomics, and proteomics technologies, high-speed cell sorting is also finding an increased number of applications.

Uses of Flow Cytometry in Immunotoxicology—Surface Marker Detection and Immunophenotypic Analysis

Flow cytometry has typically been used for assessment of cell surface phenotypes on immune cells. Immunophenotyping has been established in many different species including humans, monkeys, mice, rats, and dogs. Markers that are routinely examined in the peripheral blood of humans and other lymphoid tissues in animals are shown in table 15.1. These markers represent surface antigens commonly expressed on leukocytes (CD45), as well as those that are restricted to T cells (CD3, TCR), T cell subsets (CD4 and CD8), B cells (CD19, CD20, mIg), monocytes (CD14), NK cells (CD56, CD161), and neutrophils. Immunophenotyping procedures have generally used the mononuclear fraction of peripheral blood obtained after cell separation procedures. However, there are numerous procedures that allow for direct phenotypic analysis in whole blood. The advantage of whole-blood assays is they are easily performed with small sample vol-

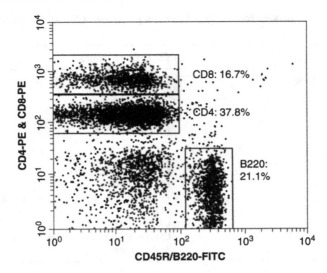

Figure 15.1. Mouse peripheral blood multiplex surface marker detection. Identification of CD4+ and CD8+ T cells and B220 positive B cells in CD45+ mouse whole blood. Whole mouse blood was simultaneously incubated with all four reagents. Data were collected using CD45 fluorescence as the trigger parameter. In this plot, data were gated on a side scatter ×CD45 PE/Cy5 plot to include only the main CD45+ population.

umes, which is particularly important in species with small blood volumes. An example of an immunophenotyping method that has been developed for the detection of all of the major lymphocyte subsets in mouse peripheral blood is illustrated later.

Application 1: Mouse Multiplex Immunophenotyping in Whole Blood

This example uses a multiplex system to allow evaluation of all three major populations of mouse lymphocytes. Figure 15.1 shows an analysis of unlysed mouse peripheral blood lymphocytes using CD45 positivity to trigger data collection. Here CD4 and CD8 are in the phycoerythrin channel, and the B-cell marker B220 (CD45R) is fluorescein isothiocyanate (FITC) labeled. This combination measures both major populations of T cells and B cells, using a single, small sample of peripheral blood. Specifically, about 40 μL whole blood was collected by tail vein puncture into 4 μL of 0.5 M EDTA and mixed immediately. An aliquot of 20 μL was mixed with 30 μL of a cocktail of CD45-PE/Cy5 (CyChrome) clone 30-F11, 1.1 μg/mL; CD4-PE clone RM4-4, 0.85 μg/mL; CD8b.2-PE clone 53-5.8, 0.42 μg/mL; and B220-FITC (CD45R) clone RA3-6B2 1.05 μg/mL. All antibodies were obtained from Pharmingen Inc. (San Diego, CA). This mixture was incubated for 60 min at room temperature. PBS was added shortly before analysis to bring the mixture up to 100 μL volume. Data were collected using triggering on the CD45-PE/Cy5 channel with a high threshold to eliminate consideration of nearly all of the red cells. Cells in the narrow CD45 peak were gated to a FITC × PE dot plot resulting in the data shown in figure 15.1. The cytometer was calibrated using Rainbow Calibration Particles (Spherotech, Inc., Libertyville, IL).

See (8) for full details of this method. This protocol can be used for a variety of applications, depending on the specific needs of the project. It is quite suitable for evaluating the effect of a chemical treatment on lymphocyte subpopulations during a repeat-dose toxicity study.

Functional Flow Cytometry in Immunotoxicology

Previous studies have attempted to correlate changes in immunophenotype with altered immune function (1, 5, 6). Xenobiotics that alter lymphoid cell survival, proliferation, or differentiation are sometimes identified through changes in the immunophenotype of peripheral cells. However, as discussed above, there are concerns that some agents may alter immune function, but that these changes may not be detected by simply performing a phenotypic analysis of subsets. In our experience, changes in immunophenotype and immune function are not always correlated and may follow different dose– and time–response relationships. Therefore, immune function tests are needed, some of which can be performed by flow cytometry.

Cell Cycle Analysis

One of the first applications of flow cytometry to experimental biology was cell cycle analysis performed using DNA staining dyes (9). As cells move through the cell cycle from G1, their DNA content changes in S phase and eventually doubles when cells reach G2. These changes are easily detected with DNA dyes such as propidium iodide, 4′,6-diamidino-2-phenylindole dihydrochloride, and viable cell dyes such as Ho-33342. More recently, bromodeoxyuridine incorporation into DNA and detection with anti-bromodeoxyuridine antibodies in permeabilized cells has become a favored method for analysis of xenobiotic-induced cell cycle changes.

Flow cytometry cell cycle assays have also recently been developed for immune function testing as an indicator of cell proliferation. An example of such a test is the local lymph node assay (LLNA). The LLNA is a test to measure chemical sensitivity in draining lymph nodes obtained from mice treated with chemical sensitizers. This test has recently been validated through ICCVAM, and its use accepted by the Food and Drug Administration as a replacement for traditional guinea pig sensitization tests that measure skin sensitization. The current approved version of the LLNA uses ^{3}H-thymidine incorporation into responding lymph node cells that are proliferating. However, bromodeoxyuridine flow cytometry assays have recently been described that have advantages in that they do not use radioactivity and they use fewer total numbers of cells (10). These assays are currently being developed and validated, and it is expected that they will provide a useful alternative to LLNA testing.

Cell Activation and Signaling

There are numerous assays and approaches that detect cell activation and signaling by flow cytometry. When activated, many cells upregulate molecules that appear on their surface. For example, surface molecules, such as MHC II proteins on B cells, are increased following treatment with cell-activating agents, and these molecules can be

Table 15.2. Flow cytometry indicators of cell function or activation

Cell Population	Marker/Indicator Example
All Cells	Cell viability (FDA, AAD, PI)
	Cell cycle (BrdU, anti-cyclins)
	Intracellular Ca^{2+} (Indo-1, Fluo-3/4, Fura-Red)
	pH (SNARF)
	Membrane potential ($DiOC_6$)
	Mitochondrial function (Rhodamine 123, Mitotracker)
	Organelle function—ER (DiO_6), lysosome (LysoTracker)
	Oxidative stress and ROS (GSH, DHE, H_2DCFDA)
	Change in phosphorylation status—(Anti-PY/PS)
	Apoptosis (Annexin-V, TUNEL, Caspase reagents)
B Cell	MHC II upregulation
	Change in mIg expression (e.g., switch to γ)
	Cytokine receptor (IL-4R)
	Adhesion molecule up or down regulation (e.g., L-selectin)
T Cell	Membrane cytokine receptor (e.g., IL-2R)
	Intracellular cytokine or chemokine expression
Monocyte	Increased phagocytosis of fluorescent beads
	Increased membrane protein levels (e.g., CD14 in mice)
	Oxidative burst measurements
	Intracelluar cytokine or chemokine
Natural killer cell	Increased levels of CD161
	Cellular cytotoxicity, cytokine secretion
Neutrophil	Increased intracellular proteins (e.g., MPO)
	Increased phagocytosis of fluorescent beads
	Oxidative burst measurements
	Increased membrane protein levels (e.g., CD11b in rats, humans

used to monitor the functionality of cell signaling pathways (11). Other markers, such as the IL-2 receptor and cell adhesion molecules, are useful markers of cell activation and can be used for this purpose. These markers are readily detected using specific antibodies and flow cytometry.

Other cell activation assays detect changes in intracellular biochemical parameters that are associated with cell function or that occur during cell signaling (12). Such changes include alterations in intracellular Ca^{2+} (13), pH, oxidative stress, changes in cell phosphorylation, cell surface molecule expression, cytokine receptor (14), intracellular cytokine expression, and others are listed in table 15.2. Many of these assays have proven useful in immunotoxicology and other toxicology applications. One of the advantages of flow cytometry is that cell activation parameters can actually be measured in defined target cells and subsets in combination with cell surface markers, such as intracellular Ca^{2+} in human peripheral blood leukocyte subsets (15).

Apoptosis—Annexin and TUNEL, and Bcl-2-Related Family Member Detection

Changes in cell viability can be detected by flow cytometry, using membrane-impermeant dyes, such as propidium iodide. Cells that fluoresce with propidium iodide have leaky membranes providing an extremely sensitive and quantitative marker of decreased membrane integrity. Cells with damaged membranes are rapidly removed from the circulation in the spleen and other reticuloendothelial system organs. There are, however, other markers of impending cell death that can be seen in circulating lymphoid cells, including markers of apoptosis (16, 17). During apoptosis, cell membranes expose phosphatidyl serine residues that bind a protein known as Annexin V. Biotinlyated Annexin V can be used to detect apoptotic cells following staining with fluorescence-tagged strept-avidin protein. Annexin V flow cytometry assays are commonly use to detected apoptotic cells in vitro and in vivo (18, 19). One caution, however, is that this assay has not been validated in several species that are used for preclinical studies, most notably the dog.

An alternative assay to detect apoptosis is the TUNEL assay (2, 20). This assay has been adapted to flow cytometry and requires fixation of cells for penetration of streptavidin into the cytoplasm to react with biotinylated-DNA that is formed during incorporation of biotin-UTP during DNA repair. The TUNEL assay has been used in both immunohistochemistry and flow cytometry studies and has been used in multiple species with good results. Alternatively, certain caspase substrates and products may be useful for identifying apoptosis in various animal species.

There are several other potential flow cytometry assays that use proteins that are expressed intracellularly during the apoptotic process and that can be detected with specific antibodies in permeabilized cells. However, caution must be observed when interpreting data with intracellular markers, and assays must be carefully controlled and validated (21). Examples of intracellular proteins that have been detected by flow cytometry and correlated with cell lysate Western blotting or other validation techniques include Bcl-2 and related protein family members (22), p53 (3, 23), cyclins (24), and phosphorylated proteins (25).

Intracellular Cytokine and Cytokine Bead Analyses

Cytokines are products of activated cells that can be detected intracellularly in activated lymphoid and nonlymphoid cells using specific antibodies. Such cytokines as IL-2 are detectable in the cytoplasm of activated cells by flow cytometry in the peripheral blood or in cells obtained from peripheral lymphoid organs (26–28). Recently, secreted cell cytokines have also been detected using antibody-coated beads and multiplex analysis (29). These assays are sensitive and specific, but their use, thus far, has been limited to analyses of in vitro production of cytokines by activated cells.

Type I Hypersensitivity Testing

Type I hypersensitivity testing may be an area of increased importance for flow cytometry detection in peripheral blood in the future. During mast cell and basophil degranulation, cytoplasmic granules containing vasoactive mediators fuse with the cell membrane during a type I allergic reaction. Granular membranes express unique antigens, such as CD63, that when fused with the cell membrane, can be detected with antibodies. Recent reports indicate that CD63 can be a useful marker for detecting drug-induced type hypersensitivity in human peripheral blood leukocytes (30–32).

NK Cell Flow Cytometry Assays

A number of laboratories have become interested of late in the development of nonisotopic methods to assess NK activity. Flow cytometry has previously been shown to be useful for measuring lysis of target cells, and assays have been developed to measure NK activity using flow cytometry (33, 34). There are also activation markers on human NK cells that may be upregulated and may be useful for monitoring function, including CD161, MHC molecules, cytokine receptors, and adhesion molecules. It is important to note that there may be important differences in the expression of surface markers on human and nonhuman species, as represented by the finding that CD56, a major marker for human NK, is not the major NK biomarker in nonhuman primates (35).

Other Toxicology Applications for Flow Cytometry

Flow cytometry can be a valuable method for evaluating both pharmacologic and toxicologic effects of xenobiotics. One major limitation is the requirement for staining and analysis of cells as single-cell suspensions that will flow through the sample chamber. For evaluation of leukocytes and bone marrow cells, this presents little problem. The method can be used with primary cell culture cells from a variety of tissues, though this significantly raises the level of technical expertise that is required. Thus, although flow cytometry is most usually applied to hematopoietic cells, there are some applications to other cell types grown in culture or obtained from dissociated tissues. In addition, there are new applications for the detection of cells that might appear in circulation as the result of a toxic response, such as circulating endothelial cells that might be released during vascular injury.

Hematopoietic Toxicology

Flow cytometry is very useful for characterizing phenotypic changes in peripheral blood that occur as a result of altered cell proliferation and differentiation in various blood dyscrasias and hematopoietic diseases. Flow cytometry has proved especially valuable in prognosis and therapeutic classification of various leukemias and lymphomas (36–39). In a preclinical setting, it is often not important to stage or classify blood disorder on the basis of cell surface phenotypes. However, such studies could be useful in instances in which an understanding of mechanisms of disease is important. In addition, it could be quite useful in evaluating an unexpected bone marrow dyscrasia.

Genotoxicity and Micronuclei Formation

The genotoxicity of small organic molecules is traditionally evaluated using several assays including the mouse micronucleus assay. Mice are treated with a single dose of the test compound and killed 24 or 48 hr later. Slides of peripheral blood and sometimes bone marrow are made and stained with acridine orange. The young erythrocytes (reticulocytes, RET) are stained with the acridine orange, which metachromatically labels both RNA and DNA. RET and RET with included micronuclei (MN-RET) are manually counted with a microscope by a trained observer. Normally 1000–2000 total RET are counted, and data are normally expressed as the percentage of MN-RET of Total RET. In addition, a separate count is taken to allow calculation of the percentage RET of total erythrocytes. Genotoxic compounds or other treatments capable of causing breaks in double-stranded DNA will cause significant increases in percentage MN-RET. Using the traditional method, data collection is quite tedious and requires a trained observer.

Recently, a series of papers has described a flow cytometric method for counting RET and MN-RET (40). This method allows objective discrimination between aneugens and clastogens (41) and has been extended to measurement in rat blood (42). This method has also recently been the subject of a successful interlaboratory validation study (43). It uses staining of the transferrin receptor (CD71) as a marker for RETs and staining with propidium iodide after RNAse treatment to label MN-containing cells. After calibration with stained malaria-containing erythrocytes, samples are analyzed, collecting enough data to include 10,000–20,000 total RET. This method uses only 50 μL of blood, which allows time-course experiments, even with mice. Research is in progress to extend this method to additional species.

Vasculitis and Circulating Endothelial Cell Detection

Flow cytometry has the possibility of being used to detect rapid toxic responses to drugs in peripheral blood. As another toxicologic example, it is known that treatment of animals with some PDEIII inhibitors causes loss of endothelial cells from vascular epithelium. Flow cytometry is being explored for the detection of changes in the number of endothelial cells in circulation as well as for identifying them to allow measurement of other changes.

Application 2: Approaches to Endothelial Cell Detection in Rat Peripheral Blood

Endothelial cells expressing cell surface markers such as CD31 (PECAM-1) (44) may also be identified by their ability to take up acetylated low-density lipoproteins (45). As an example, figure 15.2A shows the measurement of the endothelial cell marker PECAM-1/CD31-PE, and Bodipy-labeled Ac-LDLs in leukocytes from rat blood following erythrocyte lysis. Figure 15.2B shows the intentional mixture of rat leukocytes and cultured rat endothelial cells. Cells were mixed or not, and then incubated 4 hr at 37°C with Ac-LDL-Bodipy (0.067 mg/mL; Molecular Probes, Eugene, OR) and CD31-PE (1 μg/mL; Pharmingen Inc., San Diego, CA). Data are ungated. This demonstrates that discernable numbers of endothelial cells released into circulation could likely be detected. These labels could be combined with other markers such as annexin-V-PE/Cy5 or the viability marker 7-AAD to perform a comprehensive analysis of the health of circulating endothelial cells. Another marker to be considered for detecting circulating endothelial cells is CD146 (46).

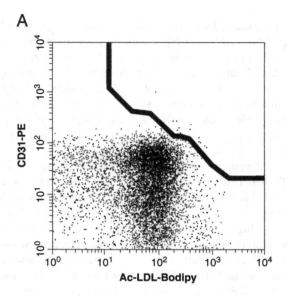

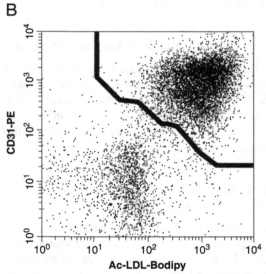

Figure 15.2. Simultaneous evaluation of rat peripheral blood leukocytes and endothelial cells. (A) Rat peripheral blood leukocytes stained with the endothelial cell markers Acetylated-LDL-Bodipy and CD31-PE. (B) Intentional mixture of rat peripheral blood leukocytes and cultured rat endothelial cells stained with Acetylated-LDL-Bodipy and CD31-PE. Data are ungated.

Applications of Flow Cytometry to Genomics

In the area of genomics and analysis of gene expression, there are at least two applications for flow cytometry in the preclinical evaluation of xenobiotic toxicology. The first application is still in development and, as described by Gusev et al. (47), requires

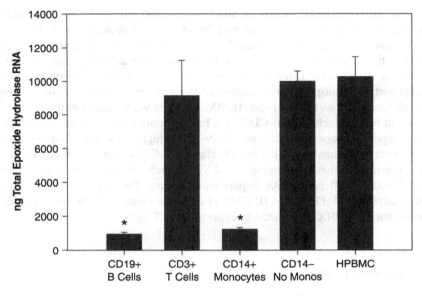

*statistically different from HPBMC

Figure 15.3. Quantitative mRNA expression detected by TaqMan for soluble epoxide hydrolase (EPHX2) expression in subsets of human peripheral blood mononuclear cells obtained by MoFlo high speed cell sorting. Results show that human peripheral blood mononuclear cells express EPHX2, but B cells (CD19+) and monocytes (CD14+) do not.

a method to amplify coding regions on genes of interest through PCR-like reactions (e.g., rolling circle amplification) and detects gene products in cells through the use of fluorescent oligonucleotide probes (decorator). This technology is not widely available or validated at this time. However, in principle, flow cytometry should allow single-cell detection of gene products and should be amenable to high-throughput screening.

High-Speed Cell Sorting and QRT-PCR Detection

The second application takes advantage of the cell sorting capability of many flow cytometers. In this application, cells can be sorted on the basis of a surface or intracellular marker of interest, and the purified cells can then be analyzed for gene expression, using traditional gene detection systems. In the example shown below, high-speed cell sorting was used to obtain enriched populations of B, T, and mononuclear cells, and each cell population was analyzed for expression of epoxide hydrolase 2 (EPHX2) expression, using quantitative real-time PCR (QRT-PCR).

Application 3: EPHX2 Detection in HPBMC

HPBMC were isolated by Ficoll-Hypaque (48), and mononuclear cells were aliquoted into 1-mL volumes at 5×10^6 cell/mL in labeling buffer (phosphate buffered saline, no calcium or magnesium, and 1% fetal bovine serum). Of the fluorochrome-conjugated monoclonal antibody (BD Biosciences, San Jose, CA), 300 μL was then added and the sample then incubated for 20 min (up to 1 hr) at 4°C and protected from light. Following incubation, cells were combined into one 15-mL conical tube, centrifuged, washed one time with labeling buffer, centrifuged for 5 min at 400 g, aspirated and resuspended at 10–20 × 10^6 cells/mL in growth media. The cell suspensions were held on ice for sorting. Multiple fluorochromes (up to four) can be used in a single sample to label different cell surface markers. HPBMC subsets were labeled with specific fluorescent antibodies such as anti-CD3-Cy-Chrome conjugated (T cells), anti-CD14-FITC conjugated (monocytes), and anti-CD19-PE conjugated (B cells). Surface marker analyses were performed on a Becton-Dickinson FACSCalibur. For high-speed cell sorting experiments, these cells were sorted by fluorochrome on the MoFlo (Cytomation) at 8,000–10,000 cells/s. An experiment showing TaqMan results for soluble EPHX2 analysis of MoFlo-sorted HPBMC is shown in figure 15.3. In this experiment, we found that the EPHX2 is limited in expression to T cells.

Current Issues

Clinical Monitoring and Biomarkers

Flow cytometry measures are potentially of great use because the method may be transferred directly to human clinical studies. This is clearly apparent where the biomarker of interest is a change in expression of a leukocyte cell surface protein. Other methods that evaluate increases or decreases in a cell population are candidates as well. The equivalent assay for humans must be validated as much as possible to ensure that it is able to detect the same type of change observed in the animal assay. Care must be taken to consider the differences in biology between the test species and humans. For example, in humans, MN-RET are scavenged by the spleen much more efficiently than in mice. This difference is of such magnitude that there is no good agreement in the literature as to the baseline level of MN-RET in circulation in humans. Therefore, failure to observe MN-RET in humans would not indicate that the test condition was nongenotoxic. In contrast, a treatment that resulted in an 80% reduction in CD4+ cells in mice would have a high probability of being immunosuppressive in humans as well. Therefore, development of a flow cytometry assay may significantly simplify the issue of monitoring for signs of toxicity in humans.

Animal Models and Reagents

Antibody reagent availability is a key issue in the detection of surface markers on lymphoid cells in rats and dogs, which are two important species that are used for preclinical evaluation of pharmaceuticals. Studies in nonhuman primates generally use human reagents that have been previously shown to detect the equivalent marker in that species. Human and murine reagents are widely available for surface marker detection, and several companies and academic labs are currently developing and validating reagents for use in other species.

Calibration of Flow Cytometers and Assay Validation

Most sponsors require that studies that will be part of an application to a regulatory agency conform to good laboratory practice standards. These standards include a requirement for calibration of the instrument. To the best of our knowledge, no regulatory body has issued specific rules for calibration of flow cytometers. However there have been recent articles that have addressed this issue (49, 50). A basic check of the instrument performance could include use of calibration beads with several intensities. It is suggested that a bead set be selected to measure fluorescence in each fluorescence channel used in the study.

The issue of assay validation is one that causes significant concern in relation to immunotoxicology studies. The problem is that nearly all immunotoxicology assays are either *in vivo* methods or require primary cell cultures. This brings in the major issue of differences among strains and species of experimental animals. An example of this problem is that CD14 is a good monocyte marker in humans, but in rodents, CD14 is expressed on activated but not resting monocytes. A similar issue is the reproducibility of reagents. Monoclonal antibodies are one of the best-defined systems in all of immunology. It is well known that clones binding to different epitopes of the same protein may show quite different performance. There are also cases in which the same clone and fluorochrome from different manufacturers will show significant differences in results. The bottom line is that an assay may be validated only for the exact set of animal strains, species, and reagents evaluated in the validation study. Substitution of assay components is acceptable, provided a comparison study shows acceptable performance by the new item as compared to the original validated component. Equivalence of assays between species cannot be assumed and must be explicitly established.

Conclusions

This chapter points out several current applications of flow cytometry that are useful in the preclinical and clinical evaluation of new chemical entities and xenobiotics for important pharmacologic and toxicologic effects. As instrumentation and reagents become more widely available for test species, the use of flow cytometry technology will certainly increase. The overall advantages of the technology are its inherent quantitative aspects and high-throughput potential. Current limitations relate to standardization of test procedures and protocols, calibration of instruments, interspecies variability in biomarkers, agreement on data analysis and interpretation, and overall assay validation between laboratories. The development of new and improved functional flow cytometry procedures and correlation with established immune function tests will also certainly move the use of the technology forward.

Acknowledgment   This article was written by the authors in their private capacity. No official support or endorsement by the Food and Drug Administration is intended or should be inferred.

References

1. Luster, M.I., Munson, A.E., Thomas, P.T., Holsapple, M.P., Fenters, J.D., White, K.L. Jr, Lauer, L.D., Germolec, D.R., Rosenthal, G.J., Dean, J.H. 1988. Development of a testing battery to assess chemical-induced immunotoxicity: National Toxicology Program's guidelines for immunotoxicity evaluation in mice. *Fundam. Appl. Toxicol.* 10:2–19.
2. Burchiel, S.W., Kerkvliet, N.L., Gerberick, G.F., Lawrence, D.A., Ladics, G.S. 1997. Assessment of immunotoxicity by multiparameter flow cytometry. *Fundam Appl. Toxicol.* 38:38–54.
3. Burchiel, S.W., Lauer, F.T., Gurulé, D., Mounho, B.J., Salas, V.M.1999. Uses and future applications of flow cytometry in immunotoxicity testing. *Methods* 19:28–35.
4. Dean, J.H., Hincks, J.R., Remandet B. 1998. Immunotoxicology assessment in the pharmaceutical industry. *Toxicol. Lett.* 102–103:247–255.
5. Burchiel, S.W., Hadley, W.M., Cameron, C.L., Fincher, R.H., Lim, T.W., Elias, L., Stewart, C.C. 1987. Analysis of heavy metal immunotoxicity by multiparameter flow cytometry: Correlation of flow cytometry and immune function data in B6C3F1 mice. *Int. J. Immunopharmacol.* 9:597–610.
6. Burchiel, S.W., Hadley, W.M., Barton, S.L., Fincher, R.H., Lauer, D., Dean, J.H. 1988. Persistent suppression of humoral immunity produced by 7,12-dimethylbenz(a)-anthracene (DMBA) in B6C3F1 mice: Correlation with changes in spleen cell surface markers detected by flow cytometry. *Int. J. Immunopharmacol.* 10:369–376.
7. ILSI Immunotoxicology Technical Committee. 2001. Application of flow cytometry to immunotoxicity testing: summary of workshop. *Toxicology* 163:39–48. Available at: http://hesi.ilsi.org/file/h1_866000flow.pdf.
8. Weaver, J.L., McKinnon, K., Broud, D.D., Germolec, D.R. 2002. Serial phenotypic analysis of mouse peripheral blood leukocytes. *Toxicol. Mech. Methods* 12:95–118.
9. Darzynkiewicz, Z, Juan, G. 1997. DNA content measurement for DNA ploidy and cell cycle analysis. Unit 7. In: Robinson et al.(Eds.), *Current Protocols in Cytometry*, Vol. 1, Wiley and Sons, New York.
10. Takeyoshi, M., Yamasaki, K., Yakabe, Y., Takatsuki, M., Kimber, I. 2001. Development of non-radio isotopic endpoint of murine local lymph node assay based on 5-bromo-2′-deoxyuridine (BrdU) incorporation. *Toxicol. Lett.* 119:203–208.
11. Davis, D.A., Burchiel, S.W. 1992. Inhibition of calcium-dependent pathways of B-cell activation by DMBA. *Toxicol. Appl. Pharmacol.* 116:202–208.
12. Rabinovitch, P.S. 2002. Studies of cell function. Chapter 9. In: J.P. Robinson, Z. Durzynkiewicz, J. Dobrucki, W. Hyun, A. Orfao, and P. Rabinovitch (Eds.), *Current Protocols in Cytometry*, Vol. 2, Wiley and Sons, New York.
13. Burchiel, S.W., Edwards, B.S., Kuckuck, F.W., Lauer, F.T., Prossnitz, E.R., Ransom, J.T., Sklar, L.A. 2000. Analysis of free intracellular calcium by flow cytometry: multiparameter and pharmacologic applications. *Methods* 21:221–230.
14. Collins, D.P. 2000. Cytokine and cytokine receptor expression as a biological indicator of immune activation: important considerations in the development of in vitro model systems. *J. Immunol. Methods* 243:125–145.
15. Mounho, B.J., Davila, D.R., and Burchiel, S.W. (1997). Characterization of intracellular calcium responses produced by polycyclic aromatic hydrocarbons in surface marker-defined human peripheral blood mononuclear cells, *Toxicol. Appl. Pharmacol.* 145:323–330.
16. Darzynkiewicz, Z., Juan, G., Li, X., Gorczyca, W., Murakami, T., Traganos, F. 1997. Cytometry in cell neurobiology: analysis of apoptosis and accidental cell death (necrosis). *Cytometry* 27:1–20.
17. Darzynkiewicz, Z. Bender, E. Milewski, P. 2001. Flow cytometry in analysis of cell cycle and apoptosis. *Semin. Hematol.* 38:179–193.
18. Koopman, G., Reutelingsperger, C.P., Kitten, G.A., Koenen, R.M., Pals, S.T., van Ores, M.H. 1994. Annexin V for flow cytometric detection of phosphatidylserine expression on B cells undergoing apoptosis. *Blood* 84:1415–1420.
19. Davis, J.W. 2nd, Melendez, K., Salas, V.M., Lauer, F.T., Burchiel, S.W. 2000. 2,3,7,8-Tetra-

chlorodibenzo-p-dioxin (TCDD) inhibits growth factor withdrawal-induced apoptosis in the human mammary epithelial cell line, MCF-10A. *Carcinogenesis* 21:881–886.

20. Gorczyca, W., Gong, J., Darzynkiewicz, Z. 1993. Detection of DNA strand breaks in individual apoptotic cells by the in situ terminal deoxynucleotidyl transferase and nick translation assays. *Cancer Res.* 53:1945–1951.

21. Koester, S.K., Bolton, W.E. 2000. Intracellular markers. *J. Immunol. Methods* 243:99–106.

22. Delia, A.A., Bordello, M.G., Bassini, D., Giard ini, R., Fontanelle, E., Puzzle, F., Fulford, K., Pyrotic, M., and Porta G.. 1992. Flow cytometric detection of the mitochondrial BCL-2 protein in normal and neoplastic human lymphoid cells. *Cytometry* 13:502–509.

23. Remivox, Y., Lauren-Puig, P., Salmon, R.J., Trélat, G., Durallium, B., Thomas, G. 1990. Simultaneous monitoring of P53 protein and DNA content of colorectal adenocarcinomas by flow cytometry. *Int. J. Cancer* 45:450–456.

24. Darzynkiewicz, Z. Gong, J. Juan, G. Ardent, B, Traganos, F. 1996. Cytometry of cyclin proteins. *Cytometry* 25:1–13.

25. Villager, F., Scott-Algera, D., CA yota, A., Sicilian, J., Nager, M.T., Dihydro, G. 1995. Flow cytometric analysis of protein-tyrosine phosphorylation in peripheral T cell subsets. Application to healthy and HIV-seropositive subjects. *J. Immunol. Methods* 185:43–56.

26. Labalette-Houache, M., Torpier, G., Capron, A., Dessaint J.P. 1991. Improved permeabilization procedure for flow cytometric detection of internal antigens. Analysis of interleukin-2 production. *J. Immunol. Methods* 138:143–153.

27. Mascher, B., Schlenke, P., Seyfarth, M. 1999. Expression and kinetics of cytokines determined by intracellular staining using flow cytometry. *J Immunol. Methods* 223:115–121.

28. Pala, P., Hussell, T., Openshaw, P.J. 2000. Flow cytometric measurement of intracellular cytokines. *J. Immunol. Methods* 243:107–124.

29. Carson, R.T., Vignali, D.A. 1999. Simultaneous quantitation of 15 cytokines using a multiplexed flow cytometric assay. *J. Immunol. Methods* 227:41–52.

30. Sainte-Laudy, J., Sabbah, A., Vallon, C., Guerin, J.C. 1998.Analysis of anti-IgE and allergen induced human basophil activation by flow cytometry. Comparison with histamine release. *Inflamm. Res.* 47:401–408.

31. Ebo, D.G., Lechkar, B., Schuerwegh, A.J., Bridts, C.H., De Clerck, L.S., Stevens, W.J. 2002. Validation of a two-color flow cytometric assay detecting in vitro basophil activation for the diagnosis of IgE-mediated natural rubber latex allergy. *Allergy* 57:706–712.

32. Sanz, M.L., Gamboa, P.M., Antepara, I., Uasuf, C., Vila, L, Garcia-Aviles, C., Chazot, M., De Weck, A.L. 2002. Flow cytometric basophil activation test by detection of CD63 expression in patients with immediate-type reactions to betalactam antibiotics. *Clin Exp Allergy* 32:277–286.

33. Callewaert, D.M., Radcliff, G., Waite, R., LeFevre, J., Poulik, M.D. 1991. Characterization of effector-target conjugates for cloned human natural killer and human lymphokine activated killer cells by flow cytometry. *Cytometry* 12:666–676.

34. Chang, L., Gusewitch, G.A., Chritton, D.B., Folz, J.C., Lebeck, L.K., Nehlsen-Cannarella, S.L. 1993. Rapid flow cytometric assay for the assessment of natural killer cell activity. *J. Immunol. Methods* 166:45–54.

35. Carter, D.L., Shieh, T.M., Blosser, R.L., Chadwick, K.R., Margolick, J.B., Hildreth, J.E., Clements, J.E., Zink, M.C. 1999. CD56 identifies monocytes and not natural killer cells in rhesus macaques. *Cytometry* 37:41–50.

36. Orfao, A., Schmitz, G., Brando, B., Ruiz-Arguelles, A., Basso, G., Braylan, R., Rothe, G., Lacombe, F., Lanza, F., Papa, S., et al. 1999. Clinically useful information provided by the flow cytometric immunophenotyping of hematological malignancies: current status and future directions. *Clin. Chem.* 45:1708–1717.

37. Campana, D., Behm, F.G. 2000. Immunophenotyping of leukemia. *J. Immunol. Methods* 243:59–75.

38. Sullivan, J.G., Wiggers, T.B. 2000. Immunophenotyping leukemias: the new force in hematology. *Clin. Lab Sci.* 13:117–122.

39. Weir, E.G., Borowitz, M.J. 2001. Flow cytometry in the diagnosis of acute leukemia. *Semin. Hematol.* 38:124–138.

40. Dertinger, S.D., Torous, D.K., Tometsko, K.R. 1996. Simple and reliable enumeration of micronucleated reticulocytes with a single-laser flow cytometer. *Mutat. Res.* 371:283–292.
41. Torous, D.K., Dertinger, S.D., Hall, N.E., Tometsko, C.R. An automated method for discriminating aneugen- vs. clastogen-induced micronuclei. 1998. *Environ. Mol. Mutagen* 31:340–344.
42. Torous, D.K., Dertinger, S.D., Hall, N.E., Tometsko, C.R. 2000 Enumeration of micronucleated reticulocytes in rat peripheral blood: a flow cytometric study. *Mutat. Res.* 465:91–99.
43. Torous, D.K., Hall, N.E., Dertinger, S.D., Diehl, M.S., Illi-Love, A.H., Celebrant, K., Mandelin, K., Bolcsfoldi, G., Ferguson, L.R., Pearson, A., et al. 2001 Flow cytometric enumeration of micronucleated reticulocytes: high transferability among 14 laboratories. *Environ. Mol. Mutagen* 38:59–68.
44. Hewett, P.W, and Murray, J.C. 1993. Human microvessel endothelial cells: isolation, culture and characterization. *In Vitro Cell Dev. Biol. Anim.* 29A:823–830.
45. Gaffney J, West D, Arnold F, Sattar A, Kumar S. 1985. Differences in the uptake of modified low density lipoproteins by tissue cultured endothelial cells. *J. Cell Sci.* 79:317–325.
46. Dignat-George, F., Sampol, J. 2000. Circulating endothelial cells in vascular disorders: new insights into an old concept. *Eur. J. Haematol.* 65:215–220
47. Gusev, Y., Sparkowski, J., Raghunathan, A., Ferguson, H. Jr, Montano, J., Bogdan, N., Schweitzer, B., Wiltshire, S., Kingsmore, S.F., Maltzman, W., et al. 2001. Rolling circle amplification: a new approach to increase sensitivity for immunohistochemistry and flow cytometry. *Am. J. Pathol.* 159:63–69.
48. Davila, D.R., Romero, D.L., and Burchiel, S.W. (1996). Human T cells are highly sensitive to suppression of mitogenesis by polycyclic aromatic hydrocarbons and this effect is differentially reversed by α-naphthoflavone. *Toxicol Appl Pharmacol* 139:333–341.
49. Owens, M.A., Vall, H.G., Hurley, A.A., Wormsley, S.B. 2000. Validation and quality control of immunophenotyping in clinical flow cytometry. *J. Immunol. Methods* 243:33–50.
50. Schwartz A, Repollet E, Vogt R, Gratama J. 1996 Standardizing flow cytometry: Construction of a standardized fluorescence calibration plot using matching spectral calibrators. *Cytometry* 26:22–31.

16

Flow Cytometry and Cell Sorting in Plant Biotechnology

DAVID W. GALBRAITH, JAN BARTOŠ,
AND JAROSLAV DOLEŽEL

Introduction

Higher plants comprise approximately 250,000 described species and represent a critical component of the planetary biomass. They contribute functions essential for life, of which the most important is photosynthesis, as it provides the means for conversion of incident solar radiation into biomass accumulation, as well as the oxygen required by aerobic life forms. Fixed carbon in the form of carbohydrate provides the basis of the food chain, and metabolic interconversions within plants provide a variety of essential dietary factors. Plants also provide biomass in the form of structural materials and are the source of many natural products with important biomedical properties. As a consequence, considerable scientific interest is invested in determining the molecular mechanisms underlying plant growth, development, metabolism, and responses to biotic and abiotic stresses. Investment has also been made in developing tools and resources for biological investigations using plants. Notable advances include the development of genetics, of means for transformation using defined DNA sequences, and most recently, of the entire nuclear genome sequences of two plant species (*Arabidopsis thaliana* and *Oryza sativa*). On the basis of information of this type and that from other sources, it is evident that higher plants share many features with other eukaryotic organisms (150). Shared features can be observed at many levels; for example, the overall method of construction of cells, in which a bilamellar plasma membrane separates the cytoplasm from the external milieu and provides primary homeostatic regulation. Eukaryotic cells of different kingdoms share organelles, as well as overall regulatory mechanisms. Shared, or highly similar, protein sequences are observed, and they perform similar functions as enzymes, regulatory molecules, or structural components.

Higher land plants have evident differences from other eukaryotes. They contain unique classes of organelles primarily devoted to energy capture from sunlight (plastids and peroxisomes). Of these, chloroplasts contain highly fluorescent pigments devoted to photosynthesis, which, particularly chlorophyll, provide unique and powerful signals that can be employed for flow cytometric analysis. Higher plants are also essentially immobile in the sporophytic stage and hence must be capable of responding to changes in environmental conditions and to biotic attack. At the cellular level, the individual cells are encased by a rigid cell wall, and cytokinesis is achieved by a mechanism involving a shared phragmoplast. This has two implications: first, a plant can be considered to largely consist of an assemblage of cells intimately connected by shared cell walls, and second, that individual plant cells are incapable of directed movement from their site of production (meristems) to their ultimate organismal location beyond that consequent to the process of cell division itself. In terms of flow cytometry, which requires single cell suspensions, the interconnection of individual plant cells offers obvious challenges.

Flow cytometry provides a means to rapidly and accurately characterize the optical properties of suspensions of cells or particles of biological or other provenance. It involves the passage of these particles serially through the focus of an intense light source. The light scattered from these particles, or emitted in the form of fluorescence, is collected using sensitive detectors screened by appropriate wavelength-specific optical filters. The amplitudes of these signals are quantified, digitized, and correlated, and the information is displayed as multiparametric frequency distributions. Cell sorting involves the automatic selection and purification of specified subpopulations contained within these frequency distributions and is most commonly achieved through the application of a high-frequency periodic mechanical disturbance to the flow stream, using a piezoelectric transducer, which causes this stream to break into droplets at a precise distance below the point of optical interrogation. Sorting is achieved by application of a potential difference to the flow stream at the point in time that the desired particle is entering into the "last-attached-droplet." The resultant charge retained on the surface of the droplet containing the desired particle allows electrodynamic displacement of the droplet via its passage through a high-voltage electric field [for a comprehensive review of flow cytometers and their operation, see (191)].

Overview and Implementation

The first reports of the application of flow cytometry and cell sorting to higher plants addressed two general areas: (1) the analysis of nuclear DNA contents, including the cell division cycle, and ploidy levels, based on use of DNA-specific fluorochromes, and (2) the selection of heterokaryons produced by fusion of protoplasts from different species. Recently, flow cytometry and cell sorting have been applied to a wide variety of different problems in plant biotechnology and genomics; for example, chromosome sorting for mapping purposes. Plant applications of flow cytometry and sorting have been particularly boosted by the development of methods for expression of intrinsically fluorescent markers within transgenic organisms, of which the green fluorescent protein (GFP) and its derivatives are the most popular and important. Flow cytometric analysis and sorting applications have also been developed for use with subcellular organelles, including nuclei, mitochondria, and chloroplasts, based on intrinsic fluorescence (chloroplasts), in vivo GFP expression and targeting, or staining with particular fluorochromes. A review of early work in this area can be found in (72).

Flow Cytometric Estimation of Nuclear DNA Contents and Ploidy Levels

The first report of the use of flow cytometry for the analysis of plant genome sizes appeared in 1983 (82). This work bypassed the problem of the incompatibility of flow cytometry with the intrinsically three-dimensional structure of plant tissues by employing a simple method of gentle homogenization to release intact nuclei. These nuclei were then stained using mithramycin as the DNA-specific fluorochrome. The intensities of the emitted fluorescence following excitation at 457 nm were determined by flow cytometry using a Coulter EPICS V instrument. Using tobacco (*Nicotiana tabacum* L.) as the model species for technique development, high-quality histograms were observed, with the major peak of fluorescence (the G_0/G_1 peak) having a coefficient of variation of around 3%. Use of mithramycin was felicitous, as the absorption and emission maxima of this fluorochrome are quite different from those of the fluorescent photosynthetic pigments, and this reduces background noise from subcellular debris and fluorescent organelles such as chloroplasts. However, other fluorochromes can be employed, such as propidium iodide, with similar results (figure 16.1A). Assignment of an absolute DNA content value to the G_0/G_1 peak can be most simply done through inclusion of cells of known DNA content as an internal standard, such as chicken red blood cells (figure 16.1B). Various caveats to these measurements are described in further sections.

In the initial report (82), these methods were found to be effective for multiple species and genera, and they have since been extended to a very large number of plant species and to other fluorochromes, of which propidium iodide (PI) and 4',6-diamidino-2-phenylindole dihydrochloride are the most popular. A searchable database of published plant nuclear DNA contents has been established at the Royal Botanical Gardens at Kew (http://www.rbgkew.org.uk/cval/homepage.html); it illustrates the extraordinary diversity of nuclear DNA contents found in flowering plants (a 2C value of ~0.1 to 254 pg). In terms of convenience, flow cytometry far outperforms other methods of nuclear DNA content measurements, including quantitative microspectrophotometry and chromosome counting, and consequently has become the method of choice for these measurements despite the high cost of the instrumentation.

In typical flow cytometric measurements of plant nuclear DNA contents, assignment of the G_0/G_1 peak can be reasonably straightforward. For example, for tobacco leaf homogenates, the flow histograms have a single predominant peak and a secondary peak at twice the DNA content of the first, with a smaller proportion of nuclei (from S-phase cells) located between the two peaks (figure 16.1). Flow histograms of the apical regions (containing the shoot meristem) show similar presumptive G_0/G_1 and G_2 peaks. Flow histograms of haploid tobacco plants, produced from anther culture, show G_0/G_1 and G_2 peaks at exactly 50% of the corresponding positions of diploid plants, as do flow histograms from *Nicotiana sylvestris*, one of the presumptive true diploid parents of tobacco (73, 193). For some tobacco tissues, for example, older leaves and roots, the proportions of G_2 nuclei are elevated with respect to those of

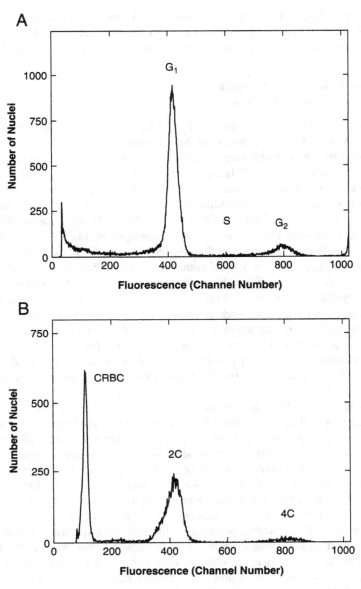

Figure 16.1. (*A*) Flow cytometric analysis of tobacco leaf homogenates stained with propidium iodide. (*B*) As for (*A*), with inclusion of chicken red blood cells as an internal standard for genome size measurement. In leaf tissue, the G1 nuclei, which have a 2C DNA content, greatly outnumber those in S-phase and in G2, which have a 4C DNA content. [Reprinted from (81) with permission from Wiley & Sons.]

G_0/G_1 nuclei. For this reason, flow cytometry can also be employed as a convenient monitor of the ploidy status of the species of interest. This is particularly useful for detecting and removing nonhaploid plants produced via anther culture, for identification of euploid plants regenerated from heterokaryons (see, e.g., 195), and for quality-

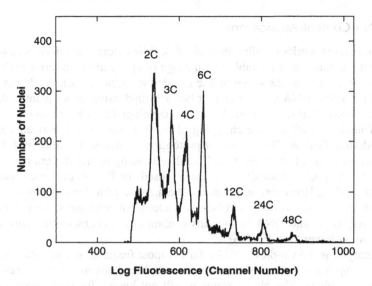

Figure 16.2. Flow cytometric analysis of nuclear DNA content distribution in maize endosperm (28 days after pollination), using propidium iodide as the fluorochrome [Reprinted from (76) with permission from Publisher Marcel Dekker.]

control assurance of ploidy level in seed production. Examples of these and further applications are discussed below.

In other species, and within particular tissues, changes in nuclear ploidy levels appear to have developmental significance within the sporophyte. This phenomenon, termed *endoreduplication*, occurs as a consequence of multiple rounds of chromosome replication in an absence of subsequent mitoses and leads to chromosomes comprising a doubling series of chromatids (42). This produces polysomatic (154) organs and tissues, which contain endoreduplicated cells. The interest in endoreduplication stems from its link to cell growth and from its frequent occurrence in storage tissues (endosperm, cotyledons, parenchyma tissue), tissues with high metabolic activity (embryo suspensor), and various other specialized cell types, such as root hairs and trichomes. Endoreduplication produces populations of nuclei having broad (and sometimes very broad) ranges of DNA content values. As a consequence, histograms of DNA content are usually shown in logarithmic scale (figure 16.2).

The occurrence of endoreduplication is not always restricted to specific cell types. For example, within *Arabidopsis thaliana*, most of the sporophyte exhibits endoreduplication (79). Similar developmental controls of endoreduplication are observed in succulents possessing small genomes (48). In these two cases, endoreduplication may solve the potential conflict between small genome sizes, scaling mechanisms linking nuclear DNA content, nuclear size, cell size, nuclear cytoplasmic information transfer, and the requirement for a minimal cell size limitations relating to sporophytic growth.

The ease of use of flow cytometry for measurement of plant nuclear DNA contents, through comparison to various standards and controls, has facilitated studies of the role of genome size in a very wide number of biological questions and problems. These are discussed in the following section.

DNA Content Applications

Since the pioneering work of Galbraith et al. (82), flow cytometric analysis of nuclear DNA content has found a remarkably broad range of applications in terms both of biological end points and of the scope of use (academic to industrial). As discussed by Doležel (55), nuclear DNA content cannot be measured using intact plant cells with sufficient precision. Various methods have been developed for the preparation of suspensions of nuclei, which include chopping of fresh tissues (82), homogenization of formaldehyde-fixed tissues (190), lysis of protoplasts in hypotonic solution (20), and release of nuclei from ethanol/acetic acid fixed cells using hydrolytic enzymes (167). A range of DNA-specific fluorochromes can be used for flow cytometric quantification of nuclear DNA. However, only some of them are suitable for estimation of DNA content in absolute units, as discussed below. Under ideal conditions, the precision of analysis is very high, and DNA histograms contain peaks characterized with coefficient of variation lower then 1% (56).

Ploidy analysis is the simplest, and by far the most frequent, use of DNA flow cytometry, with applications ranging from breeding and seed quality testing to taxonomy and population biology. The ploidy status is still not known for many accessions of cultivated species, and systematic analyses reveal surprising results. By screening alfalfa germplasm for ploidy level, Brummer et al. (29) found that many accessions had ploidy levels that were different than expected, and the authors recommended their reclassification. Unlike chromosome counting in root meristem tissue, which represents only one histological layer, flow cytometric analysis of leaf tissues representing all three histological layers (LI, LII, and LIII) enables reliable detection of mixoploidy. De Schepper et al. (49) screened the ploidy of six species and 88 cultivars of *Rhododendron*. In addition to various ploidy levels, one cultivar was found to be a cytochimera, with two histological layers being mixoploid (2x + 4x) and one layer being tetraploid. Cytochimeras and plants of different ploidy levels have also been found in *Hosta* (229).

Commercial cultivars of some crops such as sugar beet and banana are triploid, and their breeding is based on different crossing strategies to obtain the desired ploidy level. Flow cytometry is employed routinely to check the ploidy in parental plants as well as in their progenies. In some crops (e.g., sugar beet, ryegrass), ploidy is an important parameter of seed lot quality. Because of high throughput, flow cytometry has been recommended as a means for determining ploidy level in seed testing. To speed up the analysis, high numbers of individual plants can be pooled in one sample (11, 54). However, precise estimation of rare contamination by seeds or plants of undesirable ploidy may require analyses of individual plants (198). Baert et al. (7) employed flow cytometry to estimate the proportion of diploid and tetraploid perennial ryegrass in a mixed sward. In these mixtures, tetraploids are less persistent than diploids, and flow cytometry provides a convenient means to determine the proportions present.

Breeders are using interspecific hybridization to transfer desired characters from one species to another. Provided parental species differ enough in nuclear DNA content, flow cytometry can detect interspecific hybrids according to their intermediate DNA values. For instance, Keller et al. (110) detected hybrid plants obtained after crosses of onion with 19 species belonging to genus *Allium*. In other work (13), flow cytometry was successfully employed to identify F1 hybrids between two coffee species having the same chromosome number but differing in DNA content. Flow cytometry has also been used to check the ploidy levels of various *Actinia* species and their hybrids obtained by embryo rescue (103). This study also uncovered the potential for wide crosses in *Actinidia*.

Protoplast fusion is an attractive way of producing hybrids between plant species separated by crossability barriers (1). However, somatic hybrids may be karyologically unstable, and elimination of chromosomes of one or both partners can occur. Flow cytometry is an attractive way of assessing karyological status of hybrids. For example, Otoni et al. (161) obtained tetraploid somatic hybrids between diploid *Passiflora* species, but somatic hybridization between *N. tabacum* and *N. rotundifolia* led to a spectrum of plants differing in DNA content, indicating random chromosome elimination and hence genomic incompatibility between the two species (105). Asymmetric protoplast fusion provides a more targeted approach toward transfer of desired traits into breeding lines, as it avoids introduction of the whole genome of the wild species. Oberwalder et al. (159) found large variations in DNA content within hybrid lines obtained by asymmetric protoplast fusion between wild species and breeding lines of potato. Flow cytometry has also been used to characterize asymmetric somatic hybrids obtained by fusion of protoplasts of *Helianthus annuus* with microprotoplasts of perennial *Helianthus* species (26). Hybrids identified using random amplified polymorphic DNA markers were found to have a higher DNA content than the receptor species, indicating that they were addition lines; this assumption was subsequently confirmed by chromosome counting.

A growing list of publications confirms the sensitivity of DNA flow cytometry and its suitability for detection of aneuploidy. Bashir *et al.* (14) reported detection of the presence of a pair of rye chromosomes in wheat-rye addition lines. More recently, flow cytometry has been used to detect differences of as little as 1.84% in DNA content (168), which provided the ability to detect the presence of a pair of rye chromosome arms added to the wheat genome. Šesek et al. (189) were able to detect monosomics and trisomics in hops (2n = 30). It should be noted that, in all cases, the analysis involved the use of an internal reference standard, which eliminated errors resulting from variation in sample preparation and instrument drift (55). Although the authors mentioned above used plant standards, Roux et al. (182) employed nuclei isolated from chicken red blood cells to detect aneuploids in triploid banana, the flow cytometric data in this case being confirmed by chromosome counts.

In the breeding and production of dioecious crops, the ability to identify plant sex at an early stage of growth is an important goal. Provided sex is determined by a pair of chromosomes differing sufficiently in DNA content, flow cytometry can be suitable to discriminate males from females, as illustrated for *Melandrium album* (56). In date palm, sex chromosomes do not differ significantly in size. However, females have a pair of GC-rich homomorphic chromosomes, whereas the corresponding pair in males is heteromorphic. On the basis of this observation, Siljak-Yakovlev et al. (196) were able to determine sex in this species, using flow cytometric analysis of adenine thymine/guanosine cytosine ratio in isolated nuclei.

Apomixis is a form of asexual reproduction through seed. Increased interest in apomixis is stimulated both by its possible use as means of maintaining uniformity in crop cultivars and to provide heterosis stability. However, the reproductive biology of apomixis provides considerable obstacles for detailed genetic study. Flow cytometry was found to provide a convenient means to screen ploidy levels in progeny from facultatively apomictic Kentucky bluegrass (104). In combination with random amplified polymorphic DNA fingerprinting, DNA flow cytometry allowed unambiguous determination of the genetic origin of aberrant progeny. An efficient screen for reproductive pathways based on flow cytometric analysis of embryo and endosperm nuclei was developed by Matzk et al. (147). This "flow cytometric seed screen" allows rapid selection of saprophytic or gametophytic mutants in sexual species, identification of pure sexual or obligate apomictic genotypes from facultative apomictic species, and analysis of inheritance of individual reproductive processes (9, 148).

Flow cytometry offers a range of applications in plant biotechnology. The method has been used to analyze genetic stability during culture in vitro and ploidy of regenerated plants (23, 30, 162). Several studies indicated a link between karyological instability in vitro and regeneration ability (40, 116). The use of double-haploid homozygous lines has become a common practice in breeding various crops. Haploids are typically obtained from anther, microspore, and ovary culture. As a rule, not only haploid but also dihaploid, and sometimes also mixoploid and polyploid regenerants, are obtained (70). Haploid regenerants are treated with a polyploidizing agent to produce fertile homozygous diploids, whereas spontaneously occurring dihaploids are directly used in breeding programs or other applications. Flow cytometry has been useful not only in screening high numbers of regenerants (69, 179) but also in optimizing the conditions for chromosome doubling in haploids (84, 207).

In many species, polyploid cultivars outperform diploid ones in yield or quality. Chromosome doubling in ornamental plants results in larger flowers and longer flowering periods (210). In addition, polyploidization is sometimes needed to facilitate crossing between species differing in ploidy level (171). Regeneration from in vitro cultures, which have been treated with a mitotic spindle inhibitor, is an efficient way to produce polyploid plants. However, as all cultured cells are not treated in identical ways, the regenerated plants are frequently mixoploid. Flow cytometric screening permits optimization of treatment conditions, detection of mixoploid regenerants, and selection of eupolyploids (183, 214). If needed, chimeras may be dissociated by several cycles of shoot tip culture (181).

Because of its ease and speed, flow cytometric ploidy determination is an ideal tool to analyze large numbers of samples from many populations. Thanks to this method, new cytotypes were discovered in a number of species (134, 201). Using flow cytometry, Amsellem et al. (2) studied the effect of ploidy levels on the invasiveness of introduced plants in insular plant communities. The method was useful in analyzing the dynamics of populations differing in ploidy and analyzing the ploidy of in hybrid zones. Burton et al. (31) studied population cytotype structure in the polyploid *Galax urceolata* with the aim of understanding the evolutionary forces governing the establishment of polyploids and their coexistence with diploids. Although triploids occurred in zones of overlap between diploids and tetraploids, they were always in the minority. The study indicated a disruptive selection for chromosome number, as the populations were predominately diploid or tetraploid. Hardy et al. (101) found no triploid plants within mixed populations of diploid and tetraploid cytotypes of *Centaurea jacea*, despite overlapping flowering periods.

Unlike ploidy analysis, flow cytometric estimation of nuclear DNA content in absolute units (pg, bp) is methodologically more demanding and requires the use of a reliable reference standard of known DNA content (55, 82). Unfortunately, no general agreement on suitable standards has been achieved, and various groups continue using different standards (64, 106, 144). Comparison of data obtained in different laboratories should therefore be done cautiously. Although the precision of flow cytometric estimation of genome size is high, interlaboratory comparisons (60) have indicated that the differences between laboratories can exceed 10%. Thus, the reliable detection of small differences between accessions is possible only by analyzing the samples using the same protocol and the same instrument, preferably at the same time. The reliability of some flow cytometric estimates is compromised by use of DNA fluorochromes that exhibit preferential binding to AT- or GC-rich regions of DNA, and the use of DNA intercalators such as ethidium bromide or propidium iodide has been recommended (60). However, the use of AT- or GC-binding dyes can provide an estimation of the AT/GC ratio in nuclear DNA (27, 89, 178). It has been noted that the estimation of AT/CC content might be prone to errors, as it assumes a random distribution of bases in DNA (12). In some species, secondary metabolites are suspected to interfere with binding of fluorochromes to DNA, giving rise to artifacts (157, 175). The applicability of likelihood functions to extract data features from genome size measurements has recently been demonstrated (52).

Knowledge of genome size is essential in various areas of plant research. For instance, this information can be used to estimate the number of clones of a genomic library needed to achieve desired genome coverage. Although the biological significance of genome size is not clear, it has been shown to correlate with various characters at cell, tissue, and organismal levels. The correlations included cell size and cell cycle duration, minimum generation time, geographical distribution, plant phenology, biomass production, radiosensitivity, and megasporogenesis (15, 16, 21, 32). In addition to understanding its biological significance, knowledge of genome size may clarify evolutionary trends (22). Current data indicate that ancestral angiosperm genomes were small and that genome size increased during evolution (22, 128). A recent study (12) did not find a correlation between genome size and the AT:GC ratio. As genome sizes are known for only about 1% of all plant species (17), any generalization regarding the significance of genome size is difficult. It is anticipated that high-throughput flow cytometric analysis will greatly help to fill this gap in our knowledge.

The analysis of geographically isolated populations may shed light on the stability of plant genome. Despite long-term isolation and lack of gene flow between 10 such populations of a grass, *Sesleria albicans*, only negligible interpopulation differences in DNA content were found (136). Similarly, only small differences in nuclear genome size were observed between vegetatively propagated clones of banana (137) and between divergent and isolated populations of Fraser fir (6). No differences were detected in nuclear DNA contents of onion cultivars taken from four continents (18). These observations contrast with reports on intraspecific variation in genome size (35, 91, 151). Whether the variation reported in these and other studies reflects large plasticity of nuclear genome or is, in contrast, a result of various methodological and taxonomical artifacts is a matter of active discussion (95). For instance, intraspecific genome size variation described for pea (33, 34), soybean (91), and pigeonpea (160) has not been confirmed (10, 96, 97). Although clarification of the occurrence of intraspecific variation will continue to be an issue for some time, increasing numbers of taxonomists are employing flow cytometric estimation of nuclear genome size as a means to improve classification (54, 152, 230). In some cases, differences in genome size between related species can be explained, at least in part, by the presence of repetitive DNA sequences (209, 211, 219). Flow cytometric analysis can also be used to chart the occurrence of apomixis in natural populations (192).

Flow Cytometric Analysis of the Cell Cycle

Flow cytometry based on the use of DNA specific fluorochromes can further be employed for analysis of the plant cell division cycle. Because the nuclear DNA content reflects the position of the cell within the cell cycle, the distribution of cells in different phases of the cycle may be estimated by analyzing distributions of relative DNA content (55, 82). Various approaches have been developed to deconvolve DNA content histograms, ranging from simple graphical and nonparametric curve-fitting methods to sophisticated parametric methods (45, 222). The relative simplicity of this type of analysis is counterbalanced by its somewhat low resolution. A further problem is associated with the fact that the fluorescence histograms provide at best only a snapshot in time of the nuclear DNA contents of the tissues and do not provide information about whether specific nuclei are actively progressing through the cell division cycle (i.e., are cycling or quiescent). This is particularly problematic for organs containing mixtures of different tissues and for tissues containing mixtures of different cell types. Furthermore, the method, as it is applied to isolated nuclei, cannot detect M-phase cells. The reliability of the analysis further depends on the precision of DNA content estimation (the coefficient of variation of the DNA peaks), on the relative proportion of background debris, and on the presence of nuclear doublets in the sample as a result of nonspecific adhesion (8, 46, 99).

Commercial software exists, for example, ModFit (Verity Software House, Topsham, ME) and Multicycle (Phoenix Flow Systems, San Diego, CA), that can deal with problems resulting from debris background, nuclei doublets, and the presence of multiple populations differing in ploidy level. Estimates can be made of the duration of individual cell cycle phases based only on measurements of overall growth rate coupled to monoparametric flow analysis (177). However, this approach relies on a simplified assumption of equal cell cycle lengths for all cells within the population and may overestimate the duration of G_1 phase because of the presence of quiescent cells.

Despite these limitations, monoparametric DNA content analysis has been the prevailing method used to examine the cell cycle in higher plants. Cell cycles have been studied in many tissue and organ types; for example, root tips (142), leaves (65), and buds (50). An interesting application involves the analysis of cell cycle distribution in seed embryo during maturation and seed priming (199). A majority of cell cycle studies was performed with synchronized suspension-cultured cells, which facilitates the analysis of gene expression associated with cell cycle regulation (67, 153, 177). In this type of work, flow cytometry can be readily employed to determine the extent of synchronization and to ascertain the position of the synchronized population within the cell cycle. For example, Peres et al. (165) used partially synchronized maize suspension cells to monitor the levels of the transcripts of two cyclin genes, cdc2, two histone genes, and a retinoblastoma gene as a function of progression through the cell cycle. Using synchronized alfalfa cell cultures, Roudier et al. (180) studied the function of a plant A2-type cyclin (CycA2). Surprisingly, the CycA2-associated kinase activity was found to be biphasic, with the results indicating that this cyclin plays a role both in S phase and at the G_2/M transition. Other workers have monitored cyclin-dependent kinase (CDK) protein levels within synchronized tobacco BY-2 cells and were able to demonstrate that a CDK lacking the PSTAIRE motif was involved in the control of G_2/M progression (173).

Flow cytometry plays an even more important role in the elucidation of the effect of various compounds on cell cycle progression. In an attempt to understand the involvement of cyclic AMP, Ehsan et al. (66) found that indomethacin, an inhibitor of cyclic AMP synthesis, causes G_1/S arrest in synchronized tobacco BY-2 cells. The same cell line was used to investigate role of plant growth regulators. Laureys et al. (124) used lovastin to inhibit cytokinin synthesis and demonstrated that transition from G_1 to S occurs even under inhibitory conditions. In contrast, cytokinin addition was found to block cells at the G_1/S boundary, implying that the G_1/S transition is triggered by downregulation of production of zeatin type cytokinins at G_1. Abscisic acid and jasmonic acid play a crucial role in altering plant morphology in response to stress. Swiatek et al. (203) investigated their effect of on cell cycle in synchronized BY-2 cells. They found that although abscisic acid has the potential to prevent DNA synthesis and arrest cells only in G_1, jasmonic acid acts to arrest cells in both G_1 and G_2 phases. In a number of studies, the availability of flow cytometry has been found essential to evaluate the effect of various CDK inhibitors on cell cycle progression (28, 172). In some cases, the use of synchronized cells from root meristems may be advantageous over cell cultures (59). Binarová et al. (24) studied the role of CDKs in cell cycle progression, using synchronized cells from field bean root meristems. Inhibition of CDK caused transient arrest both at the G_1/S and G_2/M boundaries.

Unlike monoparametric analysis of DNA content, multiparametric methods allow more detailed analysis of cell cycle (43). For instance, it is possible to distinguish between cycling and quiescent cells and between cells in G_2 and M phase of the cell cycle. The most popular methods rely on incorporation of 5-bromo-2'-deoxyuridine (BrdU) into newly synthesized DNA in place of thymidine. Incorporated BrdU partially quenches the fluorescence of Hoechst 33258 and 33342 dyes (123), which permits flow cytometric detection of nuclei that have incorporated this analog. DNA content may be simultaneously determined in flow, using a DNA fluorochrome whose fluorescence is not quenched. In plants, Glab et al. (88) and Trehin et al. (208) used the BrdU/Hoechst 33258 technique to investigate the effect on cell cycle progression of the CDK inhibitors olomoucine and roscovitine. Olomoucine and roscovitine were shown to arrest cycling *Arabidopsis* cells in late G_1 and G_2 phases. Trehin et al. (208) also showed that selected plant growth regulators were not able to induce cell cycle progression beyond the G_1 arrest induced by roscovitine. Other workers have used the same technique to follow the cell cycle progression in tobacco BY-2 cells irradiated with ultraviolet light (164).

Incorporated BrdU can be also detected using specific antibodies (92). In contrast to the BrdU/Hoechst 33258 technique, which requires continuous incubation with BrdU, detection of BrdU incorporation by immunofluorescence can be done using both short (0.5–3 hr) pulses and longer periods of labeling. A short BrdU pulse followed by growth in BrdU-free medium permits not only detection of S-phase cells but also establishment of cell cycle length (T_C) or potential doubling times (for cultures within which not all cells are cycling), and duration of the individual cell cycle phases (T_{G1},

A   B

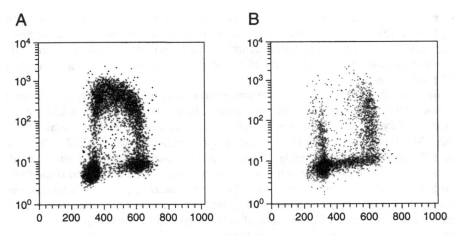

Figure 16.3. Biparametric analysis of cell cycle kinetics within meristematic root-tip cells of field bean (*Vicia faba*). Roots were incubated in a medium containing 30 μM 5-bromo-2'-deoxyuridine (BrdU) for 1 hr and then held in medium lacking BrdU. Samples were taken immediately (*a*) and 6 hr (*b*) after removal from BrdU. Root tips were fixed in 4% formaldehyde, and the nuclei were isolated by mechanical homogenization. Incorporated BrdU was detected via indirect immunofluorescence. 4'-6-diamino-2-phenylindole (DAPI) was used for determination of relative nuclear DNA content. Note the progression of the labeled (BrdU-positive) population from S to G_2/M and then to the G_1 phase of the cell cycle following BrdU removal. This type of pulse-chase analysis enables estimation of the duration of the individual phases of the cell cycle. *X* axis: relative nuclear DNA content; *Y* axis: relative BrdU content (J. Bartoš and J. Doležel, unpublished data).

T_S, T_{G2+M}; figure 16.3; 205, 224). Because the antibodies recognize BrdU only in single-stranded DNA, the protocol must be optimized to ensure a sufficient and equal amount of DNA denaturation within the samples to detect BrdU while leaving enough nondenatured DNA to stain for total DNA content. Up to now, few authors have used this method in plant cell cycle studies. Yanpaisan et al. (227), employing heat denaturation of DNA, analyzed cell cycle kinetics in suspension-cultured cells of *Solanum aviculare*. Lucretti et al. (133), in developing a protocol for cell cycle analysis within root-tip meristems, preferred DNA denaturation by acid treatment. Pelayo et al. (163), employing strong acid treatment for DNA denaturation, evaluated reinitiation of DNA synthesis in onion root meristem following treatment with hydroxyurea.

Other reports of multiparametric cell cycle analysis have included simultaneous detection of DNA and RNA (43). Together, they permit discrimination of quiescent cells, of DNA, and of proliferating cell nuclear antigen (122), which is specifically expressed in S phase, and of DNA and phosphorylated histone H3 (108), which allows discrimination between G_2 and mitotic cells. Despite the evident potential of these methods, they have rarely been applied in plant biology. Bergounioux et al. (20) analyzed DNA and RNA content in *Petunia hybrida* protoplasts during the first 48 hr of in vitro culture. RNA content increased after initiation of the cell cycle.

As previously noted, in many plant species, cell and tissue differentiation is accompanied by endoreduplication. Flow cytometry has been shown to be invaluable in identifying polysomatic tissues, determining the extent of endoreduplication, and analyzing the effect of various intrinsic and extrinsic factors. Galbraith et al. (79) discovered that the extent of endoreduplication in *A. thaliana* is tissue specific and characteristic of the stage of organ development. Similarly, endoreduplication in pod walls of temperate grain legumes is developmentally controlled (121). Flow cytometric analyses of maize endosperm showed that the extent of endoreduplication is controlled by the nuclear genome of the maternal parent and can be altered by defective kernel mutations (53, 114, 115). In all cases, the functional significance of endoreduplication remains to be elucidated. However, several studies have started to probe the underlying regulatory mechanisms. Plant hormones have been implicated in control of endoreduplication (83, 143). Sun et al. (202) have implicated downregulation of a novel B1-type cyclin in the cellular transition to endoreduplication in developing maize endosperm.

Another form of alteration to the conventional cell cycle is termed apoptosis, a process of programmed cell death accompanied by characteristic changes, originally described for mammalian cells, which include cellular shrinkage, chromatin condensation, and DNA fragmentation, followed by nuclear disintegration and formation of apoptotic bodies (111). Apoptosis plays crucial roles in development, in disease states, and in response to pathogen attack and exposure to environmental stress (36, 149). A variety of flow cytometric methods have been devised to detect apoptotic changes in animal cells (44, 217). Most of these changes can be detected in plants (223). The simplest assay for apoptosis involves detection of DNA fragmentation by measuring nuclear DNA content. Fragmentation leads to the occurrence of a so-called sub-G_1 peak. Marubashi et al. (146) and Yamada et al. (225) employed detection of a sub-G_1 peak, via laser-scanning cytometry, as a means to monitor operation of the apoptotic pathway in interspecific *Nicotiana* hybrids. Sub-G_1 peaks were also detected after induction of apoptosis in suspension-cultured tobacco cells (158). Apoptosis can also be monitored through detection of the 3′ OH ends of DNA breaks produced by endonucleolytic cleavage. This is done by attachment of biotin-tagged deoxynucleotide using terminal deoxynucleotidyl transferase, which is subsequently detected using fluorescently tagged avidin. The TUNEL (terminal deoxynucleotidyl transferase–mediated deoxyuridine triphosphate-biotin nick-end labeling) assay has been successfully modified to detect DNA fragmentation in apoptotic plant cells (158). Finally, exposure of phosphatidylserine at the outer surface of the plasma membrane precedes apoptotic nuclear changes (145). O'Brien et al. (158) were able to detect appearance of surface phosphatidylserine in apoptotic plant cells, using fluorescently labeled Annexin V.

Flow Cytometric Analysis and Sorting of Isolated Chromosomes

Compared to human and animal species, the development of procedures for analysis and sorting of mitotic chromosomes (flow cytogenetics) was delayed mainly because of the absence of plant tissues rich in the mitotic cells required for preparation of chromosome suspensions, and because of difficulties associated with releasing intact chromosomes from plant cells enclosed by rigid walls (61). Various approaches have been developed to overcome these problems, and since the first report of De Laat and Blaas (47), chromosome analysis and sorting have been reported in 12 plant species, includ-

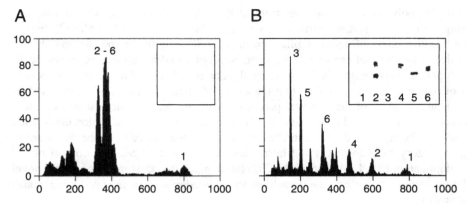

Figure 16.4. Flow karyotypes obtained after analysis of DAPI-stained chromosome suspensions prepared from two lines of field bean (*Vicia faba*). (*a*) Flow karyotype of line Inovec, a cultivar having a standard karyotype. Because of their similar sizes, chromosomes 2–6 form a composite peak and cannot be individually discriminated and sorted. Only the largest chromosome (chromosome 1) forms a separate peak and can be easily discriminated and sorted (insert: flow sorted chromosome 1). (*b*) Flow karyotype of translocation line EF. The lengths, and hence the relative DNA contents of the chromosomes, are changed as a result of reciprocal translocations between chromosomes 2 and 3 and chromosomes 4 and 5, respectively (insert: flow sorted chromosomes 1–6 were identified after labeling of the *Fok*I repeat sequence). *X* axis: relative DAPI fluorescence intensity; *Y* axis: number of events.

ing economically important legumes and cereals (62). A typical protocol for chromosome isolation and sorting consists of cell cycle synchronization and accumulation of cells in metaphase, release of chromosomes into an isolation buffer, staining of the chromosome suspension using a DNA fluorochrome, and flow analysis of the relative fluorescence intensity (63). The resultant histograms of fluorescence intensity are termed the *flow karyotype*, within which, ideally, each chromosome should be represented by a single peak (figure 16.4). Any chromosome thus represented can be sorted at high speed and purity. The suitability of flow karyotypes for sorting can also be addressed through modeling (39).

Different types of biological materials have been employed as sources of mitotic chromosomes. In their pioneering work, De Laat and Blaas (47) isolated chromosomes from suspension-cultured cells of *Haplopappus gracilis* (2n = 4). The choice of a model with only two chromosomes differing in size facilitated their discrimination within the flow karyotype and allowed sorting at purities over 95%. Analysis and sorting of chromosomes prepared from suspension-cultured cells have subsequently been reported for *Nicotiana plumbaginifolia* (213), *Lycoperson esculentum*, *Lycoperson pennelli* (3), and *Triticum aestivum* (221). As an alternative to cultured cells, which can be karyologically unstable, leaf mesophyll protoplasts synchronized by transfer to nutrient medium have been used for preparation of chromosomes of *N. plumbaginifolia* and *Petunia hybrida* (38, 39).

In 1992, Doležel et al. (58) developed a protocol for the preparation of chromosome suspensions from root-tip meristems of seedlings. A combined treatment with a DNA synthesis inhibitor and a mitotic spindle poison led to the accumulation of over 50%

of the cells within metaphase. Suspensions of intact chromosomes were prepared by mechanical homogenization of formaldehyde-fixed root tips. Because of its simplicity and reliability, the procedure has been widely adopted and used for a number of species, including *Vicia faba* (132), *Pisum sativum* (98), *Hordeum vulgare* (135), *T. aestivum* (220), and *Cicer arietinum* (184). A modified version of the protocol, omitting formaldehyde fixation, has been used in *Zea mays*, *T. aestivum*, and *Oryza sativa* (125–127). However, formaldehyde-fixed chromosomes are more stable and resistant to mechanical shearing forces during sorting. Veuskens et al. (218) introduced the idea of using *Agrobacterium rhizogenes*–transformed root cultures instead of seedlings. Cultured roots are an attractive system for maintenance of specific chromosome stocks, which are difficult to propagate by seed. Neumann *et al.* (155) used hairy root cultures to prepare chromosome suspensions of *P. sativum*.

In many plant species, chromosomes have similar sizes and DNA contents. Usually, only one or two chromosomes can be discriminated, and unfortunately, the number of sortable chromosomes does not increase with increasing chromosome number (62). Quite surprisingly, bivariate analysis of AT/GC ratio, which has been successfully used in human and animal flow cytogenetics, has not been helpful in discriminating plant chromosomes (131, 188). This is most probably a consequence of the relatively homogeneous distribution of repetitive DNA sequences across plant chromosomes. To address this problem, Macas et al. (138) developed a protocol for fluorescent labeling of repetitive DNA sequences on chromosomes in suspension, using primed in situ DNA labeling. Pich et al. (170) demonstrated the utility of this approach by discriminating all five acrocentric chromosomes of *V. faba*, which have similar DNA contents.

The most efficient strategy for chromosome discrimination, proposed by Lucretti et al. (132), involves the use of genetics to construct plant stocks containing specific chromosomal combinations. Doležel and Lucretti (57) showed that the use of *V. faba* lines, where chromosome length has been changed because of one or more translocation events, permitted sorting of all chromosome types within the karyotype. The utility of chromosome translocations for chromosome sorting was subsequently confirmed in *P. sativum* (156), *H. vulgare* (135), and *T. aestivum* (221). In a similar manner to translocations, deletions can result in changed chromosome morphology and thereby enable chromosome discrimination. Sorting of single-chromosome arms from ditelosomic lines of *T. aestivum* has been reported (figure 16.5a; 87, 120). Chromosome-addition lines are another attractive approach for chromosome discrimination: The alien chromosome can be readily sorted, provided it differs in DNA content from all of the chromosomes of the host species. Thus, Li et al. (129) sorted maize chromosome 9 from an oat–maize chromosome addition line, whereas Kubaláková et al. (118) sorted various rye chromosomes from wheat–rye addition lines (figure 16.5b).

Chromosome Sorting Applications

In contrast to traditional chromosome preparations, chromosomes flow sorted onto microscope slides are free of cytoplasmic contaminants. They therefore are invaluable for physical mapping of DNA sequences using fluorescence in situ hybridization and primed in situ synthesis methods of labeling (117, 119). Furthermore, Valárik et al. (212) demonstrated that chromosomes sorted onto microscope slides could be stretched

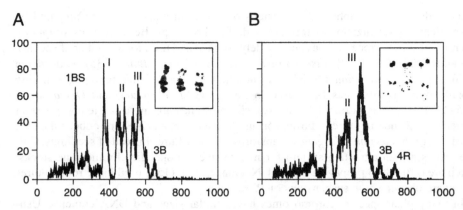

Figure 16.5. Flow karyotypes obtained after analysis of DAPI-stained chromosome suspensions prepared from two lines of wheat (*Triticum aestivum*, 2n = 6x = 42). (*a*) The flow karyotype of wheat line Pavon carrying only the short arm of chromosome 1B (1BS) consists of three composite peaks (I–III) representing groups of wheat chromosomes and a peak representing chromosome 3B. In addition, a peak of 1BS is clearly discriminated (insert: flow-sorted 1BS identified after labeling of GAA microsatellites). (*b*) The flow karyotype of a Chinese Spring/Imperial wheat/rye addition line carrying a pair of rye chromosome 4 (4R). Note that the peak of 4R is clearly discriminated, which facilitates sorting of the chromosome (insert: flow sorted chromosome 4R was identified after labeling the pSc119,2 repeat sequence). X axis: relative DAPI fluorescence intensity; Y axis: number of events.

up to 100-fold, which allows physical mapping with a spatial resolution comparable to that of pachytene chromosomes. Provided the chromosomes have been fixed by formaldehyde, they are also suitable for localization of proteins using immunofluorescent methods (25, 204). Last but not least, flow-sorted chromosomes can be used for the analysis of chromatin structure by electron microscopy (187).

Physical gene mapping using polymerase chain reaction (PCR) has been by far the most frequent application of chromosome flow sorting in plants. Tens to hundreds of chromosomes, which are sufficient for PCR, can be sorted in few minutes. It is therefore possible to employ two-step sorting (132) to considerably increase the purity of the sorted fraction. Macas et al. (139) used flow-sorted chromosomes to localize seed storage protein genes and pseudogenes to individual chromosomes of *V. faba*. Using flow-sorted autosomes and sex chromosomes, Kejnovský et al. (109) physically mapped four male reproductive organ–specific genes in the dioecious plant *Melandrium album*. Similarly, Neumann *et al.* (156) used PCR to localize genetic markers to specific chromosomes of *P. sativum* and assigned linkage groups IV and VII to chromosomes 4 and 7, respectively.

Chromosomal DNA amplified by PCR with degenerate oligonucleotide primers has been found useful for construction of chromosome-specific DNA libraries. Macas et al. (140) prepared a set of chromosome-specific DNA libraries from seven chromosome types purified from two translocation lines of *V. faba*. The set, with a mean insert size ranging between 310 and 487 bp, covers the whole *V. faba* genome more than once. Short-insert DNA libraries have been used for targeted isolation of molecular markers from defined genomic regions. Thus, Arumuganathan et al. (4) generated a chromosome 2–specific library of *Lycopersicon pennellii*, and from it isolated 11 RFLP markers, all of which mapped to chromosome 2. More recently, Požárková et al. (174) used flow-sorted chromosomes of *V. faba* for targeted retrieval of microsatellite markers specific to a part of chromosome 1. Chromosome 1–specific markers were mapped to linkage groups 1a and 1b of the composite map of *V. faba*. The work resulted in assignment of the linkage group 1b to chromosome 1 and confirmed the previous assignment of group 1a to the same chromosome.

In contrast to short-insert libraries, no long-insert DNA libraries (e.g., cloned in a BAC vector) have been constructed from flow-sorted plant chromosomes. Chromosome-specific DNA libraries with insert sizes exceeding 100 kb would be extremely useful in analyzing complex genomes, such as those of *T. aestivum* and other crops. The delay may be a result of the fact that the construction of BAC libraries requires microgram amounts of high–molecular weight DNA. Nevertheless, recent results indicated that it is possible to prepare high–molecular weight DNA from sorted plant chromosomes (197, 220). It seems thus realistic to expect the construction of chromosome-specific BAC libraries in plants in the near future (see note on page 312).

Flow Cytometric Analysis and Sorting of Plant Protoplasts

Plant protoplasts are produced by enzymatic digestion of the cellulosic cell wall, using lytic enzymes produced by wood-degrading microorganisms. Digestion is done in the presence of slightly hypertonic osmolytes, typically polyhydric alcohols such as sorbitol and mannitol. Protoplasts are spherical, fragile, single cells, but under appropriate culture conditions, many examples are seen of regeneration of the cell wall, which then can be followed by cell division, tissue differentiation, and plant regeneration. Much interest has therefore focused on the use of flow cytometry and cell sorting for selection and purification of protoplasts of defined properties for further experimentation in culture. Protoplast diameters typically range from 20 to 100 μm, and this originally presented a conflict with the sizes of standard flow cytometer tips (50–70 μm). This was resolved (102) by the use of large flow tips (100–200 μm) and the observation that the physics that governs cell sorter operation is reasonably scalable.

In brief, formation of droplets is governed by the relationship $\lambda > \pi D_j$, where λ is the wavelength of the undulation produced by the piezoelectric transducer and D_j is the diameter of the fluid jet emerging from the flow tip (102). The relationship between the piezo frequency (f), the undulation wavelength, and the flow velocity (c) is governed by the wave equation $\lambda \nu = c$. The flow velocity is evidently a function of the diameter of the flow tip and the system pressure. Combining these restrictions has the following consequences: for successful sorting using large flow tips, it is necessary to reduce both the system pressure and the piezo drive frequency, relative to the settings used for standard, smaller flow tips [examples of appropriate instrument settings are given in (101)]. This reduces the rate at which sorting can be done and therefore extends the time required for sorting large numbers of cells. As a general rule, longer sort delay settings (the time delay between detecting the desired cell and applying the voltage to the flow stream) are required, as the droplet breakoff point occurs at a lower point below the flow tip. Given the fragile nature of protoplasts, setting up for sorting (which in part involves defining the sort delay setting) requires the use of indestructible fluorescent particles having similar diameters to the protoplasts. This is because particles within the flow stream can perturb the process of droplet formation, and therefore sorting parameters optimized using small fluorescent particles may not be optimal for sorting of larger particles. We find that pollen preparations from various plant species are suitable for this purpose (102). Under appropriate conditions, protoplast viability is maintained during sorting, and these sorted protoplasts are capable of regenerating into fertile plants (1, 74).

Given their large sizes, flow analysis of protoplasts can be simply achieved using forward or 90° angle light-scatter signals, although neutral-density filters are frequently required to reduce the intensities of the scattered light. Time-of-flight measurements can also be used for accurate sizing of protoplasts (78) and for calculation of surface areas and volumes (74). The types of available experiments are greatly increased through incorporation of fluorescence into the measurements. The presence of chlorophyll within photosynthetic cells provides a naturally intense fluorescent signal, which can be both readily excited using available laser wavelengths and detected with the conventional filter sets of commercial flow cytometers. The intensity of the fluorescence emission can be used to quantify chlorophyll contents of protoplasts (78), and fluorescence can also be added either through treatment of protoplasts or their parent cells with fluorescent reagents, resulting in attachment (covalent or noncovalent) of fluorescence to the protoplasts (75, 77), or through addition of fluorescently tagged molecules that bind to the cell surface (215). This approach has been applied pairwise to different protoplast populations to allow flow sorting of heterokaryons produced by induced protoplast fusion, followed by regeneration of somatic hybrid plants (1, 68, 130). Various reports have employed naturally fluorescent cellular components (e.g., organelles and metabolites) for characterization of protoplast integrity, physiology, and metabolism (19, 74, 85, 86, 102). Fluorescent signals derived from expression of the GFP of *Aequorea victoria* have also been employed for flow analysis and sorting of protoplasts (80, 194), and this approach has been extended for the analysis of signal transduction in higher plant stress responses (100).

Flow Cytometric Analysis and Sorting of Plant Organelles

The ability of flow cytometry to analyze subcellular organelles is not limited to the DNA content and cell cycle applications outlined above. In higher plants, chloroplasts are naturally fluorescent, and other organelles can be rendered fluorescent either through targeted expression of fluorescent proteins or following treatment with organelle-specific fluorochromes.

Nuclei

Targeting of GFP to the plant nucleus can be achieved through translational fusion to nuclear localization signals (NLSs), providing the size of the fusion protein exceeds the exclusion limit of the nuclear pore (~40 kDa). This allows retention of the protein within the nucleus after NLS-mediated active transport into the nucleoplasm. We have employed an effective unipartite NLS from an orphan transcription factor of *N. tabacum* to direct a GFP-β glucuronidase chimeric protein to the nuclei of transfected protoplasts (93) and transgenic plants (37, 94). Flow cytometry can be employed to detect fluorescent nuclei within tissue homogenates, and in combination with nuclear expressed sequence tag analysis (141), leads to the strategy of flow sorting of nuclei for global characterization of gene expression.

In brief, this strategy depends on the known existence of 5' regulatory sequences that confer cell or tissue type-specific expression of transgenic markers, such as GFP. Coupling these sequences to the nuclear-targeted version of GFP results in transgenic plants with fluorescent nuclei restricted to the cells within which the regulatory sequences are active (37). Flow sorting in principle provides a means to selectively purify those nuclei. Characterization of global gene expression can be done using a modified form of amplified fragment length polymorphism analysis (141), in which the 3' ends of nuclear transcripts are selectively amplified and characterized via polyacrylamide gel electrophoresis. Alternative strategies, involving variants of serial amplification of gene expression (216) or hybridization of linearly amplified transcripts to DNA microarrays also appear promising (D.W. Galbraith, F.C. Gong, and R.P. Elumalai, unpublished data).

Chloroplasts

The first report of the application of flow cytometry for the specific analysis of chloroplasts employed forward and 90° light scatter and fluorescence emission signals to characterize intact chloroplasts and thylakoids prepared from spinach and maize (5). Subsequently, flow cytometry has been used to explore the integrity of isolated spinach chloroplasts, based on forward angle and 90° light scatter and chlorophyll autofluorescence (186). When combined with labeling with fluorescein isothiocyanate–linked lectins, intact chloroplasts could be distinguished from thylakoid membranes and from various chloroplast membrane subfractions. Flow cytometric analysis of lectin binding also provided a convenient means to determine the degree of surface exposure of specific glycosyl residues on the various organelle fractions.

Flow cytometry has also been employed for characterization of chloroplast thylakoids isolated from mesophyll and bundle sheath cells of maize (169). Differences in chlorophyll fluorescence spectra provide a convenient means to distinguish these different thylakoid types, and flow sorting can then be used to isolate them in highly purified form. The authors were unable to detect the major light-harvesting complex of photosystem II in the sorted bundle sheath chloroplasts, indicating that conventional biochemical methods for purification of these chloroplasts results in contamination by mesophyll chloroplasts.

Fluorescent tags for chloroplast function can also be employed for flow cytometric analysis and sorting. Subramanian et al. (200) produced an epitope-tagged transit peptide taken from the precursor of the small subunit of pea ribulose bisphosphate carboxylase, a chloroplast-targeted protein. Binding of this chimeric peptide to the chloroplast translocation apparatus was monitored, using FITC-conjugated S-protein as a means to detect the epitope. The peptide was found to bind to the translocation apparatus but not to be translocated across the chloroplast envelope. Binding was saturable, slightly dependent on ATP, and inhibited by addition of guanosine triphosphate. Confocal microscopy indicated patch-like labeling, implying that the translocation apparatus is localized to discrete regions on the chloroplast envelope.

Yang et al. (226) have described a method for the measurement of starch granules within chloroplasts, using flow cytometry. Although it does not provide an absolute measurement of the amounts of starch, this method might be useful as a screening tool to rapidly identify mutants, leading to altered starch levels. Flow cytometric methods have also been described for the analysis of suspensions of starch granules prepared from detergent-treated endosperm (107). These suspensions provide a rapid and accurate means for counting starch granule numbers and revealed at least three subclasses of granules. The timing of the appearance of these subclasses was regulated during kernel development, but their significance remains to be established.

Mitochondria

The only report of the application of flow cytometry for the analysis of plant mitochondria (166) employed Rhodamine 123 to monitor membrane potential. Variations were detected in membrane potential following the addition of succinate and ATP and in response to treatments with metabolic inhibitors. Changes in the light scatter properties of the mitochondria were found to be related to modifications of the inner membrane-matrix system rather than to changes in overall volume or surface modifications to the outer membrane. In further work, these flow cytometric methods were able to distinguish known subpopulations of *Arum maculatum* mitochondria.

Future Directions

High-Throughput Biology

Much current interest is focused on the development and use of high-throughput methods to provide large data sets of information concerning the molecular functions of cells, tissues, and organisms. Technological advances have made it possible to investigate these functions in a parallel, genome-wide manner. This has led to the opening of new and revolutionary fields of scientific endeavor, which, based on the class of molecule that is being investigated, are popularly designated as the various "-omics" (genomics, proteomics, metabolomics, and so on; 71). Plant biologists have participated fully in this revolution, including the completion of sequencing of the genome of one model plant (206) and, in draft form, of one crop (90, 228). DNA microarrays, one of the most important technologies of functional genomics (51), were originally developed to study gene expression in *Arabidopsis thaliana* (185). Given the inherent similarities of kingdoms of eukaryotes, it is obvious that further advances in genomics technologies will be fully applicable to the plant kingdom.

Flow cytometry and cell sorting are likely to play an important role in the development of further high-throughput applications. Sorting on the basis of transgenic expression of marker molecules has the potential to provide homogeneous populations of protoplasts for further study, which should allow dissection of transcriptional regulatory mechanisms underlying cell and tissue-type specific gene expression. The ability to target fluorescent proteins to nuclei also should allow sorting of nuclei from specific cellular subtypes, and this again should prove a crucial tool for dissecting gene expression. The ability to sort pure populations of organelles will provide a means for studying their protein complements. In this respect, further methods for flow analysis and sorting of mitochondria and chloroplasts can be anticipated on the basis of the successful transgenic expression and targeting of GFP and related molecules to these organelles (112, 113, 176). Similar methods for sorting other organelles, such as golgi and peroxisomes, should allow rapid progress in understanding how these organelles are constructed and how they function. Reports are also beginning to emerge of the use of global methods to produce large numbers of transgenic plants individually expressing fluorescent protein fusions (41). The application of fluorescence resonance energy transfer methods to transgenic plants expressing combinations of different fluorescent proteins should enable systematic analysis of protein–protein interactions in vivo. It is evident that flow cytometric methods might well have a role to play in measurements of this type.

In terms of sorting chromosomes, there is no doubt that large-insert chromosome- or chromosome-arm-specific libraries would considerably simplify the analysis and mapping of complex genomes of most crop plants. Based on the published data, it seems realistic to expect construction of the first such libraries in the very near future. The analysis of the evolution and propagation of repetitive DNA sequences within plant genomes represents what has been until now a neglected application of flow cytogenetics. Flow cytometry is the only method that can produce sufficient amounts of DNA from individual chromosomes for separate analysis. The possibility of sorting large amounts of chromosomes also opens the way to the application of proteomics approaches to systematically identify and analyze proteins of mitotic chromosomes. A final, attractive aspect of chromosome sorting would be their use for chromosome-mediated gene transfer.

The progress in the analysis of cell cycle regulation at a molecular level and an increasing number of candidate genes requires powerful methods to analyze cell cycle kinetics and gene expression. Multiparametric flow cytometry has a potential for a detailed cell cycle analysis, including endoreduplication and apoptosis. Pulse-designed BrdU experiments should be an alternative to monoparametric methods for estimation of cell cycle parameters. There is no doubt that with the availability of antibodies to various proteins, multiparametric flow cytometry will play an increasingly important role in plant cell cycle studies.

Although ploidy analysis using flow can be considered a routine application, further improvements can be expected, aimed particularly at increasing the resolution to permit routine detection of aneuploidy. Experiments aiming at screening large populations of plants outside of the laboratory environment would benefit from the development of reliable protocols for storage of plant tissues before analysis. At present, the methods for sample preparation require fresh plant materials, which restricts the broader use of flow cytometry in taxonomy and population biology. Alternatively, development of fully portable flow cytometers could obviate this problem. Clearly, flow cytometry will continue to play a major role in obtaining data on genome size, particularly for the large number of plant taxa for which this information is currently limited. Such information is essential to clarify the evolutionary trends in genome size evolution and its biological role. Finally, agreement on a unified set of standards has yet to be achieved. Such an agreement would facilitate comparison of data obtained by different research teams.

Acknowledgments   This work was supported by grants to D. Galbraith from the National Science Foundation (awards DBI 9813360, DBI 9872657, DBI 0211857, and DBI 0321663) and by research grants to J. Doležel from the Czech Science Foundation (awards 521/03/0595, 204/04/1207, and 521/04/0067).

Note   We have created three large-insert DNA libraries from flow-sorted chromosomes in wheat: a subgenomic BAC library specific for chromosomes 1D, 4D, and 6D, a BAC library specific for chromosomes 3B, and a BAC library specific for short arm of chromosome 1B (1BS). The first two have already been published: Janda, J., Bartoš, J., Šafář, J., Kubaláková, M., Valárik, M., Číhalíková, J., Šimková, H., Caboche, M., Sourdille, P., Bernard, M., Chalhoub, B., Doležel, J. (2004), Construction of a subgenomic BAC, library specific for chromosomes 1D, 4D and 6D of hexaploid wheat, *Theor. Appl. Genet.* 109:1337–1345; Šafář, J., Bartoš, J. Janda, J., Bellec, A., Kubaláková, M., Valárik, M., Pateyron, S., Weiserová, J., Tušková, R., Číhalíková, J., Vrána, J., Šimková, H., Faivre-Rampant, P., Sourdille, P., Caboche, M., Bernard, M., Doležel, J., Chalhoub, B. (2004), Dissecting large and complex genomes: flow sorting and BAC cloning of individual chromosomes from bread wheat, *Plant J.* 39:960–968).

References

1. Afonso, C.L., Harkins, K.R., Thomas-Compton, M., Krejci, A., Galbraith, D.W. 1985. Production of somatic hybrid plants through fluorescence activated sorting of protoplasts. *Biotechnology* 3:811–816.
2. Amsellem, L., Chevallier, M.H., Hossaert-Mckey, M. 2001. Ploidy level of the invasive weed *Rubus alceifolius* (*Rosaceae*) in its native range and in areas of introduction. *Plant Syst. Evol.* 228:171–179.
3. Arumuganathan, K., Slattery, J.P., Tanksley, S.D., Earle, E.D. 1991. Preparation and flow cytometric analysis of metaphase chromosomes of tomato. *Theor. Appl. Genet.* 82:101–111.
4. Arumuganathan, K., Martin, G.B., Telenius, H., Tanksley, S.D., Earle, E.D. 1994. Chromosome 2-specific DNA clones from flow-sorted chromosomes of tomato. *Mol. Gen. Genet.* 242:551–558.
5. Ashcroft, R.G., Preston, C., Cleland, R.E., Critchley, C. 1986. Flow cytometry of isolated chloroplasts and thylakoids. *Photobiochem. Photobiophys.* 13:1–14.
6. Auckland, L.D., Johnston, J.S., Price, H.J., Bridgwater F.E. 2001. Stability of nuclear DNA content among divergent and isolated populations of Fraser fir. *Can. J. Bot.* 79:1375–1378.
7. Baert, J., Reheul, D., Vanbockstaele, E., DeLoof, A. 1992. A rapid method by flow cytometry for estimating persistence of tetraploid perennial ryegrass in pasture mixtures with diploid perennial ryegrass. *Biol. Plant.* 34:381–385.
8. Bagwell, C.B., Mayo, S.W., Whetstone, S.D., Hitchcox, S.A., Baker, D.R., Herbert, D.J., Weaver, D.L., Jones, M.A., Lovett, E.J. 1991. DNA histogram debris theory and compensation. *Cytometry* 12:107–118.
9. Bantin, J., Matzk, F., Dresselhaus, T. 2001. *Tripsacum dactyloides* (*Poaceae*): a natural model system to study parthenogenesis. *Sex. Plant Reprod.* 14:219–226.
10. Baranyi, M., Greilhuber J. 1996. Flow cytometric and Feulgen densitometric analysis of genome size in *Pisum. Theor. Appl. Genet.* 92:297–307.
11. Barker, R.E., Kilgore, J.A., Cook, R.L., Garay, A.E., Warnke, S.E. 2001. Use of flow cytometry to determine ploidy level of ryegrass. *Seed Sci. Technol.* 29:493–502.
12. Barow, M., Meister A. 2002. Lack of correlation between AT frequency and genome size in higher plants and the effect of nonrandomness of base sequences on dye binding. *Cytometry* 47:1–7.
13. Barre, P., Layssac, M., D'Hont, A., Louarn, J., Charrier, A., Hamon, S., Noirot M. 1998. Relationship between parental chromosomic contribution and nuclear DNA content in the

coffee interspecific hybrid *C. pseudozanguebariae* x *C. liberica* var. "dewevrei." *Theor. Appl. Genet.* 96:301–305.

14. Bashir, A., Auger, J.A., Rayburn, A.L. 1993. Flow cytometric DNA analysis of wheat-rye addition lines. *Cytometry* 14:843–847.

15. Bennett, M.D. 1973. Nuclear characters in plants. *Brookhaven Symp. Biol.* 25:344–366.

16. Bennett, M.D., Smith, J.B. 1976. Nuclear DNA amounts in angiosperms. *Phil. Trans. R. Soc. Lond.* B274:227–274.

17. Bennett, M.D., Bhandol, P., Leitch, I.J. 2000. Nuclear DNA amounts in angiosperms and their modern uses—807 new estimates. *Ann. Bot.* 86:859–909.

18. Bennett, M.D., Johnston, S., Hodnett, G.L., Price, H.J. 2000. *Allium cepa* L. cultivars from four continents compared by flow cytometry show nuclear DNA constancy. *Ann. Bot.* 85:351–357.

19. Bergounioux, C., Brown, S.C., Petit P.X. 1992. Flow cytometry and plant protoplast cell biology. *Physiol. Plant.* 85:374–386.

20. Bergounioux, C., Perennes, C., Brown, S.C., Gadal P. 1988. Cytometric analysis of growth-regulator-dependent transcription and cell-cycle progression in *Petunia* protoplast cultures. *Planta* 175:500–505.

21. Bharathan, G. 1996. Reproductive development and nuclear DNA content in angiosperms. *Am. J. Bot.* 83:440–451.

22. Bharathan, G., Lambert, G.M., Galbraith, D.W. 1994. Nuclear DNA contents of monocotyledons and related taxa. *Am. J. Bot.* 81:381–386.

23. Binarová, P., Doležel J. 1988. Alfalfa embryogenic cell-suspension culture—growth and ploidy level stability. *J. Plant Physiol.* 133: 561–566.

24. Binarová, P., Doležel, J., Dráber, P., Heberle-Bors, E., Strnad, M., Bögre, L. 1998. Treatment of *Vicia faba* root tip cells with specific inhibitors to cyclin-dependent kinases leads to abnormal spindle formation. *Plant J.* 16:697–707.

25. Binarová, P., Hause, B., Doležel, J., Dráber P. 1998. Association of γ-tubulin with kinetochore/centromeric region of plant chromosomes. *Plant J.* 14:751–757.

26. Binsfeld, P.C., Wingender, R., Schnabl, H. 2000. Characterization and molecular analysis of transgenic plants obtained by microprotoplast fusion in sunflower. *Theor. Appl. Genet.* 101:1250–1258.

27. Blondon, F., Marie, D., Brown, S., Kondorosi, A. 1994. Genome size and base composition in *Medicago sativa* and *M. truncatula* species. *Genome* 37:264–270.

28. Bogre, L., Zwerger, K., Meskiene, I., Binarova, P., Csizmadia, V., Planck, C., Wagner, E., Hirt, H., Heberlebors E. 1997. The cdc2Ms kinase is differently regulated in the cytoplasm and in the nucleus. *Plant Physiol.* 113:841–852.

29. Brummer, E.C., Cazcarro, P.M., Luth, D. 1999. Ploidy determination of alfalfa germplasm accessions using flow cytometry. *Crop Sci.* 39:1202–1207.

30. Brutovska, R., Cellarova, E., Doležel, J. 1998. Cytogenetic variability of in vitro regenerated *Hypericum perforatum* L. plants and their seed progenies. *Plant Science* 133:221–229.

31. Burton, T.L., Husband B.C. 1999. Population cytotype structure in the polyploid *Galax urceolata* (*Diapensiaceae*). *Heredity* 82:381–390.

32. Cavalier-Smith, T. 1978. Nuclear volume control by nucleoskeletal DNA, selection for cell volume and cell growth rate, and the solution of the C-value paradox. *J. Cell Sci.* 34:247–278.

33. Cavallini, A., Natali L. 1990. Nuclear DNA variability within *Pisum sativum* (*Leguminosae*): cytophotometric analysis. *Plant Syst. Evol.* 173:179–183.

34. Cavallini, A., Natali, L., Cionini, G., Gennai D. 1993. Nuclear DNA variability within *Pisum sativum* (*Leguminosae*): nucleotypic effects on plant growth. *Heredity* 70:561–565.

35. Ceccarelli, M., Falistocco, E., Cionini, P.G. 1992. Variation of genome size and organisation within hexaploid *Festuca arundinacea*. *Theor. Appl. Genet.* 83:273–278.

36. Chandra, J., Samali, A., Orrenius, S. 2000. Triggering and modulation of apoptosis by oxidative stress. *Free Radical Biol. Med.* 29:323–333.

37. Chytilová, E., Macas, J., Galbraith, D.W. 1999. Green fluorescent protein targeted to the nucleus, a transgenic phenotype useful for studies in plant biology. *Ann. Bot.* 83:645–654.

38. Conia, J., Bergounioux, C., Perennes, C., Muller, P., Brown, S., Gadal, P. 1987. Flow cytometric analysis and sorting of plant chromosomes from *Petunia hybrida* protoplasts. *Cytometry* 8:500–508.
39. Conia, J., Muller, P., Brown, S., Bergounioux, C., Gadal, P. 1989. Monoparametric models of flow cytometric karyotypes with spreadsheet software. *Theor. Appl. Genet.* 77:295–303.
40. Coutos-Thevenot, P., Jouanneau, J.P., Brown, S., Petiard, V., Guern, J. 1990. Embryogenic and non-embryogenic cell lines of *Daucus carota* cloned from meristematic cell clusters: relation with cell ploidy determined by flow cytometry. *Plant Cell Rep.* 8:605–608.
41. Cutler, S.R., Ehrhardt, D.W., Griffitts, J.S., Somerville C.R. 2000. Random GFP :: cDNA fusions enable visualization of subcellular structures in cells of Arabidopsis at a high frequency. *Proc. Natl. Acad. Sci. USA* 97:3718–3723.
42. D'Amato, F. 1977. *Nuclear Cytology in Relation to Development.* Cambridge: Cambridge University Press.
43. Darzynkiewicz, Z. 1994. Simultaneous analysis of cellular RNA and DNA content. *Methods Cell Biol.* 41:401–420.
44. Darzynkiewicz, Z., Bedner, E., Smolewski P. 2001. Flow cytometry in analysis of cell cycle and apoptosis. *Semin. Hematol.* 38:179–193.
45. Dean, P.N. 1986. Data analysis in cell cycle kinetics research. In *Techniques in Cell Cycle Analysis*, ed. J.E. Gray, and Z. Darzynkiewicz, pp. 207–253. New York: Humana Press
46. Dean, P.N., Gray, J.W., Dolbeare, F.A. 1982. The analysis and interpretation of DNA distributions measured by flow cytometry. *Cytometry* 3:463–467.
47. De Laat, A.M.M., Blaas, J. 1984. Flow-cytometric characterization and sorting of plant chromosomes. *Theor. Appl. Genet.* 67:463–467.
48. DeRocher, E.J., Harkins, K.R., Galbraith, D.W., Bohnert H.J. 1990. Developmentally regulated systemic endopolyploidy in succulents with small genomes. *Science* 250:99–101.
49. De Schepper, S., Leus L., Mertens, M., Van, B.E., De, L.M., Debergh, P., Heursel J. 2001. Flow cytometric analysis of ploidy in *Rhododendron* (subgenus Tsutsusi). *Hortscience* 36:125–127.
50. Devitt, M.L., Stafstrom, J.P. 1995. Cell-cycle regulation during growth-dormancy cycles in pea axillary buds. *Plant Molec. Biol.* 29:255–265.
51. Deyholos, M.K., Galbraith, D.W. 2001. High-density DNA microarrays for gene expression analysis. *Cytometry* 43:229–238.
52. Diaz-Frances, E., Sprott, D.A. 2001. Statistical analysis of nuclear genome size of plants with flow cytometer data. *Cytometry* 45:244–249.
53. Dilkes, B.P., Dante, R.A., Coelho, C., Larkins, B.A. 2002. Genetic analyses of endoreduplication in *Zea mays* endosperm: evidence of sporophytic and zygotic maternal control. *Genetics* 160:1163–1177.
54. Dimitrova, D., Ebert, I., Greilhuber, J., Kozhuharov, S. 1999. Karyotype constancy and genome size variation in Bulgarian *Crepis foetida* s.l. (*Asteraceae*). *Plant Syst. Evol.* 217:245–257.
55. Doležel, J. 1991. Flow cytometric analysis of nuclear DNA content in higher plants. *Phytochem. Anal.* 2:143–154.
56. Doležel, J., Göhde W. 1995. Sex determination in dioecious plants *Melandrium album* and *M. rubrum* using high-resolution flow cytometry. *Cytometry* 19:103–106.
57. Doležel, J., Lucretti S. 1995. High-resolution flow karyotyping and chromosome sorting in *Vicia faba* lines with standard and reconstructed karyotypes. *Theor. Appl. Genet.* 90:797–802.
58. Doležel, J., Číhalíková, J., Lucretti, S. 1992. A high-yield procedure for isolation of metaphase chromosomes from root tips of *Vicia faba*. L. *Planta* 188:93–98.
59. Doležel, J., Číhalíková, J., Weiserová, J., Lucretti, S. 1999. Cell cycle synchronization in plant root meristems. *Methods Cell Sci.* 21:95–107.
60. Doležel, J., Greilhuber, J., Lucretti, S., Meister, A., Lysák, M.A., Nardi, L., Obermayer, R. 1998. Plant size estimation by flow cytometry: Inter-laboratory comparison. *Ann. Bot.* 82 (Suppl. A):17–26.

61. Doležel, J., Lucretti, S., Schubert I. 1994. Plant chromosome analysis and sorting by flow cytometry. *Crit. Rev. Plant Sci.* 13:275–309.
62. Doležel, J., Kubaláková, M., Bartoš, J., Macas, J. 2004. Flow cytogenetics and plant genome mapping. *Chrom. Res.* 12:77–91.
63. Doležel, J., Macas, J., Lucretti, S. 1999. Flow analysis and sorting of plant chromosomes. In *Current Protocols in Cytometry*, ed. J.P. Robinson, Z. Darzynkiewicz, P.N. Dean, L.G. Dressler, A. Orfao, P.S. Rabinovitch, C.C. Stewart, H.J. Tanke, and L.L. Wheeless, pp. 5.3.1–5.3.33. New York: Wiley.
64. Doležel, J., Sgorbati, S., Lucretti, S. 1992. Comparison of three DNA fluorochromes for flow cytometric estimation of nuclear DNA content in plants. *Physiol. Plant.* 85:625–631.
65. Doleželová, M., Doležel, J., Neštický, M. 1992. Relationship of embryogenic competence in maize (*Zea mays* L.) leaves to mitotic-activity, cell-cycle and nuclear-DNA content. *Plant Cell Tiss. Org. Cult.* 31:215–221.
66. Ehsan, H., Roef, L., Witters, E., Reichheld, J.P., Van Bockstaele, D., Inze, D., Van Onckelen H. 1999. Indomethacin-induced G1/S phase arrest of the plant cell cycle. *FEBS Lett.* 458:349–353.
67. Fabian, T., Lorbiecke, R., Umeda, M., Sauter, M. 2000. The cell cycle genes cycA1;1 and cdc2Os-3 are coordinately regulated by gibberellin *in planta*. *Planta* 211:376–383.
68. Fahleson, J., Eriksson, I., Landgren, M., Stymne, S., Glimelius K. 1994. Intertribal somatic hybrids between *Brassica napus* and *Thlaspi perforatum* with high content of the *T. perforatum*-specific nervonic acid. *Theor. Appl. Genet.* 87:795–804.
69. Farnham, M.W., Caniglia, E.J., Thomas, C.E. 1998. Efficient ploidy determination of anther-derived broccoli. *Hortscience* 33:323–327.
70. Farnham, M.W. 1998. Doubled-haploid broccoli production using anther culture: effect of anther source and seed set characteristics of derived lines. *J. Am. Soc. Hort. Sci.* 123:73–77.
71. Fields, S., Johnson, M. 2002. A crisis in postgenomic nomenclature. *Science* 296:671–672.
72. Galbraith, D.W. 1989. Flow cytometry and cell sorting: applications to higher plant systems. *Intl. Rev. Cytol.* 116:165–227.
73. Galbraith, D.W. 1990. Flow cytometric analysis of plant genomes. *Methods Cell Biol.* 33:549–562.
74. Galbraith, D.W. 1990. Isolation and flow cytometric characterization of plant protoplasts. *Methods Cell Biol.* 33:527–547.
75. Galbraith, D.W., Galbraith, J.E.C. 1979. A method for the identification of fusion of plant protoplasts derived from tissue cultures. *Zeits. Pflanzenphysiol.* 93:149–158.
76. Galbraith, D.W., Lambert G.M. 1995. Advances in the flow cytometric characterization of plant cells and tissues. In *Flow Cytometric Applications in Cell Culture*, ed. M. Al-Rubeai and A.N. Emery, pp. 311–326. New York: Marcel Dekker.
77. Galbraith, D.W., Mauch T.J. 1980. Identification of fusion of plant protoplasts II. *Zeits. Pflanzenphysiol.* 98:129–140.
78. Galbraith, D.W., Harkins, K.R., Jefferson R.A. 1988. Flow cytometric characterization of the chlorophyll contents and size distributions of plant protoplasts. *Cytometry* 9:75–83.
79. Galbraith, D.W., Harkins, K.R., Knapp S. 1991. Systematic endopolyploidy in *Arabidopsis thaliana*. *Plant Physiol.* 96:985–989.
80. Galbraith, D.W., Herzenberg, L.A., Anderson M.T. 1999. Flow cytometric analysis of transgene expression in higher plants: green fluorescent protein. *Methods Enzymol.* 302:296–315.
81. Galbraith, D.W., Doležel, J., Lambert, G., Macas, J. 1998. DNA and ploidy analyses in higher plants. In *Current Protocols in Cytometry*, ed. J.P. Robinson, Z. Darzynkiewicz, P.N. Dean, L.G. Dressler, A. Orfao, P.S. Rabinovitch, C.C. Stewart, H.J. Tanke, and L.L. Wheeless. Wiley, New York, pp 7.6.1–7.6.22.
82. Galbraith, D.W., Harkins, K.R., Maddox, J.R., Ayres, N.M., Sharma, D.P., Firoozabady, E. 1983. Rapid flow cytometric analysis of the cell cycle in intact plant tissues. *Science* 220:1049–1051.
83. Gendreau, E., Orbovic, V., Höfte, H., Traas, J. 1999. Gibberellin and ethylene control endoreduplication levels in the *Arabidopsis thaliana* hypocotyls. *Planta* 209:513–516.

84. Geoffriau, E., Kahane, R., Bellamy, C., Rancillac, M. 1997. Ploidy stability and *in vitro* chromosome doubling in gynogenic clones of onion (*Allium cepa* L). *Plant Sci.* 122:201–208.

85. Giglioli Guivarch, N., Pierre, J.N., Brown, S., Chollet, R., Vidal, J., Gadal, P. 1996. The light-dependent transduction pathway controlling the regulatory phosphorylation of C4 phosphoenolpyruvate carboxylase in protoplasts from *Digitaria sanguinalis*. *Plant Cell* 8:573–586.

86. Giglioli Guivarch, N., Pierre, J.N., Vidal, J., Brown, S. 1996. Flow cytometric analysis of cytosolic pH of mesophyll cell protoplasts from the crabgrass *Digitaria sanguinalis Cytometry* 23:241–249.

87. Gill, K.S., Arumuganathan, K., Lee J.H. 1999. Isolating individual wheat (*Triticum aestivum*) chromosome arms by flow cytometric analysis of ditelosomic lines. *Theor. Appl. Genet.* 98:1248–1252.

88. Glab, N., Labidi, B., Qin, L.X., Trehin, C., Bergounioux, C., Meijer, L. 1994. Olomoucine, an inhibitor of the Cdc2/Cdk2 kinases activity, blocks plant-cells at the G1 to S and G2 to M cell-cycle transitions. *FEBS Lett.* 353:207–211.

89. Godelle, B., Cartier, D., Marie, D., Brown, S.C., Siljak-Yakovlev, S. 1993. Heterochromatin study demonstrating the non-linearity of fluorometry useful for calculating genomic base composition. *Cytometry* 14:618–626.

90. Goff et al. 2002. A draft sequence of the rice genome (*Oryza sativa* L. ssp. *japonica*). *Science* 296:92–100.

91. Graham, M.J., Nickell, C.D., Rayburn, A.L. 1994. Relationship between genome size and maturity group in soybean. *Theor. Appl. Genet.* 88:429–432.

92. Gratzner, H.G. 1982. Monoclonal antibody to 5-bromodeoxyuridine and 5-iododeoxy-uridine: a new reagent for detection of DNA-replication. *Science* 218:474–475.

93. Grebenok, R.J., Pierson, E.A., Lambert, G.M., Gong, F.C., Afonso, C.L., Haldeman-Cahill, R., Carrington, J.C., Galbraith D.W. 1997. Green-fluorescent protein fusions for efficient characterization of nuclear localization signals. *Plant J.* 11:573–586.

94. Grebenok, R.J., Lambert, G.M., Galbraith, D.W. 1997. Characterization of the targeted nuclear accumulation of GFP within the cells of transgenic plants. *Plant J.* 12:685–696.

95. Greilhuber, J. 1998. Intraspecific variation in genome size. A critical reassessment. *Ann. Bot.* 82 (Suppl. A):27–35.

96. Greilhuber, J., Obermayer, R. 1997. Genome size and maturity group in *Glycine max* (soybean). *Heredity* 78:547–551.

97. Greilhuber, J., Obermayer, R. 1998. Genome size variation in *Cajanus cajan* (*Fabaceae*): a reconsideration. *Pl. Syst. Evol.* 212:135–141.

98. Gualberti, G., Doležel, J., Macas, J., Lucretti S. 1996. Preparation of pea (*Pisum sativum* L.) chromosome and nucleus suspensions from single root tips. *Theor. Appl. Genet.* 92:744–751.

99. Haag, D., Feichter, G., Goerttler, K., Kaufman, M. 1987. Influence of systematic errors on the evaluation of the S phase proportions from FNA distributions of solid tumors as shown for 328 breast carcinomas. *Cytometry* 8:377–385.

100. Hagenbeek, D., Rock, C.D. 2001. Quantitative analysis by flow cytometry of abscisic acid-inducible gene expression in transiently transformed rice protoplasts. *Cytometry* 45:170–179.

101. Hardy, O.J., Vanderhoeven, S., De L.M., Meerts P. 2000. Ecological, morphological and allozymic differentiation between diploid and tetraploid knapweeds (*Centaurea jacea*) from a contact zone in the Belgian Ardennes. *New Phytologist* 146:281–290.

102. Harkins, K.R., Galbraith, D.W. 1987. Factors governing the flow cytometric analysis and sorting of large biological particles. *Cytometry* 8:60–71.

103. Hirsch, A.M., Testolin, R., Brown, S., Chat, J., Fortune, F.D., Bureau, J.M., De Nay, D. 2001. Embryo rescue from interspecific crosses in the genus *Actinidia* (kiwifruit). *Plant Cell Rep.* 20:508–516.

104. Huff D.R., Bara J.M. 1993. Determining genetic origins of aberrant progeny from facultative apomictic Kentucky bluegrass using a combination of flow cytometry and silver-stained RAPD markers. *Theor. Appl. Genet.* 87:201–208.

105. Ilceva, S., San, L.H., Zagorska, N., Dimitrov, B. 2001. Production of male sterile interspecific somatic hybrids between transgenic *N. tabacum* (bar) and *N. rotundifolia* (nptII) and their identification by AFLP analysis. *In Vitro Cell Dev. Biol. Plant* 37:496–502.
106. Johnston, J.S., Bennett, M.D., Rayburn, A.L., Galbraith, D.W., Price, H.J. 1999. Reference standards for determination of DNA content of plant nuclei. *Am. J. Bot.* 86: 609–613.
107. Jones, R.J., Srienc, F., Roessler, J. 1992. Flow cytometric determination of light-scattering and absorbency properties of starch granules. *Starch* 44:243–247.
108. Juan, G., Traganos, F., Darzynkiewicz Z. 2001. Methods to identify mitotic cells by flow cytometry. *Methods Cell Biol.* 63:343–354.
109. Kejnovský, E., Vrána, J., Matsunaga, S., Souček, P., Široký, J., Doležel, J., Vyskot, B. 2001. Localization of male-specifically expressed *MROS* genes of *Silene latifolia* by PCR on flow-sorted sex chromosomes and autosomes. *Genetics* 158:1269–1277.
110. Keller, E.R.J., Schubert, I., Fuchs, J., Meister A. 1996. Interspecific crosses of onion with distant *Allium species* and characterization of the presumed hybrids by means of flow cytometry, karyotype analysis and genomic *in situ* hybridization. *Theor. Appl. Genet.* 92:417–424.
111. Kerr, J.F.R, Wyllie, A.H., Currie, A.R. 1972. Apoptosis: a basic biological phenomenon with wideranging implications in tissue kinetics. *Brit. J. Cancer* 26:239–257.
112. Kohler, R.H., Zipfel, W.R., Webb, W.W., Hanson, M.R. 1997. The green fluorescent protein as a marker to visualize plant mitochondria in vivo. *Plant J.* 11:613–621.
113. Kohler, R.H., Cao, J., Zipfel, W.R., Webb, W.W., Hanson, M.R. 1997. Exchange of protein molecules through connections between higher plant plastids. *Science* 276:2039–2042.
114. Kowles, R.V., McMullen, M.D., Yerk, G., Phillips, R.L., Kraemer, S., Scrienc, F. 1992. Endosperm mitotic activity and endoreduplication in maize affected by defective kernel mutations. *Genome* 35:68–77.
115. Kowles, R.V., Yerk, G.L., Haas, K.M., Phillips, R.L. 1997. Maternal effects influencing DNA endoreduplication in developing endosperm of *Zea mays*. *Genome* 40:798–805.
116. Kubaláková, M., Doležel, J., Lebeda, A. 1996. Ploidy instability of embryogenic cucumber (*Cucumis sativus* L.) callus culture. *Biol. Plant.* 38:475–480.
117. Kubaláková, M., Lysák, M.A., Vrána, J., Šimková, H., Číhalíková, J., Doležel, J. 2000. Rapid identification and determination of purity of flow-sorted plant chromosomes using C-PRINS. *Cytometry* 41:102–108.
118. Kubaláková, M., Valarik, M., Vrána, J., Valárik, M., Číhalíková, J., Molnar-Lang, M., Doležel, J. 2003. Analysis and sorting of rye (*Secale cereale* L.) chromosomes using flow cytometry. *Genome* 46:893–905.
119. Kubaláková, M., Vrána, J., Číhalíková, J., Lysák, M.A., Doležel, J. 2001. Localisation of DNA sequences on plant chromosomes using PRINS and C-PRINS. *Methods Cell Sci.* 23:71–82.
120. Kubaláková, M., Vrána, J., Číhalíková, J., Šimková, H., Doležel, J. 2002. Flow karyotyping and chromosome sorting in bread wheat (*Triticum aestivum* L.). *Theor. Appl. Genet.* 104:1362–1372.
121. Lagunes-Espinoza, L.D.C., Huyghe, C., Bousseau, D., Barre, P., Papineau J. 2000. Endoreduplication occurs during pod wall development in temperate grain legumes. *Ann. Bot.* 86:185–190.
122. Larsen, J.K., Landberg, G., Roos, G. (2001). Detection of proliferating cell nuclear antigen. *Methods Cell Biol.* 63:419–431.
123. Latt, S.A. 1973. Microfluorometric detection of deoxyribonucleic acid replication in human metaphase chromosomes. *Proc. Natl. Acad. Sci. USA* 70:3395–3399.
124. Laureys, F., Smets, R., Lenjou, M., Van, B.D., Inze, D., Van, O.H. 1999. A low content in zeatin type cytokinins is not restrictive for the occurrence of G(1)/S transition in tobacco BY-2 cells. *FEBS Lett.* 460:123–128.
125. Lee, J.-H., Arumuganathan, K. 1999. Metaphase chromosome accumulation and flow karyotypes in rice (*Oryza sativa* L.) root tip meristem cells. *Mol. Cells* 9:436–439.
126. Lee, J.-H., Arumuganathan, K., Kaeppler, S.M., Kaeppler, H.F., Papa, C.M. 1996. Cell synchronization and isolation of metaphase chromosomes from maize (*Zea mays* L.) root tips for flow cytometric analysis and sorting. *Genome* 39:697–703.

127. Lee, J.-H., Arumuganathan, K., Yen, Y., Kaeppler, S. Kaeppler, H., Baezinger, P.S. 1997. Root tip cell cycle synchronization and metaphase-chromosome isolation suitable for flow sorting in common wheat (*Triticum aestivum* L.). *Genome* 40:633–638.

128. Leitch, I.J., Chase, M.W., Bennett M.D. 1998. Phylogenetic analysis of DNA C-values provides evidence for a small ancestral genome size in flowering plants. *Ann. Bot.* 82 (Suppl. A):85–94.

129. Li, L.J., Arumuganathan, K., Rines, H.W., Phillips, R.L., Riera-Lizarazu, O., Sandhu, D., Zhou, Y., Gill, K.S. 2001. Flow cytometric sorting of maize chromosome 9 from an oat-maize chromosome addition line. *Theor. Appl. Genet.* 102:658–663.

130. Liu, J.H., Dixelius, C., Eriksson, I., Glimelius, K. 1995. *Brassica napus* (+) *B. tournefortii*, a somatic hybrid containing traits of agronomic importance for rapeseed breeding. *Plant Science* 109:75–86.

131. Lucretti, S., Doležel J. 1997. Bivariate flow karyotyping in broad bean (*Vicia faba*). *Cytometry* 28:236–242.

132. Lucretti, S, Doležel, J., Schubert, I., Fuchs, J. 1993. Flow karyotyping and sorting of *Vicia faba* chromosomes. *Theor. Appl. Genet.* 85:665–672.

133. Lucretti, S., Nardi, L., Nisini, P.T., Moretti, F., Gualberti, G., Doležel, J. 1999. Bivariate flow cytometry DNA/BrdUrd analysis of plant cell cycle. *Methods Cell Sci.* 21:155–166.

134. Lysák, M.A., Doležel J. 1998. Estimation of nuclear DNA content in *Sesleria* (Poaceae). *Caryologia* 51:123–132.

135. Lysák, M.A., Číhalíková, J., Kubaláková, M., Šimková, H., Künzel, G., Doležel, J. 1999. Flow karyotyping and sorting of mitotic chromosomes of barley (*Hordeum vulgare* L.). *Chrom. Res.* 7:431–444.

136. Lysák, M.A., Rostková, A., Dixon, J.M., Rossi, G., Doležel, J. 2000. Limited genome size variation in *Sesleria albicans. Ann. Bot.* 86:399–403.

137. Lysák, M.A., Doleželová, M., Horry, J.P., Swennen, R., Doležel, J. 1999. Flow cytometric analysis of nuclear DNA content in *Musa. Theor. Appl. Genet.* 98:1344–1350.

138. Macas, J., Doležel, J., Gualberti, G., Pich, U., Schubert, I., Lucretti, S. 1995. Primer-induced labelling of pea and field bean chromosomes in situ and in suspension. *BioTechniques* 19:402–408.

139. Macas, J., Doležel, J., Lucretti, S., Pich, U., Meister A., Fuchs J., Schubert I. 1993. Localization of seed protein genes on flow-sorted field bean chromosomes. *Chrom. Res.* 1:107–115.

140. Macas, J., Gualberti, G., Nouzová, M., Samec, P., Lucretti, S., Doležel, J. 1996. Construction of chromosome-specific DNA libraries covering the whole genome of field bean (*Vicia faba* L.). *Chrom. Res.* 4:531–539.

141. Macas, J., Lambert, G.M., Doležel, D., Galbraith, D.W. 1998. NEST (nuclear expressed sequence tag) analysis: a novel means to study transcription through amplification of nuclear RNA. *Cytometry* 33:460–468.

142. Majewska, A., Furmanowa, M., Sliwinska, E., Glowniak, K., Guzewska, J., Kuras, M., Zobel A. 2000. Influence of extract from shoots of *Taxus baccata* var. "Elegantissima" on mitotic activity of meristematic cells of *Allium cepa* L. roots. *Acta Soc. Botan. Polon.* 69:185–192.

143. Mambelli, S., Setter, T.L. 1998. Inhibition of maize endosperm cell division and endoreduplication by exogenously applied abscisic acid. *Physiol. Plant.* 104:266–272.

144. Marie, D., Brown, S.C. 1993. A cytometric exercise in plant DNA histograms, with 2C values for 70 species. *Biol. Cell* 78:41–51.

145. Martin, S.J., Reutelingsperger, C.P.M., McGahon, A.J., Rader, J.A., van Schie, R.C.A.A., LaFace, D.M., Green, D.R. 1995. Early redistribution of plasma-membrane phosphatidylserine is a general feature of apoptosis regardless of the initiating stimulus: inhibition by overexpression of Bcl-2 and Abl. *J. Exp. Med.* 182:1545–1556.

146. Marubashi, W., Yamada, T., Niwa, M. 1999. Apoptosis detected in hybrids between *Nicotiana glutinosa* and *N. repanda* expressing lethality. *Planta* 210:168–171.

147. Matzk, F., Meister, A., Schubert I. 2000. An efficient screen for reproductive pathways using mature seeds of monocots and dicots. *Plant J.* 21:97–108.

148. Matzk, F., Meister, A., Brutovská, R., Schubert, I. 2001. Reconstruction of reproductive diversity in *Hypericum perforatum* L. opens novel strategies to manage apomixis. *Plant J.* 26:275–282.

149. Meier, P., Finch, A., Evan, G. 2000. Apoptosis in development. *Nature* 407:796–801.

150. Meyerowitz, E.M. 2002. Plants compared to animals: the broadest comparative study of development. *Science* 295:1482–1485.

151. Michaelson, M.J., Price, H.J., Johnston, J.S., Ellison, J.R. 1991. Variation of nuclear DNA content in *Helianthus annuus* (*Asteraceae*). *Am. J. Bot.* 78:1238–1243.

152. Mishiba, K.I., Ando, T., Mii M., Watanabe H., Kokubun H., Hashimoto G., Marchesi E. 2000. Nuclear DNA content as an index character discriminating taxa in the genus *Petunia sensu* Jussieu (*Solanaceae*). *Ann. Bot.* 85:665–673.

153. Nagata, T, Nemoto, Y, Hasezawa, S. 1992. Tobacco BY-2 cell line as the "HeLa" cells in the cell biology of higher plants. *Intl. Rev. Cytol.* 132:1–30.

154. Nagl, W. 1978. *Endopolyploidy and Polyteny in Differentiation and Evolution*. Amsterdam: North-Holland.

155. Neumann, P., Lysák, M., Doležel, J., Macas, J. 1998. Isolation of chromosomes from *Pisum sativum* L. hairy root cultures and their analysis by flow cytometry. *Plant Sci.* 137:205–215.

156. Neumann, P., Požárková, D., Vrána, J., Doležel, J., Macas, J. 2002. Chromosome sorting and PCR-based physical mapping in pea (*Pisum sativum* L.). *Chrom. Res.* 10:63–71.

157. Noirot, M., Barre, P., Louarn, J., Duperray, C., Hamon, S. 2000. Nucleus-cytosol interactions—a source of stoichiometric error in flow cytometric estimation of nuclear DNA content in plants. *Ann. Bot.* 86:309–316.

158. O'Brien, I.W., Reutelingsperger, C.M., Holdaway, K.M. 1997. Annexin-V and TUNEL use in monitoring the progression of apoptosis in plants. *Cytometry* 29:28–33.

159. Oberwalder, B., Schilde-Rentschler, L., Ruoss, B., Wittemann, S., Ninnemann H. 1998. Asymmetric protoplast fusions between wild species and breeding lines of potato—effect of recipients and genome stability. *Theor. Appl. Genet.* 97:1347–1354.

160. Ohri, D., Jha, S., Kumar, S. 1994. Variability in nuclear DNA content within pigeonpea, *Cajanus cajan* (*Fabaceae*). *Pl. Syst. Evol.* 189: 211–216.

161. Otoni, W.C., Blackhall, N.W., d'Utra Vaz, F.B., Casali, V.W., Power, J.B., Davey, M.R. 1995. Somatic hybridization of the *Passiflora* species, *P. edulis* f. *flavicarpa* Degener. and *P. incarnata* L. *J. Exp. Bot.* 46:777–785.

162. Palomino, G., Doležel, J., Cid, R., Brunner, I., Mendez, I., Rubluo, A. 1999. Nuclear genome stability of *Mammillaria san-angelensis* (Cactaceae) regenerants induced by auxins in long-term *in vitro* culture. *Plant Sci.* 141:191–200.

163. Pelayo, H.R., Lastres, P., De la Tore, C. 2001. Replication and G(2) checkpoints: their response to caffeine. *Planta* 212:444–453.

164. Perennes, C., Glab, N., Guglieni, B., Doutriaux, M.P., Phan, T.H., Planchais, S., Bergounioux, C. 1999. Is arcA3 a possible mediator in the signal transduction pathway during agonist cell cycle arrest by salicylic acid and UV irradiation? *J. Cell Sci.* 112: 1181–1190.

165. Peres, A., Ayaydin, F., Nikovics, K., Gutierrez, C., Horvath, G.V., Dudits, D.N., Feher, A. 1999. Partial synchronization of cell division in cultured maize (*Zea mays* L.) cells: differential cyclin, cdc2, histone and retinoblastoma transcript accumulation during the cell cycle. *J. Exptl. Bot.* 50:1373–1379.

166. Petit, P.X. 1992. Flow cytometric analysis of Rhodamine123 fluorescence during modulation of the membrane potential in plant mitochondria. *Plant Physiol.* 98:279–286.

167. Pfosser, M. 1989. Improved method for critical comparison of cell-cycle data of asynchronously dividing and synchronized cell-cultures of *Nicotiana tabacum*. *J. Plant Physiol.* 134:741–745.

168. Pfosser, M., Amon, A., Lelley, T., Heberle-Bors, E. 1995. Evaluation of sensitivity of flow cytometry in detecting aneuploidy in wheat using disomic and ditelosomic wheat-rye addition lines. *Cytometry* 21:387–393.

169. Pfundel, E., Meister, A. 1996. Flow cytometry of mesophyll and bundle sheath chloroplast thylakoids of maize (*Zea mays* L.). *Cytometry* 23:97–105.

170. Pich, U., Meister, A., Macas, J., Doležel, J., Lucretti, S., Schubert, I. 1995. Primed in situ labelling facilitates flow sorting of similar sized chromosomes. *Plant J.* 7:1039–1044.

171. Pinheiro, A.A., Pozzobon, M.T., do Valle, C.B., Penteado, M.I.O., Carneiro V.T.C. 2000. Duplication of the chromosome number of diploid *Brachiaria brizantha* plants using colchicine. *Plant Cell Rep.* 19:274–278.

172. Planchais, S., Glab, N., Trehin, C., Perennes, C., Bureau, J.M., Meijer, L., Bergounioux, C. (1997). Roscovitine, a novel cyclin-dependent kinase inhibitor, characterizes restriction point and G2/M transition in tobacco BY-2 cell suspension. *Plant J.* 12:191–202.

173. Porceddu, A., Stals, H., Reichheldt, J.P., Segers, G., De Veylder, L., Barroco, R.D., Casteels, P., Van Montagu, M., Inze, D., Mironov, V. 2001. A plant-specific cyclin-dependent kinase is involved in the control of G(2)/M progression in plants. *J. Biol. Chem.* 276:36354–36360.

174. Požárková, D., Koblížková, A., Roman, B., Torres, A.M., Lucretti, S., Lysák, M., Doležel, J., Macas, J. 2002. Development and characterization of microsatellite markers from chromosome 1-specific DNA libraries of *Vicia faba*. *Biol. Plant.* 45:337–345.

175. Price, H.J., Hodnett, G., Johnston, J.S. 2000. Sunflower (*Helianthus annuus*) leaves contain compounds that reduce nuclear propidium iodide fluorescence. *Ann. Bot.* 86: 929–934.

176. Reed, M.L., Wilson, S.K., Sutton, C.A., Hanson, M.R. 2001. High-level expression of a synthetic red-shifted GFP coding region incorporated into transgenic chloroplasts. *Plant J.* 27:257–265.

177. Richard, C., Granier, C., Inze, D., De Veylder L. 2001. Analysis of cell division parameters and cell cycle gene expression during the cultivation of *Arabidopsis thaliana* cell suspensions. *J. Exptl. Bot.* 52:1625–1633.

178. Ricroch, A., Brown, S.C. 1997. DNA base composition of *Allium* genomes with different chromosome numbers. *Gene* 205:255–260.

179. Rokka, V.M., Ishimaru, C.A., Lapitan, N.V., Pehu E. 1998. Production of androgenic dihaploid lines of the disomic tetraploid potato species *Solanum acaule* ssp. acaule. *Plant Cell Rep.* 18: 89–93.

180. Roudier, F., Fedorova, E., Gyorgyey, J., Feher, A., Brown, S., Kondorosi, A., Kondorosi, E. 2000. Cell cycle function of a *Medicago sativa* A2-type cyclin interacting with a PSTAIRE-type cyclin-dependent kinase and a retinoblastoma protein. *Plant J.* 23:73–83.

181. Roux, N., Doležel, J., Swennen, R., Zapata-Arias, F.J. 2001. Effectiveness of three micropropagation techniques to dissociate cytochimeras in *Musa* spp. *Plant Cell Tiss. Organ Cult.* 66:189–197.

182. Roux, N., Toloza, A., Radecki, Z., Zapata-Arias, F.J., Doležel, J. 2002. Rapid detection of aneuploidy in *Musa* using flow cytometry. *Plant Cell Rep.* 21:483–490.

183. Roy, A.T., Leggett, G., Koutoulis, A. 2001. In vitro tetraploid induction and generation of tetraploids from mixoploids in hop (*Humulus lupulus* L.). *Plant Cell Rep.* 20:489–495

184. Rychtarová, K., Ohri, D., Vrána, J., Číhalíková, J., Kahl, G., Doležel, J. 2001. Development of flow cytogenetics for chickpea (*Cicer arietinum*). In: Abstracts of the Symposium *Plant Molecular Biology for the New Millennium*, p. 89. Třeboň: Institute of Plant Molecular Biology, Academy of Sciences of the Czech Republic.

185. Schena, M., Shalon, D., Davis, R.W., Brown, P.O. 1995. Quantitative monitoring of geneexpression patterns with a complementary-DNA microarray. *Science* 270:467–470.

186. Schroeder, W.P., Petit, P.X. 1992. Flow cytometry of spinach chloroplasts: determination of intactness and lectin-binding properties of the envelope and the thylakoid membranes. *Plant Physiol.* 100:1092–1102.

187. Schubert, I., Doležel, J., Houben, A., Scherthan, H., Wanner, G. 1993. Refined examination of metaphase chromosome structure at different levels made feasible by new isolation methods. *Chromosoma* 102:96–101.

188. Schwarzacher, T., Wang, M.L., Leitch, A.R., Miller, N., Moore, G., Heslop-Harrison J.S. 1997. Flow cytometric analysis of the chromosomes and stability of a wheat cell-culture line. *Theor. Appl. Genet.* 94:91–97.

189. Šesek, P., Šuštar-Vozlič, J., Bohanec, B. 2000. Determination of aneuploids in hop (*Humulus lupulus* L.) using flow cytometry. *Eur. J. Physiol.* 439 (Suppl):R16–R18.

190. Sgorbati, S., Levi, M., Sparvoli, E., Trezzi, F., Lucchini, G. 1986. Cytometry and flow cytometry of 4′,6-diamidino-2-phenylindole (DAPI)-stained suspensions of nuclei released from fresh and fixed tissues of plants. *Physiol. Plant.* 68:471–476.

191. Shapiro, H.M. 2002. *Practical Flow Cytometry*, 4th ed. Wiley-Liss, New York.

192. Sharbel, T.F., Mitchell-Olds, T. 2001. Recurrent polyploid origins and chloroplast phylogeography in the *Arabis holboellii* complex (Brassicaceae). *Heredity* 87:59–68.

193. Sharma, D.P., Firoozabady, E., Ayres, N.M., Galbraith D.W. 1983. Improvement of anther culture in Nicotiana: media, culture conditions and flow cytometric determination of ploidy levels. *Zeits. fur Pflanzen.* 111:441 450.

194. Sheen, J., Hwang, S., Niwa, Y., Kobayashi, H., Galbraith D.W. 1995. Green fluorescent protein as a new vital marker in plant cells. *Plant J.* 8:777–784.

195. Sigareva, M.A., Earle, E.D. 1997. Direct transfer of a cold-tolerant Ogura male sterile cytoplasm into cabbage (*Brassica oleracea* ssp. *capitata*) via protoplast fusion. *Theoret. Appl. Genet.* 94:213–220,

196. Siljak-Yakovlev, S., Benmalek, S., Cerbah, M., Coba de la Pena, T., Bounaga, N., Brown, S.C., Sarr, A. 1996. Chromosomal sex determination and heterochromatin structure. *Sex. Plant Reprod.* 9:127–132.

197. Šimková, H., Číhalíková, J., Vrána, J., Lysák M.A., Doležel, J. 2003. Preparation of high molecular weight DNA from plant nuclei and chromosomes isolated from root tips. *Biol. Plant.* 46:369–373.

198. Sliwinska, E., Steen, P. 1995. Flow cytometry estimation of ploidy in anisoploid sugarbeet populations. *J. Appl. Genet.* 36:111–118.

199. Sliwinska, E., Jing, H.C., Job, C., Job, D., Bergervoet, J.H.W., Bino, R.J., Goot, S.P.C. 1999. Effect of harvest time and soaking treatment on cell cycle activity in sugarbeet seeds. *Seed Sci. Res.* 9:91–99.

200. Subramanian, C., Ivey, R., Bruce B.D. 2001. Cytometric analysis of an epitope-tagged transit peptide bound to the chloroplast translocation apparatus. *Plant J.* 25:349–363.

201. Suda, J., Lysak, M.A. 2001. A taxonomic study of the *Vaccinium* sect. *Oxycoccus* (Hill) W.D.J. Koch (*Ericaceae*) in the Czech Republic and adjacent territories. *Folia Geobot.* 36:303–320.

202. Sun, Y.J., Flannigan, B.A., Setter T.L.1999. Regulation of endoreduplication in maize (*Zea mays* L.) endosperm. Isolation of a novel B1-type cyclin and its quantitative analysis. *Plant Mol. Biol.* 41:245–258.

203. Swiatek, A., Lenjou, M., Van Bockstaele, D., Inze, D., Van Onckelen, H. 2002. Differential effect of jasmonic acid and abscisic acid on cell cycle progression in tobacco BY-2 cells. *Plant Physiol.*128:201–211.

204. Ten Hoopen, R., Manteuffel, R., Doležel, J., Malysheva, L., Schubert, I. 2000. Evolutionary conservation of kinetochore protein sequences in plants. *Chromosoma* 109:482–489.

205. Terry, N.A., White, R.A. 2001. Cell cycle kinetics estimated by analysis of bromodeoxyuridine incorporation. *Methods Cell Biol.* 63:355–374.

206. The Arabidopsis Genome Initiative (2000). Analysis of the genome sequence of the flowering plant *Arabidopsis thaliana*. *Nature* 408:796–815.

207. Tosca, A., Pandolfi, R., Citterio, S., Fasoli, A., Sgorbati, S. 1995. Determination by flow cytometry of the chromosome doubling capacity of colchicine and oryzalin in gynogenetic haploids of *Gerbera*. *Plant Cell Rep.* 14:455–458.

208. Trehin, C., Planchais, S., Glab, N., Perennes, C., Tregear, J., Bergounioux, C. 1998. Cell cycle regulation by plant growth regulators: involvement of auxin and cytokinin in the re-entry of Petunia protoplasts into the cell cycle. *Planta* 206:215–224.

209. Uozu, S., Ikehashi, H., Ohmido, N., Ohtsubo, H., Ohtsubo, E., Fukui, K. 1997. Repetitive DNA sequences. Cause for variation in genome size and chromosome morphology in the genus *Oryza*. *Plant Mol. Biol.* 35:791–799.

210. Väinölä, A. 2000. Polyploidization and early screening of *Rhododendron* hybrids. *Euphytica* 112: 239–244.

211. Valárik, M., Šimková, H., Hřibová, E., Šafář, J., Doleželová, M., Doležel, J. 2002. Isolation, characterization and chromosome localization of repetitive DNA sequences in bananas (*Musa* spp.). *Chrom. Res.* 10:89–100.

212. Valárik, M., Bartoš, J., Kovářová, P., Kubaláková, M., de Jong, H., Doležel, J. 2004. High-resolution FISH on super-stretched flow-sorted plant chromosomes. *Plant J.* 37:940–950.
213. Van der Valk, H.C.P.M., Verhoeven, H.A. 1988. Application of flow cytometry for the isolation of metaphase chromosomes from protoplasts of *Nicotiana plumbaginifolia*. In *Progress in Plant Protoplast Research*, ed. K.J. Puite, J.J.M. Dons, H.J. Huizing, A.J. Kool, M. Koornneef, and F.A. Krens, pp. 303–304. Dordrecht: Kluwer Academic.
214. Van Duren, M., Morpurgo, R., Doležel J. Afza R. 1996. Induction and verification of autotetraploids in diploid banana (*Musa acuminata*) by in vitro techniques. *Euphytica* 88:25–34.
215. Vankesteren, W.J.P., Tempelaar M.J. 1993. Surface labeling of plant-protoplasts with fluorochromes for discrimination of heterokaryons by microscopy and flow cytometry. *Cell Biol. Int.* 17:235–243.
216. Velculescu, V.E., Zhang, L., Vogelstein, B., Kinzler, K.W. 1995. Serial analysis of gene expression. *Science* 270:484–487.
217. Vermes, I., Haanen, C., Reutelingsperger C. 2000. Flow cytometry of apoptotic cell death. *J. Immunol Meth.* 243:167–190.
218. Veuskens, J., Marie, D., Hinnisdaels, S., Brown, S.C. 1992. Flow cytometry and sorting of plant chromosomes. In *Flow Cytometry and Cell Sorting*, ed. A. Radbruch, ed., pp. 177–188. Berlin: Springer.
219. Vicient, C.M., Suoniemi, A., Anamthamat-Jonsson, K., Tanskanen, J., Beharav, A., Nevo, E., Schulman, A.H. 1999. Retrotransposon BARE-1 and its role in genome evolution in the genus *Hordeum*. *Plant Cell* 11:1769–1784.
220. Vrána, J., Kubaláková, M., Šimková, H., Číhalíková, J., Lysák, M.A., Doležel, J. 2000. Flow-sorting of mitotic chromosomes in common wheat (*Triticum aestivum* L.). *Genetics* 156:2033–2041.
221. Wang, M.L., Leitch, A.R., Schwarzacher, T., Heslop-Harrison J.S., Moore, G. 1992. Construction of a chromosome-enriched HpaII library from flow-sorted wheat chromosomes. *Nucleic Acids Res.* 20:1897–1901.
222. Watson, J.V. 1992. Flow cytometry data analysis. – Basic concepts and statistics. Cambridge: Cambridge University Press.
223. Weir, I.E. 2001. Analysis of apoptosis in plant cells. *Methods Cell Biol.* 63:505–526.
224. White, R.A., Terry, N.H., Meistrich, M.L. 1990. New methods for calculating kinetic properties of cells in vitro using pulse labelling with bromodeoxyuridine. *Cell Tissue Kinet.* 23:561–573.
225. Yamada, T., Marubashi, W., Niwa, M. 2000. Apoptotic cell death induces temperature-sensitive lethality in hybrid seedlings and calli derived from the cross of *Nicotiana suaveolens* x *N. tabacum*. *Planta* 211:614–622.
226. Yang, Y.P., Juang, Y.S., Hsu, B.D. 2002. A quick method for assessing chloroplastic starch granules by flow cytometry. *J. Plant Physiol.* 159:103–106.
227. Yanpaisan, W., King, N.J., Doran, P.M. 1998. Analysis of cell cycle activity and population dynamics in heterogeneous plant cell suspensions using flow cytometry. *Biotechnol. Bioeng.* 58:515–528.
228. Yu, et al. 2002. A draft sequence of the rice genome (*Oryza sativa* L. ssp. *indica*). *Science* 296:79–92.
229. Zonneveld, B.M., Van Iren, F. 2000. Flow cytometric analysis of DNA content in *Hosta* reveals ploidy chimeras. *Euphytica* 111:105–110.
230. Zonneveld, B.J.M., Van Iren, F. 2001. Genome size and pollen viability as taxonomic criteria: application to the genus *Hosta*. *Plant Biol.* 3:176–185.

17

Analysis of GTP-Binding Protein–Coupled Receptor Assemblies by Flow Cytometry

PETER SIMONS, CHARLOTTE M. VINES, T. ALEXANDER KEY,
ROSS M. POTTER, MEI SHI, LARRY A. SKLAR,
AND ERIC R. PROSSNITZ

Overview

GTP-binding protein–coupled receptors (GPCRs) represent the largest family of integral membrane signal-transducing molecules in the human genome, with estimates of at least 600 members (23, 53). As such, they represent the targets of approximately 30%–50% of the prescription drugs on the market. They are involved in virtually every physiological process in the human body, with ligands including light, odorants, amines, peptides, proteins, lipids, and nucleotides. Binding of these ligands on the extracellular surface of the receptor leads to conformational changes within the receptor, resulting in a multitude of cellular responses. GPCRs, as their name implies, function through the actions of heterotrimeric GTP-binding proteins (G proteins). These G proteins then couple to a diverse array of effector molecules at the cell surface and inside the cell. GPCRs contain a common structural motif, with seven transmembrane alpha helices. With the recent description of the three-dimensional crystal structure of rhodopsin in its inactive state, a greater, though still incomplete, understanding of the functions of this receptor family has been achieved (30). In addition to the activation of G proteins, GPCRs undergo extensive regulation mediated primarily by a variety of kinases, including second messenger kinases and the family of G protein–coupled receptor kinases (GRKs) (36). Following receptor phosphorylation by GRKs, additional proteins named arrestins associate with GPCRs (13). The traditional role of these molecules has been to serve as desensitizing agents, preventing further association of the receptor with G proteins. However, recent studies have demonstrated that arrestins can serve as adapters in the process of receptor internalization as well as scaffolds in the activation of numerous kinase pathways. Interactions between GPCRs and cellular proteins such

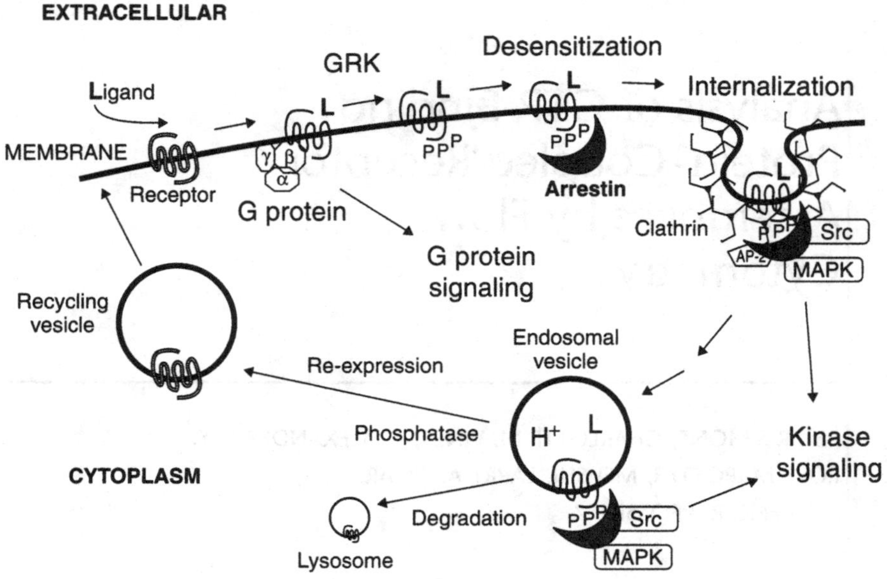

Figure 17.1. Schematic of GTP-binding protein–coupled receptor pathway.

as adaptins, rab GTPases, phosphatases, and ion channels have also been described (16). Thus, it has become apparent that understanding the interactions between GPCRs and their associated proteins is critical for any detailed understanding of receptor function. An overview of the activation and regulation of GPCRs is shown in figure 17.1 to provide a context for the approaches to be described in the remainder of this chapter.

Most GPCRs are expressed at low levels on the plasma membrane, on the order of 10,000–100,000 receptors per cell. For this reason, most studies of GPCR function use heterologous (over)expression of the receptor of interest in cells that do not endogenously express the receptor. Frequently used cell lines include HEK (human embryonic kidney), CHO (Chinese hamster ovary), and COS (African green monkey kidney fibroblast) cells. Studies of receptor function in cellular environments include the determination of effector system activation, the colocalization of proteins with receptors by fluorescence microscopy, and the association of proteins with receptors as determined by co-immunoprecipitation. Such approaches allow for the most part qualitative determinations of protein–protein interactions to be made. To make detailed measurements of specific protein–protein interactions, access to individual molecules must be attained. This can be accomplished in two ways: either through the purification of one or both of the proteins under study, or through the specific labeling of one or both of the proteins under study. In the case of GPCRs, the situation is further complicated by the primary protein of interest being an integral membrane protein that can only be isolated from other membrane proteins through detergent solubilization of the membrane (3). Thus, the choice of detergent and solubilization conditions adds to the complexity of the system.

In this chapter, we describe numerous approaches that we have developed to allow for the study of molecular assemblies involving GPCRs. The prototypical system we will describe is that of the human N-formyl peptide receptor (FPR) (41). The FPR is a member of the chemoattractant/chemokine subfamily of GPCRs that is involved in leukocyte trafficking and inflammatory responses throughout the body in response to bacterially produced peptides. The ligands for this receptor are short hydrophobic peptides (of bacterial origin) that begin with an N-formyl group on the amino-terminal methionine, followed by two to five hydrophobic amino acids such as leucine, phenylalanine, norleucine, and tyrosine. The prototypical ligand is N-formyl-met-leu-phe, although ligands of greater length are fully functional (54). The advantage of using this ligand-receptor system is the availability of a collection of ligands of varying affinity that can be fluorescently labeled to assess the presence and functional state of the FPR. Such fluorescently labeled ligands display great specificity for the FPR, establishing one of the criteria for gaining access to individual molecules. Small molecules such as these ligands can also be covalently attached to surfaces, potentially providing for the display of specific proteins.

With the ability to modify proteins through molecular biological techniques comes the potential to incorporate novel epitopes and generate novel chimeric proteins that facilitate the study of protein complexes. In the former case, the addition of short sequences can allow for the rapid isolation or display of the protein of interest. Examples of such sequences or tags include the hexahistidine tag, allowing binding to Ni-chelate surfaces, and epitopes recognized by widely available antibodies, such as the FLAG, myc, or HA tags. In the latter case, chimeric proteins can be generated that fuse fluorescent proteins, such as GFP and its variants, to the protein of interest, allowing the presence of a single protein to be determined. Together, these modifications allow for the specific display of particular proteins on appropriate surfaces as well as for the detection of binding interactions of the fluorescently tagged protein to the surface-bound component.

As an alternate approach to using entire proteins, individual domains or peptides can also be used to characterize molecular assemblies. In the former case, molecular biological techniques can be used to create fusion proteins of glutathione S-transferase (or a host of other partners, such as maltose binding protein) and the protein domain of interest. Commercial endeavors currently use this approach to define the specificity of domain–domain interactions (e.g., AxCell Biosciences, Newtown, PA, www.axcellbio.com). In the latter case, small portions of proteins can be chemically synthesized by solid-phase techniques. The advantage of this approach is that one has complete control over the composition of the peptide. This can be of particular importance in the case of amino acid modifications such as phosphorylation, methylation, sulfation, and so forth. The disadvantage is that one is somewhat constrained to relatively short peptides, on the order of 50 or so amino acids. Longer peptides are possible though often beyond the capabilities of "standard" syntheses. In the case of membrane proteins, either type of construct is largely constrained to portions of the protein that are excluded from the membrane, with the goal that the final product will be soluble. In this chapter, we will describe recent studies that use many of these approaches.

Receptor Biology

In any attempt to understand complex biological systems, it is crucial to remember the underlying biology of the components involved. GPCRs are a diverse family of receptors that respond to an immense collection of extracellular signals. Typically, cellular responses are initiated in the millisecond to second time frame.

GPCRs transduce extracellular stimuli into intracellular messages via coupling to heterotrimeric G proteins. GPCR signaling fundamentally involves a balance between signal propagation and termination within the cell (see figure 17.1). In that sense, GPCR-mediated signaling reflects the coordination between mechanisms governing ligand binding, receptor activation, signal propagation, receptor phosphorylation, endocytosis, and downregulation (13, 37). The timescales of these processes range from subseconds (activation and signaling), to seconds (phosphorylation), to minutes (endocytosis), to hours (downregulation).

The signaling of GPCRs has been defined in terms of a ternary complex model (42, 50). In short, agonist (L) binding initiates or stabilizes a receptor (R) conformational change, leading to binding and activation of a heterotrimeric G protein (G). The resultant ligand, receptor, and G protein (LRG) complex has been shown to display high affinity for ligand, as well as sensitivity to the addition of guanine nucleotides. Activation of the α subunit by receptor initiates exchange of GDP for GTP and the proposed dissociation of the $\alpha\beta\gamma$ heterotrimer (17). For many GPCRs, the uncoupled $\beta\gamma$ subunit remains tethered to the plasma membrane via lipid anchors and initiates signaling from that position. In contrast, α subunits can traffic cytoplasmically, initiating activation either at the membrane or within the cytosol. Differences in G protein–dependent signaling across GPCRs is in part a function of the specific members of the α, β, and γ subunit families to which a given GPCR binds. The classical effectors of GPCRs include adenylyl cyclases, phospholipases, and ion channels (15).

Homologous desensitization has been hypothesized to proceed in terms of an alternative ternary complex model (26). In short, activation leads to covalent modification of the intracellular domains of the receptor by G protein–coupled receptor kinases, which specifically phosphorylate serine and threonine residues. For some GPCRs, phosphorylation partially uncouples the receptor from heterotrimeric G proteins (28, 29). Complete uncoupling is accomplished through the physical capping of an arrestin molecule (A) onto the phosphorylated GPCR (R^P). Ultimately, receptor complexes internalize into intracellular compartments, thus limiting access of extracellular ligands. Phosphorylation presumably stabilizes a conformation required to promote interaction with cellular elements that directly promote internalization (8). Downregulation of the total cellular complement of receptors as a result of reduced mRNA and protein synthesis, as well as lysosomal and plasma membrane degradation, can further "desensitize" stimulated cells over extended timescales.

Arrestins are at the crossroads of processes governing cellular responsiveness, with well-characterized roles in the signaling, desensitization, internalization, and resensitization of many GPCRs (13). Arrestins translocate to activated, GRK-phosphorylated GPCRs and mediate their desensitization effect by physically uncoupling them from heterotrimeric G proteins. Arrestin binding subsequently targets some GPCRs for endocytosis through the direct recruitment of endocytic machinery, including clathrin heavy chain and the β2-adaptin subunit of the AP2-adaptor complex (14, 24). The ultimate fate of receptor–arrestin complexes differs among receptors (12). It appears that information governing how GPCR–arrestin complexes behave is contained within the C-terminal tail of many GPCRs (56).

There are various physiological consequences of heptahelical receptor stimulation that are not solely mediated by G protein activation. It is now known that some GPCRs act as scaffolds promoting the formation and compartmentalization of G protein-independent signal transduction complexes (37). Arrestins have been implicated in this process, acting as scaffolds for mitogen-activated protein kinase, Src-related kinases, the Dissheveled protein, and the PI3 kinase. The manner in which arrestins confer spatial and temporal specificity to G protein–independent signaling pathways is a topic of increasing interest. Agonist-activated receptors are ultimately internalized via sequestration into intracellular membrane compartments of the cell. Intense efforts over the last decade have underscored important diversity in the patterns of GPCR endocytosis and intracellular trafficking (13). It is apparent that different mechanisms of internalization exist for different GPCRs. Also, different mechanisms of internalization exist for some GPCRs within different cell lines (57).

The intracellular trafficking of GPCRs following arrestin-mediated desensitization is necessary for receptor resensitization. The molecular mechanisms governing the rate at which receptors recycle and reestablish agonist responsiveness are poorly understood. Some internalized GPCRs recycle rapidly back to the plasma membrane fully resensitized, whereas others are retained inside the cell and recycle slowly or not at all (2). The dephosphorylation of GRK-phosphorylated receptors in endosomes appears to be a critical step in the resensitization pathway. An event presumably necessary for the dephosphorylation of GPCRs is their dissociation from arrestin either at or near the plasma membrane or internalized into endosomes. The ability of arrestin to remain associated with some receptors but not others suggests that arrestin may regulate the cellular trafficking and dephosphorylation of GPCRs and, ultimately, their kinetics of resensitization (33). Thus, variations across GPCR intracellular sequences may have significant divergent effects on arrestin coupling and, correspondingly, diverse functional consequences.

We have begun to describe the characteristics of G protein– and arrestin-based ternary complexes in both cell-based and cell-free assays with the FPR using flow cytometry, confocal microscopy, and spectrofluorometry. The ability to discriminate receptor complexes in vitro is predicated on two facts. First, ligand dissociation rates vary on the basis of receptor assemblies. Ternary complexes of both arrestins and G_i proteins with the phosphorylated and nonphosphorylated FPR, respectively, display high affinity for agonist. Second, addition of guanine nucleotides causes the rapid dissociation of G proteins, but not arrestins, from receptors. This dissociation, in turn, results in the rapid dissociation of ligand, as the uncoupled receptor exhibits low affinity for agonist. In general, our findings support the notion that discrete differences in the phosphorylation state of the FPR are sufficient to modulate its affinity for heterotrimeric G proteins, activated states of arrestins, and concomitantly, for ligand.

Surprisingly, our work has suggested that phosphorylation of the wild-type FPR is sufficient to prevent G protein binding in vitro. In contrast, wild-type arrestin binding has been demonstrated to be dependent on phosphorylation in two serine/threonine clusters in the receptor carboxyl terminus. Furthermore, we have recently shown that these domains differentially regulate arrestin and agonist affinity. At this time, our work is focused on elucidating specific effector and receptor determinants underlying ternary complex formation, as well as the downstream consequences of these interactions.

Flow Cytometric Approaches to Assess GPCR Function

Flow cytometry is a commonly used approach for studying cell-based functions of GPCRs. Through the use of flow cytometry on intact cells, the contribution of all proteins that contribute to the functioning of a GPCR can be assessed in vivo. Analysis of real-time events and fixed time points can be used to assess ligand/receptor interactions as well as receptor processing and receptor-mediated cell activation. In addition, flow cytometry is a powerful technique to identify specific populations of cells, such as those transiently transfected with GFP-based chimerae, in which the GPCR is to be analyzed (4). In general, there are three methods of tracking the behavior of a GPCR that is expressed on the surface of a cell: binding of a fluorescinated ligand, binding of fluorescently tagged antibodies and expression of a GPCR–GFP fusion.

Ligand Binding

When a fluorescinated ligand is available, equilibrium and kinetic fluorescent measurements can be extremely useful tools for examining receptor–ligand binding interactions and their regulation (46, 48). In the case of the FPR, a wide variety of fluorescent ligands are available, displaying affinities from μM to pM (54). Because of the small size of many GPCR agonists, fluorescent derivatives of ligands are available mostly for the peptide-based ligands, although even these are not abundant. Ligand binding can be measured in real time on the cytometer, as well as for fixed or stabilized samples (45). In these assays, cells are harvested by centrifugation, washed once in serum-free medium (SFM) and resuspended in SFM. (Use of PBS [phosphate-buffered saline] and Tris-buffered saline can have varying results depending on the cell type and the length of time the cells are maintained at 37°C or have been left on ice.) Binding can be carried out on ice to prevent ligand-dependent cellular activation and receptor internalization. An important parameter is the affinity of the derivatized ligand, as poor affinities (K_d values greater than approximately 50 nM) can lead to excessive background (nonspecific) fluorescence signals. Nonspecific binding is determined by incubating the cells with excess nonfluorescent ligand. Alternatively, the nonspecific binding can be analyzed by incubation of the parental cells, which lack the transfected receptor, in the presence of fluoresceinated peptide alone. We have found that both methods produce comparable results, unless the concentration of the fluorescent ligand enables it to compete with the blocking ligand.

Ligand affinity measurements can be made by determining the level of labeled ligand binding as a function of concentration. We often use concentrations of fluorescent ligand varying from 10^{-11} to 10^{-7} M. One critical factor in simple affinity measurements is to avoid depletion of the ligand through binding to the cell surface receptors. The cell density in this case needs to be reduced to 10^5/mL or less, depending on the level of receptor expression per cell, to prevent depletion of ligand at the lowest ligand concentrations. Nonspecific binding is again assessed in the presence of excess unlabeled ligand. Alternatively, ligand affinities can be determined by measuring association and dissociation rates from cells (47). Both of these parameters can be accomplished in real time. Further information regarding the states of the receptor can be obtained through the study of permeabilized cells, in which the states of the receptor can be manipulated (48). Absolute numbers of receptors expressed on cells can be determined by comparing the fluorescence intensity to reference beads containing known amounts of fluorescein or antibody binding sites, such as Simply Cellular calibration beads (Bangs Labs, Fishers, IN).

Receptor Internalization

Receptor internalization can be measured using either fluorescinated ligands or antibodies to the GPCR in the absence of available labeled ligands. In recent years, this approach, using for the most part amino-terminally epitope-tagged receptors, such as the β_2 adrenergic receptor, has gained wide acceptance compared to the traditional approaches using radiolabeled ligands (10). In the case of the FPR, similar results are obtained for N-formyl-methionyl-leucyl-phenylalanine, fluorescent ligand, or antibody-based approaches for determining receptor internalization (40). Whereas radioligand uptake measures ligand accumulation over time, flow cytometry-based receptor internalization assays measure the agonist-dependent depletion of GPCRs from the cell surface. In such an assay, cells are harvested, washed, resuspended in SFM, and allowed to equilibrate in a 37°C water bath for 10 min. The cells are then treated with a saturating concentration of low-affinity peptide ($100 \times K_d$) and allowed to internalize the ligand for varying time intervals. Internalization of the receptor is arrested at each time point by the transfer of an additional aliquot of cells to ice-cold SFM. Once all time points have been obtained, the cells are rinsed four times in ice-cold SFM to remove the remaining low-affinity peptide, and the receptor remaining on the surface is labeled with either antibody or fluorescinated ligand. The overall internalization rates are measured by dividing the mean channel fluorescence (MCF) level of the internalized sample by the MCF of an untreated control (all values being corrected for nonspecific binding). As the mechanisms involved in receptor internalization begin to be elucidated, simple, accurate methods to evaluate these processes become all the more critical (56).

Receptor Recycling

Receptor recycling is a critical part of normal cellular function. The rapid reexpression of internalized receptors to the cell surface, following removal of ligand and receptor dephosphorylation, allows for the continued responsiveness of cellular processes. Although certain receptors are not recycled at all [e.g., thrombin receptors (35)], others recycle quickly (33), although this can depend on many additional factors such as cell type and conditions of stimulation. Receptor recycling kinetics can be measured in real time or by isolating time points. In this assay, the receptor is allowed to optimally internalize in the presence of excess ligand ($100 \times K_d$). Following the internalization step the cells are pelleted, quickly rinsed, and allowed to incubate at 37°C in the absence of external peptide. The recycling of the receptor is arrested by transferring an aliquot of 10^6 cells to two volumes of ice-cold SFM. The extent of receptor recycling at each time point can be measured using either antibodies or fluorescenated ligands by flow cytometry.

Other Cellular Functions Measured by Flow Cytometry

Flow cytometry can also be used to measure numerous additional cellular responses, often allowing the determination of multiple parameters simultaneously. These responses include monitoring intercellular calcium fluxes (5), polymerization of filamentous-actin [F-actin (22)], cellular adhesion and cellular aggregation (32, 44), phagocytosis (27), oxygen radical production (51), apoptosis/cell cycle (34), and degranulation (1, 11). For a more extensive review on these applications, see (52).

Assemblies of Molecular Complexes on Beads

The biophysical analysis of molecular complexes involving G protein–coupled receptors and associated proteins is a particularly daunting task. Not only are the components expressed at low levels within cells but they are also for the most part integral or peripheral membrane proteins, requiring solubilization with detergents to isolate the proteins of interest (6, 7, 20, 49). The traditional view of GPCR signaling involves the formation of a ternary LRG complex. In this scheme, the ligand stabilizes the receptor in an active conformation, which then acts as the catalyst, serving to initiate GDP release from the α subunit of the heterotrimeric G protein (42). An alternate ternary complex involving ligand, receptor, and arrestin has been described that likely serves to inactivate G protein–mediated signaling while simultaneously activating numerous kinases—potentially independently of G protein activation (31). To study the interactions between two or more components of a signaling complex by flow cytometry, one member must be anchored to a solid surface. Furthermore, depending on the approach used, the other component must be labeled in some way to allow its binding interaction to be detected. Flow cytometry is a well-established technique that has been used extensively to analyze either cell-based or bead-based associations (see chapter 9). To examine interactions of molecular assemblies with this approach requires that one of the interacting components be physically associated, either covalently or with high affinity, to the beads, while the other component should be labeled with a fluorophore, permitting detection in the cytometer. Many approaches exist to accomplish each of these tasks.

Fluorescent Labeling of Proteins

In general, there are two approaches to label a protein of interest with a fluorophore. The first approach again requires the protein to be present in a pure form. If this is achievable, fluorophores can be directly and covalently attached to the protein. The most common approaches involve the labeling of amino groups (with succinimidyl moieties) or sulfhydryl groups (with maleimidyl moieties). This approach is only feasible if such labeling does not interfere with the functions of the protein. However, because of the large and ever-growing collection of fluorophores available with virtually any desired fluorescent properties, such approaches are of great value. As an alternative approach, fluorescent chimeric proteins can be generated using molecular biological techniques through the fusing of the protein of interest to proteins such as green fluorescent protein of *Aequoria victoria* or one its variants with emission peaks between about 425 and 525 nm. Suitable red fluorescent proteins with longer emission wavelengths are also now available (9). Although much larger in size (~27 kDa) than the chemical fluorescent probes (typically ~1 kDa), the addition of a GFP to either terminus of a protein often has little effect on protein structure or function and makes them useful for both in vivo and in vitro applications of molecular assemblies. In the case of GPCRs, the use of carboxy-terminally fused GFP chimerae has become widespread mostly for microscopy-based studies. However, the fact that virtually all of these receptor chimerae appear to maintain normal function with respect to signaling and processing indicates their usefulness for the synthesis of relevant molecular assemblies.

Attachment Schemes

The most important consideration in this area is that the immobilized molecule retain its biological activity and be displayed in such a way that it can be recognized by its binding partner. The simplest approach, if the protein to be displayed exists in a pure state, is to adsorb the purified proteins to polystyrene beads. To achieve covalent attachment, microspheres that possess reactive groups such as carboxyl, succinimidyl, or maleimidyl groups can be used (18, 55). If the protein cannot be obtained in a pure state, antibodies to the protein of interest can be attached to the beads to allow for an immunopurification step that results in the display of the protein of interest on antibody-containing beads. If a specific antibody is not available, a more general approach using anti-epitope antibodies can be used. This approach is predicated on the ability to use genetic approaches to add the epitope to the protein of interest. These epitopes may consist of only a few amino acids (examples include HA [hemagglutinin], myc, and FLAG epitopes), or they may consist of an entire protein or domain sequences. An example of the latter is Glutathione S-transferase, which can be used to immobilize GST fusion proteins on glutathione-containing beads. Alternatively, short hexahistidine (6His) tags can also be engineered into the protein to immobilize the protein directly onto nickel-chelate beads (49). Finally, beads derivatized with streptavidin, which are commercially available, can be used to bind biotinylated proteins (or other biological compounds such as oligonucleotides). It should be clear that the many of these approaches are adaptations of purification techniques with the modification that the "resin" is replaced by a bead of compatible size for flow cytometry (typically 5–25 μ).

GPCR-Based Assemblies

The ternary complex of a GPCR with a G protein and a ligand (LRG) can theoretically be attached to beads by the receptor, the G protein, or the ligand. We have attempted to couple each of these moieties to various beads in an effort to detect LRG assemblies on beads by flow cytometry with a robust specific to nonspecific binding ratio. Familiarity with affinity chromatography and immunoprecipitations led us to believe that such attachments would be simple, but attaching a sufficiently high concentration of the first molecule to a bead in an orientation that allowed the subsequent partners to bind was more difficult to achieve than we initially expected. In the sections that follow, our experiences—both successful and not—with the N-formyl peptide receptor are described.

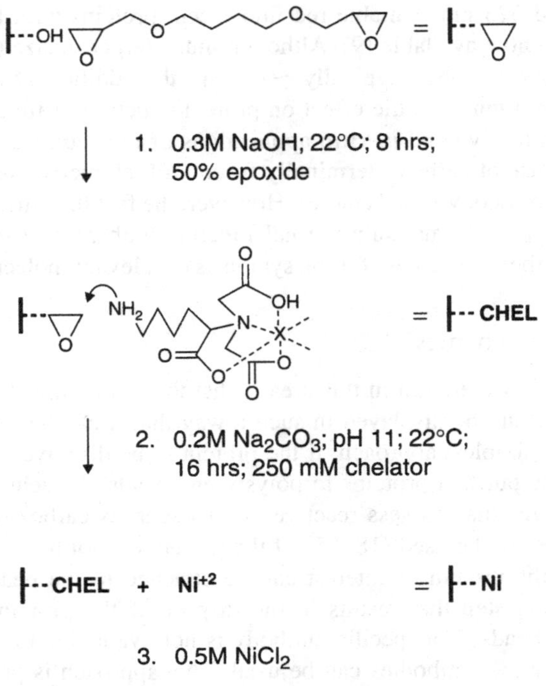

1. 0.3M NaOH; 22°C; 8 hrs;
   50% epoxide

2. 0.2M Na$_2$CO$_3$; pH 11; 22°C;
   16 hrs; 250 mM chelator

|--CHEL + Ni^{+2} = |--Ni

3. 0.5M NiCl$_2$

Figure 17.2. Synthesis of dextran-chelate-nickel beads.

Synthesis of Dextran-Chelate-Nickel (DCNi) Beads

The availability of hexahistidine-tagged proteins prompted us to synthesize beads that would bind such proteins and work well with a flow cytometer. Recently, Nolan and coworkers have also described the generation of flow cytometry–compatible nickel-chelate beads (25). The approaches used polystyrene microspheres, silica microspheres, or lipid-linked metal chelator complexes adsorbed to silica microspheres forming self-assembled bilayer membranes where the metal chelators have lateral mobility. In our approach, Superdex Peptide beads, a cross-linked agarose/dextran matrix with an exclusion limit of 7000 Da and an average size of 13 μm, were removed from a packed column purchased from Amersham Pharmacia Biotech. We have also been able to analyze the far less expensive Superdex 30 Prep Grade beads (average size 34 μm) by flow cytometry. Beads were activated with a bis-epoxide and then coupled to a chelator that contained an amino group, as shown in figure 17.2 (19, 38). Three mL of a 50% slurry of the beads was reduced to a wet cake by vacuum filtration and washed three times with 50 mL of water to remove the ethanol the beads were supplied with. The wet cake was transferred to a flask, and 5 mL of 0.6 M NaOH, 10 mg NaBH$_4$, and 5 mL 1,4-butanediol diglycidyl ether were added. The flask was rotated to keep the beads in suspension and incubated for 8 hr at 37°C; some bubbling occurred in the first hour. The beads were washed by vacuum filtration twice with water, twice with PBS, and twice with water again, and then were stored for up to 1 week at 4°C, or for at least 2 months dried, also at 4°C. One settled volume of these epoxy-activated beads was coupled with one volume of the chelator N_α,N_α-bis(carboxymethyl)-L-lysine (Fluka) in 0.2 M Na_2CO_3, pH 11, adjusting the pH again after addition to the beads; we used 2.5, 25, and 250 mM chelator in three different reactions to obtain different substitution levels on the beads. The coupling proceeded at 22°C overnight with mixing to keep the beads in suspension. The beads were washed four times and then packed into a column. $NiCl_2$ (0.1 M) was passed through the column; the two highest-substituted batches became visibly blue/green, while the lightly substituted batch remained white. The column was rinsed with water and subsequently PBS. Atomic absorption analysis of the three samples after wet-ashing showed the content of Ni to be 1.5, 16, and 30 mM for the settled beads; substitution appears to be proportional to the concentration of amino compound up to 25 mM and then begins to saturate.

Characterization of DCNi Beads

Hexahistidine-tagged enhanced green fluorescent protein (H6GFP; generously supplied by Dr. John Nolan, Los Alamos National Laboratory) was used to determine the suitability of the DCNi beads for display of proteins in a flow cytometer. This H6GFP was found to have a molar fluorescence equal to 60% of our standard fluoresceinated formyl peptide, formyl-MLFK-FITC (L^F; Penninsula Labs/Bachem, San Carlos, CA), within experimental error. The least-substituted DCNi beads, at a concentration of 50,000 DCNi/ml in PBS $\pm$ 10 mM EDTA, were mixed with 10 nM H6GFP on ice, and samples were taken at intervals to determine the MCF of the beads. Figure 17.3A shows a dot plot of the forward and side scatter of the DCNi beads, which indicates that the beads range in size over a factor of about four, similar to the size range of cells. Figure 17.3B shows two histograms of these beads, the lower of which represents autofluorescence and the higher of which represents H6-GFP incubated with the beads, both of which spread over a range of fluorescence similar to a population of cells. These distributions are wider than for commercial polystyrene beads but narrower than the range exhibited by the silica-based nickel chelate particles we have used previously (49). Figure 17.3C shows that in the absence of EDTA, H6-GFP bound the beads quickly, reaching a maximum at about 20 min. By comparison, with standardized microspheres (Flow Cytometry Standards Corporation/Bangs Laboratories, Fishers, IN), about 3.3×10^6 H6GFP/bead were bound at the maximum level after 20–60 min, corresponding to 0.3 nM H6-GFP on beads, with 9.7 nM H6-GFP left in solution. (This was already more than twice the fluorescence of the highest standard bead population in the standards kit and was approaching the level at which some fluorescence quenching might occur. Assuming that bovine serum albumin has a "parking area" of 38 nm^2, there are 14 million parking spaces on a 13-μm bead, and if half of these sites can be occupied by random fall, the H6GFP has occupied 3 million of 7 million expected sites; the site density of the more highly substituted beads should therefore lead to slightly higher occupancy, limited by physical contact.)

At 30 min, in figure 17.3C, a portion of the beads was treated with 10 mM EDTA, and the H6-GFP was eluted off the beads rather slowly, leaving about 20% of the H6-GFP on the beads after 20 min. In figure 17.3D, H6-GFP remained on the beads through five PBS washes over 2 hr, showing the high stability of this platform. Addition of EDTA again brought the bead fluorescence down slowly. Unfortunately, the addition of 1 mg/mL bovine serum albumin or 0.01% Tween-20 dramatically reduced

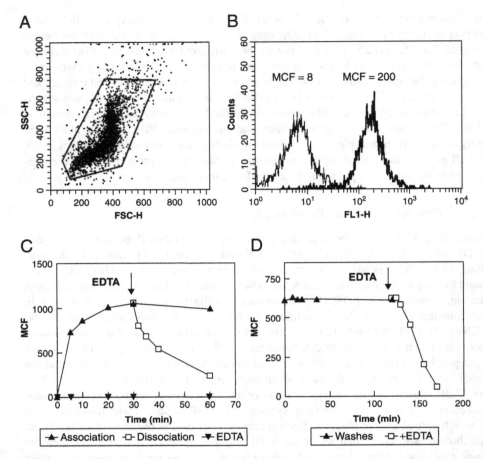

Figure 17.3. Characterization of dextran-chelate-nickel (DCNi) beads for flow cytometry using hexahistidine-tagged enhanced green fluorescent protein (H6-GFP). (A) Dot plot of forward scatter versus side scatter. (B) Histograms of bead fluorescence versus number of beads. (C) Rates of association and dissociation of H6-GFP to DCNi beads with stirring a 4°C. (D) Stability of DCNi-H6-GFP bond to washes.

the binding of H6-GFP to the beads (data not shown), demonstrating the high non-specific binding of nickel chelate beads in general. This correlated with low binding of the hexahistidine-tagged FPR in crude solubilized membrane preparations.

LRG Assembly Approaches

Receptor Assemblies in Solution

As a preface to the generation of receptor assemblies on beads, it was first necessary to study the properties of solubilized receptors and their binding partners in solution to confirm that the observed interactions reflected the known properties of the proteins in vivo. Furthermore, the prior determination of optimal detergents and reconstitution conditions in solutions would be essential for any attempts to reconstitute the complexes on solid surfaces (beads). Our early work with solubilized receptors demonstrated that reconstitution with G proteins was feasible and allowed us to determine optimal conditions (7). Subsequent studies examined the properties of the phosphorylated FPR as well as assemblies with arrestins (6, 20, 21). In conjunction with the reconstitution studies, we examined cellular complex formation by confocal fluorescence microscopy to confirm that the ability to reconstitute assemblies in vitro accurately reflected the complexes formed in vivo.

Receptors on Beads

DCNi-H6FPR Protein as a Docking Site for LR Assembly

Previous work in this laboratory (49) has demonstrated the display of an LR assembly on silica-based nickel chelate beads, using detergent-solubilized formyl peptide receptor with a hexahistidine tag on the carboxyl terminus (FPR-CH6). Briefly, FPR-CH6 was expressed in U937 cells, the cells were lysed by nitrogen cavitation, and crude postnuclear membrane preparations were stored at $-80°C$ in aliquots of 0.5 mL, corresponding to 10^8 cell equivalents. These aliquots were then thawed and solubilized in 1% dodecyl maltoside and diluted in solutions containing 0.1% dodecyl maltoside. We used both the FPR-CH6 and a second construct with a hexahistidine tag on the amino terminus of FPR (MAHHHHHHETN-receptor; NH6-FPR). The NH6-FPR would be potentially better for allowing unrestricted binding to the cytoplasmic sites on the receptor. To determine the concentration of soluble receptor and its K_d for fluorescent ligand, we took advantage of the fact that the fluorescence of the ligand is largely unaffected by binding to the receptor but is quenched about 95% by binding to a high-affinity antibody to FITC (49). From such experiments, we found that the sFPR-CH6 preparation contained approximately 37 nM sFPR-CH6 at 10^8 cell equivalents/mL, with a K_d of 6–10 nM, while the sNH6-FPR preparation had 10 nM sNH6-FPR at 10^8 cell equivalents/mL with a K_d of 3–7 nM for the fluorescent ligand.

However, as the silica particles are highly heterogeneous and shatter easily, receptor preparations were also examined on DCNi beads for the purpose of RG reconstitution. Receptor preparations (25 or 50 μL) plus 50 μL of DCN at 5×10^7 beads/mL were brought to 200 μL with DHPS (0.1% dodecyl maltoside, 30 mM HEPES, pH 7.5, 20 mM potassium chloride, 100 mM sodium chloride, an intracellular buffer mimic), mixed for 3 hr, then centrifuged. The supernatants were removed and the beads resuspended in 50 μL DHPS. DCNi-H6FPR beads (5×10^4) were shaken with the indicated concentrations of L^F in 200 μL DHPS in Facscan tubes for 1 hr at 4°C, and then their fluorescence was determined by flow cytometry. Control beads, incubated with solubilized membranes, which contained no FPR, defined nonspecific binding. Results showed that as many as 200,000 sFPR-CH6/bead and 40,000 sNH6-FPR/bead were bound. Thus, doubling both the concentration of the receptor and the time of mixing gave an average of 70% more receptor on the beads. The K_d values averaged 7.5 nM for the sFPR-CH6 and 4 nM for the sNH6-FPR, which is experimentally indistinguishable from the soluble values.

Previous work had shown that the addition of 0.5 μM bovine brain G proteins (a mixture of types; Calbiochem, La Jolla, CA) to a solubilized FPR preparation with ligand led to the formation of a soluble LRG complex over a time period of 2 hr at 4°C, as shown by a decrease in the rate of dissociation of the fluorescent ligand from the complex compared to the rate of dissociation from the FPR alone or in the presence of GTPγS (49). In the case of DCNi beads charged with NH6-FPR, one would expect that addition of fluoresceinated bovine brain G protein to the beads would lead to some nonspecific binding and fluorescence and that the addition of nonfluorescent ligand would lead to an increase in binding and fluorescence, which could be attributed to LRG assembly on the bead. We fluoresceinated bovine brain G protein to about three fluoresceins per heterotrimer and found that the nonspecific binding was 350,000 fluoresceins/bead, which was 10 times the fluorescence shown by beads loaded with NH6-FPR at the low level. Similarly, a poor ratio of specific to nonspecific binding was obtained with the silica-based nickel chelate beads. This prompted us to find systems with higher specific binding or lower nonspecific binding.

Goat-Anti-Mouse (GAM) Beads with Anti-H6 Antibodies

Polystyrene beads coated with GAM antibodies suitable for flow cytometry have a capacity of about 650,000 antibody molecules per bead (Bangs Laboratories, Fisher, IN). We doubled the manufacturer's amount and concentration of monoclonal antibody, otherwise following their coating procedure, and coated the beads with six different commercial antihexahistidine antibodies to determine whether any would bind FPR-CH6 or NH6-FPR. None bound NH6-FPR, possibly because the amino acids flanking that H6 site were not optimal for any of the antibodies, and only one (Roche) bound FPR-CH6 (data not shown), which has the H6 site at the very carboxy terminus of the protein sequence. We titrated these beads with FPR-CH6 plus saturating fluorescent ligand (50 nM) to construct a binding curve and obtained a K_d for the FPR-CH6 to the anti-H6 beads of 3.6 nM, with a B_{max} of 37,000 FPR-CH6/bead. This was again a disappointingly small number of receptors per bead, compared to 1,300,000 sites per bead.

The specificity of the interaction was shown by the fact that the FPR-CH6 could be blocked from binding to the bead by preincubation with soluble anti-H6 antibody. Most important, these GAM-anti-H6-FPR-CH6 beads were unable to show the presence of an LRG complex because soluble anti-H6 antibody competed with the binding of G protein (data not shown). We were, however, able to show the presence of a ligand-FPR-arrestin complex (L^{F}RA). Incubations were carried out in 10-μL volumes in a 96-V-well plate (Costar), including 30 nM FPR-CH6, 50 nM fluorescent ligand, various concentrations of truncated arrestin-2, and 50,000 beads per assay. The plate was shaken on a vortex mixer for 2 hr at 4°C and then diluted to 200 μL with buffer for Facscan measurement. The added arrestin resulted in higher binding than in the arrestin-lacking sample, indicating that a high ligand-affinity form of the receptor had been formed. The increase over the receptor-only binding was plotted against added arrestin and demonstrated a saturable response. Despite the demonstration of an L^{F}RA complex, the lack of a demonstrable L^{F}RG complex made this approach less than ideal as a universal technique.

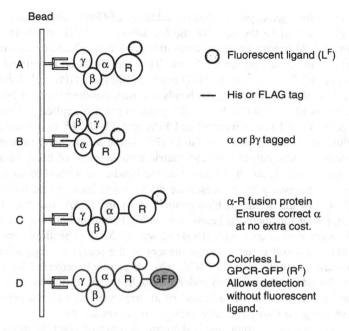

Figure 17.4. Ligand, receptor, and G protein assembly schematic diagrams.

G Protein on Beads

DCNi-G Protein as a Docking Site for LRG Assembly

Because hexahistidine-tagged proteins can bind tightly to DCNi beads through the interaction of histidines and nickel, the availability of pure H6-β1γ2 G protein heterodimer made it possible for us to detect the reconstitution of GPCR and its cognate G proteins on beads on ligand activation by using a flow cytometric method. The H6-β1γ2 was generously provided by Dr. R.R. Neubig (University of Michigan). It was genetically constructed in such a way that the hexahistidine tag was covalently attached to the amino terminus of the γ subunit and projected away from its partners in three-dimensional reconstructions (J. Garrison, personal communication). Four potential schemes of LRG assembly on beads are shown in figure 17.4. Scheme A shows H6-β1γ2 bound to the bead, and soluble αi3, wild-type FPR, and a fluorescent ligand forming an L^{F}RG assembly on the bead. Scheme B shows a similar assembly, but with the α subunit tagged. Scheme C shows the same H6-β1γ2 as in A, but this time a fusion protein of FPR with the αi2 subunit, and a fluorescent ligand, combine to form an L^{F}R-αi2β1γ2 assembly on the bead. Scheme D uses the same H6-β1γ2 and soluble αi3 as in A, but this time a fusion protein of FPR with enhanced green fluorescent protein (R^F) is used with a nonfluorescent ligand to form an LR^{F}G assembly on the bead.

Assembly with Wild-type Receptor on DCNi-H6-β1γ2αi3 beads (Scheme A)   Because the endogenous G protein concentration in our solubilized membrane preparation was relatively low, we expected that the addition of Gαi3, which has the highest affinity with FPR, would be the best for the formation of L^FRG on beads. To make DCNi-H6-$\beta1\gamma2\alpha$i3 (G beads), equal concentrations of H6-$\beta1\gamma2$ and Gαi3 were mixed (32 μM each) and incubated on ice for 5 min. This was followed by mixing with the DCNi beads (18 pmol G protein per 600,000 beads in 100 μL DHPS with 1 mM $MgCl_2$ (G buffer) for 1 hr, then centrifuging the beads and resuspending them in 50 μL of G buffer. These beads (2 μL, 24,000 beads, 0.7 pmol G protein applied per assay well) were mixed in 96-V-well plates with 60 nM FPR and 75 nM L^F in a total of 10 μL for 2 hr to allow assembly to take place in DHPS. These assays were brought to 200 μL in buffer immediately before flow cytometric measurement of bead fluorescence. To provide evidence that L^FRG had formed on the beads, we wished to show that L^F, R, and G were all necessary for fluorescence above some background level and that GTPγS would return the signal to background. In figure 17.5A, bar one, the background binding found for uncoated beads was 10 MCF, and in bar two, the assembly obtained with beads coated only with H6-$\beta1\gamma2$ was 17 MCF. The signal above background was likely a result of αi subunits present in the receptor preparation. In bar three, the standard assembly gave 27 MCF, with all elements provided for L^FRG assembly. Next, the addition of GTPγS indeed brought the bead fluorescence down to the background level, whereas the addition of an irrelevant fluorescent peptide (specific for the α4 integrin) also gave only background fluorescence.

In bar six, the lack of receptor resulted in an increase of bead fluorescence over background, but this was because without receptor, there was increased free fluorescent ligand, and this free ligand was sufficient to result in the higher background observed (data not shown). Specific assembly was defined as the difference between total assembly and assembly in the presence of GTPγS.

Assembly Using a Receptor–Gα$_i$2 Fusion Protein on DCNi-H6-$\beta1\gamma2$ Beads (Scheme C) To obviate the need for adding purified α subunits, an FPR-Gαi2 fusion protein (FPR-α) was constructed and transfected into U937 cells, where the PCR-generated cDNAs of the FPR carboxy terminus and Gαi2 amino terminus were ligated together, resulting in a spacer consisting of three alanine residues. The transfected cells were normal in culture in terms of cell morphology, proliferation, calcium mobilization, receptor internalization, and recycling (43). The chimeric receptor also colocalized with arrestin upon ligand activation. The fusion protein created a locally high concentration of Gαi2 and a 1:1 ratio of receptor and G protein. This unique feature makes the fusion protein ideal for macromolecular assembly experiments and eliminated the need for exogenous Gα protein.

We anticipated that the endogenous $\beta\gamma$ in the solubilized fusion protein membrane preparation might reconstitute with the FPR-α to form an LRG complex in solution and prevent the FPR-α from binding the $\beta\gamma$ on beads. Therefore, the addition of certain amount of GTP might be beneficial by causing the dissociation of FPR-α from endogenous $\beta\gamma$, and when GTP was gradually hydrolyzed by the Gα, the FPR-α would then be able, at least in part, to bind the $\beta\gamma$ on the beads. L^F was used for both activating the receptor and detecting for the L^FRG assembly on beads by flow cytometry. The actual assembly was carried out in a 96-well plate by mixing 29–38 nM FPR-α, 20 nM of L^F and 5×10^4 DCNi-H6-$\beta1\gamma2$ beads, with 0, 1, 5, 10, 25, 50, or 100 μM GTP or 3 μM GTPγS in a 10-μL reaction. This assembly gave a specific signal sim-

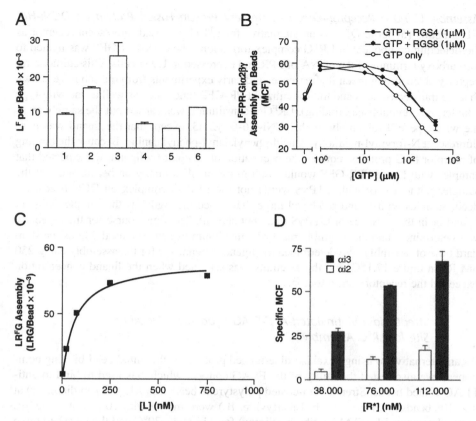

Figure 17.5. Representative assembly behavior. (A) Assembly of $L^F RG$. (B) Effect of RGS isozymes on assembly of the fusion protein on beads to form $L^f R$-$\alpha G\beta\gamma$. (C) Time of assembly versus $L^F RG$. (D) Assembly of $LR^F G$ on streptavidin-biotinylated-anti-FLAG beads with $\alpha i2$ and $\alpha i3$. The nonspecific binding was about 50 mean channel fluorescence.

ilar to the data shown in figure 17.5A, and the signal was indeed made larger by the addition of 1–10 μM GTP and smaller by yet larger amounts of GTP, as predicted (data not shown); 5 μM GTP was chosen as the standard addition for this assembly.

We took advantage of the high affinity with which $\beta\gamma$ subunits reassemble with the receptor–Giα fusion protein (figure 17.5B) to probe the activity of members of the RGS (regulators of G protein signaling) family of proteins (provided by Dr. R. Neubig). Because the assembly depended on the high-affinity α to $\beta\gamma$ interaction ($\sim 10^{-8}$ M), rather than the lower-affinity interaction between R and G in detergent ($\sim 10^{-6}$ M), complexes were formed at near stoichiometric ratios of the components. The data were consistent with the idea that RGS, by enhancing the cleavage of GTP, affected the GTP dose–response for the formation of the bead-based complex. The shift in the dose–response was consistent with the consumption of 10 μM or more GTP in 10 μL over 2 hr in the presence of R-α, RGS, and L. Assuming that 24-nM R-α (the total added) consumed 10 μM GTP, the turnover number was ~ 5 min^{-1} for the receptor-Giα fusion protein at 7°C, implying ~ 40 min^{-1} at 37°C.

Assembly Using a Receptor–Green Fluorescent Protein Fusion Protein on DCNi-H6-β1γ2 Beads (Scheme D)   As an alternative for L^{F}RG, we used fluorescent receptor as the detecting component in LR^{F}G complex formation. The cDNA of GFP was ligated to the carboxy terminus of the cDNA of FPR and expressed in U937 cells. This chimeric receptor yielded strong green fluorescence, and early experiments from our lab have shown that the transfected cells and the solubilized FPR-GFP functioned as well as the wild-type counterparts (unpublished finding). LR^{F}G reconstitution was carried out the same way as the wild-type FPR assembly, with DCNi-H6-β1γ2αi3, except that the ligand was non-fluorescent N-forymyl-methionyl-leucyl-phenylalanyl-phenylalanine. Because the binding of receptor to G proteins requires the occupation of receptor by ligand, we expected that samples with L and FPR-GFP would result in maximal assembly on beads, and that the samples without L or with GTPγS would not form LR^{F}G complex on DCN beads. Indeed, samples with ligand produced more than twice the signal as the samples without ligand or in the presence of GTPγS (data not shown). The time course for the assembly was determined, and the assembly reached a maximum in 1 hr. We used 2 hr as the standard time of assembly. We determined the ligand dependence for the assembly using 250 nM R^F in figure 17.15C, and the assembly was saturated when the ligand concentration exceeded the receptor concentration.

Streptavidin-Biotinylated-Anti-FLAG Beads as a Docking Site for LR^{F}G Assembly

As an alternative to using hexahisitidine-tagged proteins as the initial bead-binding component, we also tested the utility of the FLAG epitope, which was used to bind to anti-FLAG coated beads. Streptavidin-coated polystyrene beads (20 μL, 6.2 μm diameter) at 4×10^7 beads/mL (Spherotech, Libertyville, IL) were mixed with 20 μL of 1 mg/mL biotinylated anti-FLAG M2 antibody (Sigma) for 2 hr at 4°–20°C and then washed three times in buffer to give about 9×10^6 FLAG-FITC binding sites per bead at 4000 beads/μL. Beads (50 μl) were first mixed with 1 μL of 3.4 μM β4γ2-FLAG-H6 (Dr. James Garrison, University of Virginia) for 1 hr at 4°C, washed, mixed with 1 μL of 7.7 μM αi3 or αi2 subunit (Calbiochem) for 1 hr, spun, and then resuspended in buffer to give αβγ heterotrimer on the surface of the beads. These beads (20,000) were mixed in a 10-μL volume with the indicated concentrations of FPR-GFP with 120 nM L, ±0.1 mM GTPγS, for 2 hr at 4°C for LR^{F}G assembly as before. In figure 17.5D, it can be seen that the specific assembly increased with increasing receptor concentration for both the αi2 and αi3 subunit experiments. This was expected, as previous data for LRG assembly in solution showed an EC$_{50}$ of 1 μM for assembly with αi3, with an even higher EC$_{50}$ for assembly with αi2, and our highest concentration of receptor here reached only 0.12 μM. In later experiments, these beads achieved a signal to background ratio of 4 to 1, with only 0.2 pmol G protein used per assembly reaction, which lowers reagent use compared to the 0.7 pmol per assembly reaction used with the DCNi beads.

Receptor-Derived Peptides on Beads

As an alternative approach to the attachment of entire GPCRs to the bead surface, as described above, we sought to determine whether synthetic peptides derived from the known amino acid sequence of the FPR might serve as docking sites for GPCR-associated proteins. Most of what is known about receptor domains involved in arrestin interaction is centered on the carboxyl terminus of the receptor (18, 55). Phosphorylation of serine and threonine residues in the carboxyl terminus is required for arrestin binding, and it has also been demonstrated that the pattern of phosphorylation in this domain affects the rates of arrestin-dependent receptor internalization and recycling (33). Because of the importance of the GPCR carboxyl terminus in arrestin interactions, we wanted to design a system whereby interactions of the GPCR carboxyl terminus with arrestin could be studied outside of the influences of the remainder of the receptor or other intracellular molecules.

To this end, we designed a synthetic polypeptide that corresponded to the amino acid sequence of the 47 residues of the carboxyl terminus of the FPR. The use of synthetic peptides confers several advantages to the experimental design. First, the polypeptide fits the requirement of isolation from other receptor domains and from other cellular molecules. The synthetic polypeptide is more easily purified than products obtained from other methods, such as by proteolytic cleavage of solubilized receptor or by bacterial expression of receptor cDNA. Synthetic products are also more uniform than the products obtained by these other methods. Although the synthesis of custom polypeptides may be initially costly, there is an overall savings of time and monetary expense, as the culture of cell lines is not required and each experiment can be carried out in less than 1 day. Last, only a small amount of polypeptide is required for each experiment.

Peptide synthesis was done by the solid-phase method using Fmoc chemistry on an Applied Biosystems 433 peptide synthesizer (by New England Peptide, Inc., Gardner, MA). Because receptor phosphorylation is required for arrestin binding, the polypeptide was synthesized in both an unphosphorylated state (hereafter called 47mer) and a phosphorylated state (hereafter called phospho47mer), in which four serine or threonine residues within the polypeptide were replaced with phosphoserine or phosphothreonine. In the synthetic process, the phosphorylated residues were incorporated as Fmoc-Ser[PO(OBzl)OH] and Fmoc-Thr[PO(OBzl)OH]. Both polypeptides were biotinylated at the amino terminus so that they could be bound to commercially available streptavidin-coated polystyrene beads manufactured by Spherotech, which are of a size and uniformity suitable for analysis by flow cytometry. In this approach, excess biotinylated polypeptide was incubated with the streptavidin-coated polystyrene beads to allow for binding via a biotin-streptavidin linkage. The beads were washed thoroughly and incubated with FITC-labeled arrestins (purified arrestins were generously provided by Dr. Seva Gurevich of Vanderbilt University). The beads were then pelleted, resuspended to a volume of 200 μL, and analyzed by flow cytometry. According to this approach, an increase in the average fluorescence of the beads would indicate that the labeled arrestin was associated with the bead-bound polypeptides.

To confirm the presence of the biotinylated peptides on the beads, staining with antibodies directed against the C-terminal 10 amino acids of the FPR followed by flow cytometry revealed that both phosphorylated and unphosphorylated biotinylated peptides stably bound to the surface of streptavidin-coated beads. We were also able to discriminate between phosphorylated and unphosphorylated bead-bound peptide by the use of FITC-labeled antiphosphoserine or antiphosphothreonine antibodies. Subsequently, we demonstrated that FITC-labeled arrestin2 bound specifically to the phospho47mer, but not 47mer, with K_d values in the low micomolar range. This result is in keeping with the previous observations that phosphorylation of the carboxyl terminus is required for arrestin binding (55). To verify this result, we enzymatically dephosphorylated bead-bound phospho47mer using alkaline phosphatase and found that this treatment prevented arrestin2–FITC binding (39). Last, because the process of FITC labeling can interfere with arrestin's ability to bind the FPR carboxyl terminus, we employed competition assays, using a FITC-labeled arrestin mutant as a reporter molecule. The competition assay provided similar data to the direct binding results.

Our results using bead-bound FPR C-terminal peptides in an arrestin-binding assay provide a strong proof of principle for bead-based assays. We have demonstrated that bead-bound synthetic peptides can interact with signaling proteins in a manner consistent with whole-receptor studies while allowing study of specific receptor domains. In addition to studying receptor–arrestin interactions, this approach could easily be adapted for the study of ligand binding, receptor–G protein interactions, or other non-receptor protein interactions. The specificity of bead-based assays, combined with their speed and relative low cost in time and materials, make them an attractive research method for receptor signaling studies and drug discovery.

Future Perspectives

There are many advantages to using flow cytometry for the analysis of macromolecular assemblies. There are also a number of obstacles to its acceptance as a standard platform for such analyses. However, with the recent advances in assay miniaturization, bead chemistry, and labeling techniques, the number and magnitude of these hurdles is decreasing. In this review, we have described novel techniques to examine the molecular complexes formed by the largest class of signaling molecules, the G protein–coupled receptors. Investigating assemblies involving a component that is an integral membrane protein provides additional challenges resulting from the requirement for detergent solubilization. Purification of the components can also add to the complexity of the system but can be circumvented through the use of highly specific labels, such as fluorescent ligands. However, some receptors, such as the β_2 adrenergic receptor, have not had until recently a suitable fluorescent ligand, which made it impossible to study $L^F RG$ assembly using fluorescent ligand as the detecting agent. Such assemblies have recently been detected by using chimeric GFP receptors as the detecting agent (44b). In the future, it may be possible to use these approaches to screen for prospective ligands of orphan receptors by using orphan receptor–GFP constructs. Moreover, we can use the GFP-receptor combined with a different-colored fluorescent ligand to address significant molecular assembly questions, which are difficult to answer using single-color flow cytometry. An example is the disassembly of the LRG complex caused by GTP binding to G proteins, where the actual sequence of the dissociation events is still unclear and clarification should come from examining each component of the molecular assembly individually (44a). These approaches are also likely to play a role in GPCR proteomics. For example, color-coded particle arrays could display different subunits of G protein heterotrimers, and it would be possible to assess the ability of a single GPCR to assemble with multiple G proteins simultaneously. One can envision mechanisms to display multiple receptors or ligands simultaneously in arrays to provide other assays of assembly specificity.

We have struggled unsuccessfully to display ligands for the formyl peptide receptor on particles. Two approaches have been attempted. The first employed covalent attachment of ligand to epoxy-activated dextran beads, whereas the second used a biotinylated ligand as a bridge for binding the FPR to streptavidin beads. Both approaches, for as-yet-unknown reasons, were unsuccessful in generating a ligand-mediated FPR-GFP complex on the bead. More recently, however, we have used a GFP fusion of the β_2-adrenergic receptor to generate not only G protein–mediated assemblies on beads using $G\alpha_s\beta\gamma$ but also receptor complexes on beads derivatized with the adrenergic antagonist, dihydroalprenolol (44b). These results further indicate the feasibility of particle-based ligand display as a practical means for assaying assemblies of known and orphan receptors.

References

1. Abrams, W. R., Diamond, L. W., Kane, A. B. 1983. A flow cytometric assay of neutrophil degranulation. *J. Histochem. Cytochem.* 31: 737–744.
2. Anborgh, P. H., Seachrist, J. L., Dale, L. B., Ferguson, S. S. 2000. Receptor/beta-arrestin complex formation and the differential trafficking and resensitization of beta2-adrenergic and angiotensin II type 1A receptors. *Mol. Endocrinol.* 14: 2040–2053.
3. Baldwin, J. M., Bennett, J. P., Gomperts, B. D. 1983. Detergent solubilisation of the rabbit neutrophil receptor for chemotactic formyl peptides. *Eur. J. Biochem.* 135: 515–518.
4. Barak, L. S., Ferguson, S. S., Zhang, J., Martenson, C., Meyer, T., Caron, M. G. 1997. Internal trafficking and surface mobility of a functionally intact beta2-adrenergic receptor-green fluorescent protein conjugate. *Mol. Pharmacol.* 51: 177–184.
5. Barten, M. J., Gummert, J. F., van Gelder, T., Shorthouse, R., Morris, R. E. 2001. Flow cytometric quantitation of calcium-dependent and -independent mitogen-stimulation of T cell functions in whole blood: inhibition by immunosuppressive drugs in vitro. *J. Immunol. Methods* 253: 95–112.
6. Bennett, T. A., Foutz, T. D., Gurevich, V. V., Sklar, L. A., Prossnitz, E. R. 2001. Partial phosphorylation of the N-formyl peptide receptor inhibits G protein association independent of arrestin binding. *J. Biol. Chem.* 276: 49195–49203.
7. Bennett, T. A., Key, T. A., Gurevich, V. V., Neubig, R., Prossnitz, E. R., Sklar, L. A. 2001. Real-time analysis of G protein-coupled receptor reconstitution in a solubilized system. *J. Biol. Chem.* 276: 22453–22460.
8. Bouvier, M., Hausdorff, W. P., De Blasi, A., O'Dowd, B. F., Kobilka, B. K., Caron, M. G., Lefkowitz, R. J. 1988. Removal of phosphorylation sites from the beta 2-adrenergic receptor delays onset of agonist-promoted desensitization. *Nature* 333: 370–373.
9. Campbell, R. E., Tour, O., Palmer, A. E., Steinbach, P. A., Baird, G. S., Zacharias, D. A., Tsien, R. Y. 2002. A monomeric red fluorescent protein. *Proc. Natl. Acad. Sci. USA* 99: 7877–7882.
10. Claing, A., Perry, S. J., Achiriloaie, M., Walker, J. K., Albanesi, J. P., Lefkowitz, R. J., Premont, R. T. 2000. Multiple endocytic pathways of G protein-coupled receptors delineated by GIT1 sensitivity. *Proc. Natl. Acad. Sci. USA* 97: 1119–1124.
11. Demo, S. D., Masuda, E., Rossi, A. B., Throndset, B. T., Gerard, A. L., Chan, E. H., Armstrong, R. J., Fox, B. P., Lorens, J. B., Payan, D. G., Scheller, R. H., Fisher, J. M. 1999. Quantitative measurement of mast cell degranulation using a novel flow cytometric annexin-V binding assay. *Cytometry* 36: 340–348.
12. Evans, N. A., Groarke, D. A., Warrack, J., Greenwood, C. J., Dodgson, K., Milligan, G., Wilson, S. 2001. Visualizing differences in ligand-induced beta-arrestin-GFP interactions

and trafficking between three recently characterized G protein-coupled receptors. *J. Neurochem.* 77: 476–485.

13. Ferguson, S. S. 2001. Evolving concepts in G protein-coupled receptor endocytosis: the role in receptor desensitization and signaling. *Pharmacol. Rev.* 53: 1–24.

14. Goodman, O. B., Jr., Krupnick, J. G., Santini, F., Gurevich, V. V., Penn, R. B., Gagnon, A. W., Keen, J. H., Benovic, J. L. 1996. Beta-arrestin acts as a clathrin adaptor in endocytosis of the beta2-adrenergic receptor. *Nature* 383: 447–450.

15. Gutkind, J. S. 1998. The pathways connecting G protein-coupled receptors to the nucleus through divergent mitogen-activated protein kinase cascades. *J. Biol. Chem.* 273: 1839–1842.

16. Hall, R. A., Lefkowitz, R. J. 2002. Regulation of G protein-coupled receptor signaling by scaffold proteins. *Circ. Res.* 91: 672–680.

17. Hamm, H. E. 1998. The many faces of G protein signaling. *J. Biol. Chem.* 273: 669–672.

18. Han, M., Gurevich, V. V., Vishnivetskiy, S. A., Sigler, P. B., Schubert, C. 2001. Crystal structure of beta-arrestin at 1.9 A: possible mechanism of receptor binding and membrane Translocation. *Structure (Camb.)* 9: 869–880.

19. Hochuli, E., Dobeli, H., Schacher, A. 1987. New metal chelate adsorbent selective for proteins and peptides containing neighbouring histidine residues. *J. Chromatogr.* 411: 177–184.

20. Key, T. A., Bennett, T. A., Foutz, T. D., Gurevich, V. V., Sklar, L. A., Prossnitz, E. R. 2001. Regulation of formyl peptide receptor agonist affinity by reconstitution with arrestins and heterotrimeric G proteins. *J. Biol. Chem.* 276: 49204–49212.

21. Key, T. A., Foutz, T. D., Gurevich, V. V., Sklar, L. A., Prossnitz, E. R. 2002. Identification of N-formyl peptide receptor phosphorylation domains that differentially regulate arrestin and agonist affinity. *J. Biol. Chem.* (in press).

22. Klut, M. E., Whalen, B. A., Hogg, J. C. 1997. Activation-associated changes in blood and bone marrow neutrophils. *J. Leukoc. Biol.* 62: 186–194.

23. Lander, E. S., Linton, L. M., Birren, B., Nusbaum, C., Zody, M. C., Baldwin, J., Devon, K., Dewar, K., Doyle, M., FitzHugh, W., et al. 2001. Initial sequencing and analysis of the human genome. *Nature* 409: 860–921.

24. Laporte, S. A., Oakley, R. H., Zhang, J., Holt, J. A., Ferguson, S. S., Caron, M. G., Barak, L. S. 1999. The beta2-adrenergic receptor/betaarrestin complex recruits the clathrin adaptor AP-2 during endocytosis. *Proc. Natl. Acad. Sci. USA* 96: 3712–3717.

25. Lauer, S. A., Nolan, J. P. 2002. Development and characterization of Ni-NTA-bearing microspheres. *Cytometry* 48: 136–145.

26. Lefkowitz, R. J. 1998. G protein-coupled receptors. III. New roles for receptor kinases and beta-arrestins in receptor signaling and desensitization. *J. Biol. Chem.* 273: 18677–18680.

27. Lehmann, A. K., Sornes, S., Halstensen, A. 2000. Phagocytosis: measurement by flow cytometry. *J. Immunol. Methods* 243: 229–242.

28. Lohse, M. J., Benovic, J. L., Caron, M. G., Lefkowitz, R. J. 1990. Multiple pathways of rapid beta 2-adrenergic receptor desensitization. Delineation with specific inhibitors. *J. Biol. Chem.* 265: 3202–3211.

29. Lohse, M. J., Benovic, J. L., Codina, J., Caron, M. G., Lefkowitz, R. J. 1990. beta-Arrestin: a protein that regulates beta-adrenergic receptor function. *Science* 248: 1547–1550.

30. Luecke, H., Schobert, B., Lanyi, J. K., Spudich, E. N., Spudich, J. L. 2001. Crystal structure of sensory rhodopsin II at 2.4 angstroms: insights into color tuning and transducer interaction. *Science* 293: 1499–1503.

31. Luttrell, L. M., Lefkowitz, R. J. 2002. The role of beta-arrestins in the termination and transduction of G-protein-coupled receptor signals. *J. Cell Sci.* 115: 455–465.

32. Neeson, P. J., Thurlow, P. J., Jamieson, G. P. 2000. Characterization of activated lymphocyte-tumor cell adhesion. *J. Leukoc. Biol.* 67: 847–855.

33. Oakley, R. H., Laporte, S. A., Holt, J. A., Barak, L. S., Caron, M. G. 1999. Association of beta-arrestin with G protein-coupled receptors during clathrin-mediated endocytosis dictates the profile of receptor resensitization. *J. Biol. Chem.* 274: 32248–32257.

34. Ormerod, M. G. 2002. Investigating the relationship between the cell cycle and apoptosis using flow cytometry. *J. Immunol. Methods* 265: 73–80.

35. Paing, M. M., Stutts, A. B., Kohout, T. A., Lefkowitz, R. J., Trejo, J. 2002. beta-Arrestins

regulate protease-activated receptor-1 desensitization but not internalization or down-regulation. *J. Biol. Chem.* 277: 1292–1300.

36. Penn, R. B., Pronin, A. N., Benovic, J. L. 2000. Regulation of G protein-coupled receptor kinases. *Trends Cardiovasc. Med.* 10: 81–89.

37. Pierce, K. L., Premont, R. T., Lefkowitz, R. J. 2002. Seven-transmembrane receptors. *Nat. Rev. Mol. Cell Biol.* 3: 639–650.

38. Porath, J. 1974. The Uppsala school in separation science and the development of bioaffinity chromatography. *Ann. Ist. Super. Sanita* 10: 95–102.

39. Potter, R. M., Key, T. A., Gurevich, V. V., Sklar, L. A., Prossnitz, E. R. 2002. Arrestin variants display differential binding characteristics for the phosphorylated N-formyl peptide receptor carboxyl terminus. *J. Biol. Chem.* 277: 8970–8978.

40. Prossnitz, E. R., Gilbert, T. L., Chiang, S., Campbell, J. J., Qin, S., Newman, W., Sklar, L. A., Ye, R. D. 1999. Multiple activation steps of the N-formyl peptide receptor. *Biochemistry (Mosc).* 38: 2240–2247.

41. Prossnitz, E. R., Ye, R. D. 1997. The N-formyl peptide receptor: a model for the study of chemoattractant receptor structure and function. *Pharmacol. Ther.* 74: 73–102.

42. Samama, P., Cotecchia, S., Costa, T., Lefkowitz, R. J. 1993. A mutation-induced activated state of the beta 2-adrenergic receptor. Extending the ternary complex model. *J. Biol. Chem.* 268: 4625–4636.

43. Shi, M., Bennett, T. A., Cimino, D. F., Maestas, D. C., Foutz, T. D., Gurevich, V. V., Sklar, L. A., Prossnitz, E. R. 2003. Functional Capabilities of an N-Formyl Peptide Receptor-G(alpha)(i)(2) Fusion Protein: Assemblies with G Proteins and Arrestins. *Biochemistry* 42: 7283–7293.

44. Simon, S. I., Chambers, J. D., Butcher, E., Sklar, L. A. 1992. Neutrophil aggregation is beta 2-integrin- and L-selectin-dependent in blood and isolated cells. *J. Immunol.* 149: 2765–2771.

44a. Simons, P., Shi, M., Foutz, T., Lewis, J., Buranda, T., Lim, W.K., Neubig, R., Garrison, J., Prossnitz, E.R., Sklar, L.A. 2003. Beads coated with epitope tagged G proteins assemble complexes of ligand-receptor-G protein for detection by flow cytometry. *Mol. Pharm.* 64:1227–1238.

44b. Simons, P., Biggs, S.M., Waller, A., Foutz, T., Cimino, D.F., Guo, Q., Neubig, R.R., Tang, W.J., Prossnitz, E.R., Sklar, L.A. 2004. Real-time analysis of ternary complex on particles: direct evidence for partial agonism at the agonists-receptor.G protein asssembly step of signal transduction. *J. Biol. Chem.* 279:13514–21.

45. Sklar, L. A. 1987. Real-time spectroscopic analysis of ligand-receptor dynamics. *Annu. Rev. Biophys. Biophys. Chem.* 16: 479–506.

46. Sklar, L. A., Bokoch, G. M., Button, D., Smolen, J. E. 1987. Regulation of ligand-receptor dynamics by guanine nucleotides. Real-time analysis of interconverting states for the neutrophil formyl peptide receptor. *J. Biol. Chem.* 262: 135–139.

47. Sklar, L. A., Finney, D. A., Oades, Z. G., Jesaitis, A. J., Painter, R. G., Cochrane, C. G. 1984. The dynamics of ligand-receptor interactions. Real-time analyses of association, dissociation, and internalization of an N-formyl peptide and its receptors on the human neutrophil. *J. Biol. Chem.* 259: 5661–5669.

48. Sklar, L. A., Mueller, H., Omann, G., Oades, Z. 1989. Three states for the formyl peptide receptor on intact cells. *J. Biol. Chem.* 264: 8483–8486.

49. Sklar, L. A., Vilven, J., Lynam, E., Neldon, D., Bennett, T. A., Prossnitz, E. 2000. Solubilization and display of G protein-coupled receptors on beads for real-time fluorescence and flow cytometric analysis. *Biotechniques* 28: 976–985.

50. Stadel, J. M., DeLean, A., Lefkowitz, R. J. 1980. A high affinity agonist beta-adrenergic receptor complex is an intermediate for catecholamine stimulation of adenylate cyclase in turkey and frog erythrocyte membranes. *J. Biol. Chem.* 255: 1436–1441.

51. Sureda, F. X., Gabriel, C., Comas, J., Pallas, M., Escubedo, E., Camarasa, J., Camins, A. 1999. Evaluation of free radical production, mitochondrial membrane potential and cytoplasmic calcium in mammalian neurons by flow cytometry. *Brain Res. Brain Res. Protoc.* 4: 280–287.

52. van Eeden, S. F., Klut, M. E., Walker, B. A., Hogg, J. C. 1999. The use of flow cytometry to measure neutrophil function. *J. Immunol. Methods* 232: 23–43.

53. Venter, J. C., Adams, M. D., Myers, E. W., Li, P. W., Mural, R. J., Sutton, G. G., Smith, H. O., Yandell, M., Evans, C. A., Holt, R. A., et al. 2001. The sequence of the human genome. *Science* 291: 1304–1351.

54. Vilven, J. C., Domalewski, M., Prossnitz, E. R., Ye, R. D., Muthukumaraswamy, N., Harris, R. B., Freer, R. J., Sklar, L. A. 1998. Strategies for positioning fluorescent probes and crosslinkers on formyl peptide ligands. *J. Recept. Signal Transduct. Res.* 18: 187–221.

55. Vishnivetskiy, S. A., Paz, C. L., Schubert, C., Hirsch, J. A., Sigler, P. B., Gurevich, V. V. 1999. How does arrestin respond to the phosphorylated state of rhodopsin? *J. Biol. Chem.* 274: 11451–11454.

56. Zhang, J., Barak, L. S., Anborgh, P. H., Laporte, S. A., Caron, M. G., Ferguson, S. S. 1999. Cellular trafficking of G protein-coupled receptor/beta-arrestin endocytic complexes. *J. Biol. Chem.* 274: 10999–11006.

57. Zhang, J., Ferguson, S. S., Barak, L. S., Menard, L., Caron, M. G. 1996. Dynamin and beta-arrestin reveal distinct mechanisms for G protein-coupled receptor internalization. *J. Biol. Chem.* 271: 18302–18305.

18

Applications of Flow Cytometry to Cell Adhesion Biology: From Aggregates to Drug Discovery

RICHARD S. LARSON, ALEXANDRE CHIGAEV,
BRUCE S. EDWARDS, SERGIO A. RAMIREZ,
STUART S. WINTER, GORDON ZWARTZ,
AND LARRY A. SKLAR

Introduction

Flow cytometry represents a powerful and evolving methodologic approach to cell adhesion biology. In its beginnings, flow cytometry was used solely to measure the expression of receptors on cellular surfaces and to correlate that expression with biologic function in non-flow-cytometry-based assays. From this primitive beginning, applications have proliferated and now include methodologies that measure real-time aggregation, receptor activity, and the downstream biologic consequences of cell adhesion. These biologic applications have led to platforms that are easily employed as drug screening and target validation tools.

Functional assays that measure cell aggregation were initially developed to measure cell–cell interactions in the immune system, especially between cytotoxic cells and various cell types targeted as the focus of their cytotoxic activity (9, 60, 65, 118). The cytotoxic "effector" cells and the "target" cells were stained with spectrally distinct fluorescent dyes, gently sedimented together into a cell pellet, and allowed to interact under static conditions for designated intervals of time. When resuspended and introduced into the flow cytometer, effector cells adherent to target cells were detected as "conjugate" particles emitting the fluorescence spectra of both dyes. Nonadherent effector and target cells were detected as monochromatically fluorescent particles. By using ion concentration–sensitive cytoplasmic fluorescent probes as the effector cell labels, it was also possible to detect physiological changes in intracellular ionized calcium and pH elicited by adhesion to target cells and to correlate these responses with cytotoxic function (33, 37; figure 18.1).

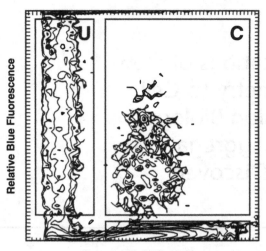

Forward Light Scatter

Figure 18.1. Flow cytometric conjugate assay to quantify natural killer (NK) cells bound to tumor target cells. NK cells loaded with the blue/violet fluorescent calcium probe indo-1 were mixed with K562 erythroleukemia cells and briefly centrifuged to facilitate cell–cell adhesion. Unbound singlet NK cells (rectangular gate labeled U) were distinguished from singlet target cells (rectangular gate labeled T) by virtue of their smaller size, resulting in a relatively small forward-light scattering signature, in combination with their blue indo-1 fluorescence. NK cells bound to K562 target cells formed conjugates (rectangular gate labeled C) that had the large light-scatter signature of K562 cells in combination with the blue fluorescence attribute of NK cells. This assay also enabled demonstration of a significant elevation of intracellular ionized calcium in target-bound NK cells as compared to unbound singlet NK cells. [Reprinted from (33) with the permission of the publisher.]

Later, methods were developed for continuously measuring ("real-time") cell adhesive interactions as they progressed over time in a fluid shear environment (112). A limitation of early adhesion kinetics analyses was that the fluid shear was generated with a magnetic stir bar and was thus neither homogeneous nor amenable to precise quantification (44, 111, 113). Subsequent refinement of these methods has enabled flow cytometric analysis of cell mixtures subjected to a more uniform and quantifiable fluid shear environment generated in a cone-plate viscometer. Cell mixtures are sampled periodically from the viscometer into a formalin fixative solution for subsequent off-line flow cytometric analysis (76, 121). These experiments have been able to demonstrate a remarkable potentiation of adhesion efficiency through the combined action of two sets of adhesion molecules and a progression of adhesion molecule use from one class to another over time (48, 75, 77). Once these methods for measuring cell adhesion were developed, applications of this technology to measure downstream biologic events such as cell signaling, activation, and growth were created and have been employed successfully in both basic science and clinical settings.

With the development of plug flow cytometry sample-handling technology (34), an online analysis approach is also possible in which the cell mixtures are transported within seconds directly from the viscometer to the flow cytometer under computer con-

Table 18.1. Adhesion molecules used in leukocyte-"target" cell interaction in blood

Target Cell Adhesion Molecules	Leukocyte Adhesion Molecules
E-selectin (CD62E)	sLe^x, sLe^a, CLA, ESL-1
P-selectin (CD62P)	PSGL-1, sLe^x, sLe^a
ICAM-1	$\alpha_M\beta_2$ (CD11b/CD18 (Mac-1), $\alpha_L\beta_2$ (CD11a/CD18 (LFA-1))
ICAM-2	$\alpha_L\beta_2$ (CD11a/CD18 (LFA-1))
VCAM-1 (CD106)	$\alpha_4\beta_1$ (CD49d/CD29 (VLA-4)), $\alpha_4\beta_7$
MadCAM-1	$\alpha_4\beta_7$, L-selectin (CD62L)
GlyCAM-1	L-selectin (CD62L)
CD34	L-selectin (CD62L)

trol (32, 35), allowing for real-time measurement of cell adhesion. The most recent methodologies to affect the study of cell–cell adhesion are techniques and reagents to study real-time changes in receptor affinity (23, 137). The combination of technologies that measure both cell adhesion and receptor affinity in real time now allows for the comparison of molecular affinity changes with cellular avidity changes. This chapter outlines the development and applications of flow cytometry in each of these technologic developments, as well as its application to adhesion biology.

Cell Adhesion Biology

Cell–cell adhesion is important in many biologic processes, including embryogenesis, inflammation, tissue repair, lymphocyte function and homing, hematopoiesis, and tumor metastasis, to name a few (57). The avidity, or adhesive activity between these cells and their targets, is regulated by the expression levels of the receptors on the cell surface (site density) and the affinity of the receptor for its ligand. There are a large number of adhesion molecule families, but three in particular have been extensively studied using flow cytometry techniques: integrins, immunoglobulin (Ig) family members, and selectins (table 18.1). The principles and technologies applied to the study of these adhesion families are readily applied to others. We will briefly summarize three areas in which cell adhesive interactions are most physiologically critical: cell adhesive interaction in the bone marrow under static conditions, cell adhesive interactions in the blood stream under flow conditions, and cell adhesive interactions during migration in the tissue. The flow cytometric applications that have been applied to each of these areas will then be later described.

Adhesive Interaction in Bone Marrow Microenvironment

In the venules and sinusoids of the bone marrow microenvironment, hematopoietic and stromal cells provide the adhesive contacts and cell-secreted molecules that maintain hematopoiesis. The adhesion molecules expressed by these cells play an essential role in hematopoiesis and work in concert with a number of soluble factors to provide a steady-state supply of mature blood elements (89). The biologic importance of integrins in disease pathogenesis is highlighted by the severe hematopoietic defects observed in mice that lack the β1-integrin subunit (55). Binding of these receptors allows for hematopoietic cell interaction with stromal cells that is in turn associated with positive and negative downstream effects on intracellular signaling. In addition to supporting normal hematopoiesis, adhesive events on stromal cells have been shown to play an important role in promoting the growth and proliferation of a wide array of acute and chronic leukemias. As a result, integrin-mediated binding has been a longstanding focus in stem cell research (4, 6). Conversely, abnormal stromal cell function has been implicated as being linked to the development of osteopetrosis, aplastic anemia, myelodysplasia, and a number of other bone marrow failure disorders (47, 103, 120).

All members of the integrin family are noncovalently associated α and β heterodimeric subunits (57). Integrins are subclassified according to β subunits, of which eight have been identified so far (β1–β8). Several α subunits (α1, α2, α3, α4, α5, α6, α7, α8, α9, α10, α11, αv, αl, αm, αx, α IIb, αE, and αd) can associate with each β subunit, leading to the generation of at least 26 heterodimers (43). Several members of the β1 and β2 subfamilies, notably VLA-4 (α 4β1, CD49d/CD29), VLA-5 (α 5β1, CD49e/CD29), and LFA-1 (αLβ2, CD11a/CD18), play key roles in mediating cell–cell adhesion in progenitor myeloid cells, monocytes, B cells, and T cells (3, 57, 95, 115). VLA-4 binds vascular cell adhesion molecule 1 (VCAM-1, CD106); VLA-5 binds VCAM-1 and fibronectin; and LFA-1 may bind three ligands, intracellular adhesion molecule (ICAM)-1, ICAM-2, or ICAM-3 (CD54, CD102, and CD50, respectively). The VCAM-1 and ICAM species are integrin ligands belonging to the Ig family (30, 79, 116, 117).

To participate in binding events, integrins must be activated, or "turned on," from cytoplasmic events through a process that has been termed "inside-out signaling." Inside-out signaling is, in turn, accomplished through changes in receptor conformation (affinity) and aggregation (avidity) (52, 57). The molecular mechanisms that control inside-out signaling and subsequent integrin activation are yet to be well understood, but protein kinase C, cytohesins, and R-ras have been shown to play important roles in mediating cytoplasmic signaling events (99). Importantly, the engagement of several types of receptors by chemokines, chemoattractants, and antigens can induce "inside-out" signaling, which play an important role in mediating cell–cell and cell–extracellular matrix interaction (127). Activation of integrins can also be induced by the binding of "activating" mAbs or by the divalent cation manganese. Although they are nonphysiologic, mAb and manganese stimulation have been invaluable for studying receptor physiology.

After activation and ligand binding at the cell surface, integrins communicate with the interior of the cell in a process that has been termed "outside-in signaling" (52). This process leads to changes in intracellular pH, calcium fluxes, tyrosine phosphorylation, and reorganization of the cytoskeleton (99). These intracellular events precede lymphocyte activation, differentiation, and transmigration. In addition, "outside-in" signaling confers a survival message to cells that are adherent to endothelial or stromal cells. Some of the signaling pathways are common to a number of integrins, such as focal adhesion kinase (FAK), whereas others appear to be specified to individual molecules (99). Through "outside-in" signaling, cell populations are mobilized to a number of tissue compartments, and are the subject of a number of studies in leukocyte biology.

The selectin family is comprised of three proteins: E- (endothelial, CD62E), P- (platelet, CD62P), and L- (leukocyte, CD62L) selectins. E- and P-selectin are expressed on stromal cells and activated endothelial cells, and L-selectin is constitutively expressed on leukocytes and leukocyte progenitors. In addition to recognizing sialylated and fucosylated lactosamines, L-selectin also binds to sialomucins, including the leukocyte-specific differentiation-associated antigen, CD34. L-selectin also appears to be important for regulating the mobilization of progenitor cells in and out of the bone marrow compartment, as well as the homing of differentiated lymphocytes to peripheral lymph nodes (43, 46).

Selectin-mediated binding is highlighted by rapid association (k_{on}) and dissociation (k_{off}) rate constants that facilitate leukocyte rolling (52, 127). The selectins also contribute to hematopoiesis, as demonstrated by a knockout murine model. In instances in which the P- and E-selectins were genetically deleted, affected mice had an extreme leukocytosis, elevated cytokine levels, and alterations in hematopoiesis, which are characterized by increased granulocytopoiesis in the bone marrow and spleen and a partial translocation of erythropoiesis to the spleen (40). In addition, a deficiency in selectin expression may interfere with leukocyte survival signals in the bone marrow compartment (133).

Over the last decade, the stromal cell coculture assay has emerged as an important and reliable tool, although an arduous one, for investigating the cellular and molecular aspects of hematopoiesis, disease pathogenesis in hematological malignancies, and the treatment strategies related to bone marrow transplantation. Stromal cells are usually obtained from fresh donor tissues or established cell lines. Initially, many studies using stromal cells were based a variety of donor-derived tissues, as early fibroblastoid cell lines such as W18Va2, M2-10B4, and KM102 did not fully support hematopoiesis (68). However, donor-derived stromal cells are not always readily available and subject to biological variability, which led a number of investigators to develop stromal cell lines that could more reliably support hematopoiesis. The recent introduction of several new cell lines, including the human cell lines HS-5, HS23, and Str-5 (96, 97) and the murine cell line HESS (124), which function as surrogates of the bone marrow microenvironment, have helped to standardize observations that are dependent on stromal cell monolayers. Depending on the cell populations being studied, the time period for which stromal cells can support cell survival varies from days to weeks. In cases in which a malignant cell population is being studied, this period of support can be much longer, and it is during these long periods of coculture that many cell–cell, cell–matrix, and soluble factor interactions take place and can be studied for their biological effects or treatment application. Flow cytometry–based assays have been developed that greatly assist in the quantification of the survival of hematopoietic cells as well as adhesive and signaling events (See Emerging Clinical Applications).

Adhesive Interactions in the Peripheral Blood under Flow Conditions

One of the important functions of vascular endothelium is its role in regulating leukocyte emigration from blood into tissues. Leukocyte trafficking and emigration distributes leukocytes to tissue-specific sites and targets immune cells to sites of antigenic

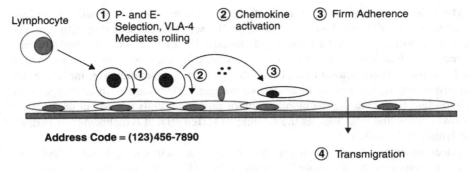

Figure 18.2. Schematic of leukocyte localization and address code hypothesis. Each molecular event can be represented as a digit of a telephone number. Only the correct number (i.e., receptors) on the leukocyte matching the corresponding numbers on the EC leads to tissue specific localization.

stimulus or microbial invasion. Recent observations show that multistep, sequential binding of adhesion and chemokine receptors may lead to exquisite tissue or organ specificity in leukocyte localization (figure 18.2). The coordinated use of integrins, selectins, and immunoglobulin family counterstructures form the molecular basis of adhesive interactions that occur between leukocytes and endothelium in blood vessels, leading to organ localization (12,16,17,57,59,66,114). This multistep adhesion model consists of four major steps:

1. Attachment (tethering) of free flowing leukocytes to endothelial cells: This first step is mediated by selectins and their ligands. In some cases, the integrin VLA-4 may be also involved.
2. Activation (triggering) of additional adhesion molecules on cell surface: Tethering of resting leukocytes on activated endothelium leads to cell activation via outside-in signaling to leukocytes. The most important activating signals are provided by chemokines and cytokines produced on the surface of endothelium or at the site of inflammation or immune response.
3. Arrest and spreading: Arrest is mediated by integrins binding to ICAMs and VCAM-1. Leukocyte integrins are functionally inactive on circulating cells, but with cellular activation, they rapidly convert to a high-affinity state that readily binds ICAMs and VCAM-1. VCAM-1 or ICAMs need to be expressed on endothelium for leukocyte binding to occur. VCAM-1 and ICAM-1 expression is induced by a variety of cytokines such as interleukin 1, interleukin 4, lipopolysaccharide, interferon α, and tumor necrosis factor α and lasts from hours to days (57; H. Tsuji, D.C. Brown, and R.S. Larson, unpublished data).
4. Migration (diapedesis) of the leukocytes into the subendothelial space: The adhesive receptors and cytokines involved in this step are not as well defined as previous steps. However, platelet endothelial cell adhesion molecule (CD31) has been shown to be important in diapedesis. This molecule is widely expressed on leukocytes and is found on cell–cell junctions on endothelial cells. CD31 is homophilic adhesion molecule, and anti-CD31 Ab and soluble CD31 have been shown to block monocyte transmigration in vitro. Shear stress is possibly a factor that regulates cell transmigration as well (24).

Each type of leukocyte (monocyte, neutrophil, eosinophil, basophil, T-lymphocyte, and B-lymphocyte) uses this sequence of events (i.e., an "address code"), but each type of leukocyte may use different combinations of selectins, chemokines, and integrins for these interactions with the endothelium under physiologic flow conditions (12, 17, 57, 59, 66, 114). Adding complexity to these events are several factors: first, chemokines have different affects on each leukocyte type. Second, adhesion receptors such as VLA-4 or LFA-1 on the surface of a leukocyte are converted from a low- to high-avidity state by chemokines. A conformational change of LFA-1 and VLA-4 is in part responsible for the avidity state change (14, 23, 57, 98). The high-avidity state of LFA-1 or VLA-4 is transient, lasting seconds to minutes (11, 23). In general, the cytokines that induce ICAM-1 or VCAM-1 are distinct from those endothelial cell–derived chemokines that induce LFA-1 avidity changes. Third, different leukocyte types require different site densities of adhesion receptors. For instance, eosinophils require a lower site density of P-selectin than do neutrophils for effective tethering. These multiple mechanisms of regulation provide for precise modulation of the recruitment of different types of leukocytes to sites of inflammation or immune response.

Adhesive Interactions in Tissues during Posttransmigratory Events

Once a leukocyte has extravasated from blood into tissue, it must migrate ultimately binding to cells, extracellular matrix components, or microbes to deliver an immune or inflammatory cell response. These cells will migrate up chemotactic gradients. These chemotactic factors and chemokines will activate integrin molecules, alone or in combination, to bind to their targets. In addition, mesenchymal cells in the tissue, particularly fibroblasts, synthesize a number of matrix proteins such as collagens, elastin, proteoglycans, and extracellular glycoproteins fibronectin, laminin, tenascin, and vitronectin. Leukocytes use integrins activated by chemokines to adhere and migrate through this microenvironment.

Growth and cell differentiation of fibroblasts in the tissue microenvironment is also affected by cell adhesion. Fibroblasts attach to ECM in special sites on their surface (i.e., focal adhesions; FAs) that are enriched by β_1 and β_3 integrins. Integrins in FAs form clusters with cytoskeletal, structural, and signaling proteins (talin, vinculin, α-actinin, c-Srk, FAK, p130cas, and paxillin; 15, 26, 54, 101). FAs serve two cellular functions: they act as a centers of cellular signaling and play a mechanical role—they maintain cell attachment to ECM and transmit force or tension at adhesion sites (2, 49, 102). The signal transduction and signal modulation by cell adhesion receptors have been previously reviewed (2). The type of ECM protein may determine the specific integrins expressed in FAs. For example, fibroblasts grown on serum or vitronectin-coated surfaces express $\alpha_v\beta_3$ integrin, whereas fibroblasts grown on fibronectin-coated surfaces express $\alpha_5\beta_1$ integrin (15).

Studies demonstrating the migratory potential of fibroblasts indicate that fibroblasts, analogous to leukocytes, are capable of directed migration toward specific chemoattractants. Different fibroblast chemoattractants have been identified in transmigration assay, using polycarbonate filters coated with gelatin or other proteins. Cytokines, chemokines, and growth factors, like tumor nerosis factor α, interleukin 4, transforming growth factor β, platelet derived growth factor, basic fibroblast growth factor, epidemal growth factor, insulin-like growth factor, Regulated upon Activation Normal T-cell Expressed and Secreted, eotaxin, monocyle chemoattractant protein, and IL-8, have been shown to induce chemotactic migration of different fibroblasts (81, 84–88, 105–107, 126). The biophysical aspects of the fibroblast-like amoeboid cell migration have been modeled as well and involve chemokine-regulation of integrins (13, 70).

Measurements of Cell Activation and Adhesion

As detailed above, adhesion molecule expression and function are regulated in physiologic environments. Chemokines, shear stress, site density, and many other factors affect the expression and function of adhesion molecules. Each factor can alter the binding strength of adhesion molecules and, consequently, that of bound cells as well, as it is the binding strength that determines how an adhering cell will behave. Cell adhesion investigations using video techniques show the detailed mechanics of individual cell–cell interactions. As a result, shear stress studies can be done using micropipette (22), centrifugation (83), parallel plate flow video (31), Couette flow video (42), Poiseuille flow (122), cone and plate flow (74), and parallel ring flow (137) techniques. One of the advantages of flow cytometry–based technologies is being able to study a significantly large population of cells rapidly. In addition, flow cytometry is also an extremely versatile tool because it can be incorporated with other apparati such as viscometers (137) and mixing and delivery devices (35). In this section, flow cytometric techniques will be discussed that allow for rapid quantification of the effects of temperature, shear stress, cell concentrations, and receptor and ligand site density on cell adhesion.

Cell Activation and Quantification of Receptor Numbers

The expression of adhesion molecules is modulated on leukocytes, endothelium, and other cells after cellular stimulation. The appearance or disappearance of these molecules can be used to measure the activation state of a cell. For instance, L-selectin is expressed on all naïve lymphocytes, a fraction of memory T cells, neutrophils, monocytes, and eosinophils (62). On activation, L-selectin on both lymphocytes and neutrophils undergoes cleavage by a membrane metalloprotease (shedding) (39, 45, 100, 134). Accordingly, increased soluble L-selectin is coincidentally detected in the plasma of patients suffering from numerous inflammatory diseases. The role of L-selectin shedding and the mechanism that regulates L-selectin shedding is poorly understood. Nonetheless, many investigators have used L-selectin shedding as a marker of cell activation and a determinant of the "resting" state of a leukocyte. In addition to L-selectin, the expression of integrins such as Mac-1 on neutrophils is upregulated with stimulation. Finally, a number of investigators have described conformationally specific antibodies that recognize the "active" state of integrins including LFA-1 (CD11a/CD18), VLA-4 $\alpha_4\beta_1$ (CD49d/CD29), $\alpha_4\beta_7$ (LPAM-1), and Mac-1 (CD11b/CD18). These conformationally specific antibodies are not widely distributed but have also been used to detect cellular activation on a broad number of cell types (51, 130).

In addition to measuring the activation state of leukocytes, the adhesion molecule repertoire expressed on the surface of the endothelium determined the subset of leukocytes that are captured. A critical role of endothelial cells is their ability to respond to cytokines and chemokines, growth factors, biomechanical forces, complement components, nitric oxide, and others. These responses are also readily measured by flow cytometric analysis. For instance, exposure of endothelial cells to endotoxins or proinflamatory cytokines like tumor necrosis factor α, IL-1, IL-6, IL-8, and interferon γ results in endothelial activation and massive expression of ICAM-1, VCAM-1, and E-selectin (61, 69, 71). Some of the endothelial responses are very rapid and do not require de novo gene transcription or protein synthesis. For instance, endothelial cell hypoxia induces an inflammatory and prothrombotic response (1). This leads to exocytosis of Weibel–Palade bodies, secretion of vWF, and P-selectin expression followed by recruitment of leukocytes and platelets (82). Exocytosis of Weibel–Palade bodies is accompanied by the surface exposure of P-selectin, cytoskeletal reorganization, activation of NO secretion, and increase in vascular permeability (126, 128).

Recently, several investigators have shown that quantification of the receptor number can be performed by flow cytometry, using fluorescently labeled beads as standards (figure 18.3; 23, 32) Through these measurements, it is now clear that relatively small changes in site density (two- to fourfold) of adhesion receptors on endothelium can dramatically alter leukocyte recruitment (32). Calibration standards such as Quantum Simple Cellular beads (Bangs Laboratories, Fishers, IN) contain a mixture of microbeads that bind specific amounts of mouse IgG antibodies. These standards are calibrated in terms of the antibody binding capacity (ABC). Cell samples and calibration beads are stained in parallel with a fluorescent antibody directed against the cell receptor of interest. A calibration plot is constructed that relates the fluorescence intensity of the beads to their ABC. From this plot, the mean or median fluorescence intensity of each cell sample is used to estimate the cell ABC. At high receptor membrane densities, the ABC may reflect as little as one-half of the total number of membrane receptors, as these are conditions in which binding of a single antibody to a pair of receptors may be favored. At low levels of membrane receptor expression, the ABC may more closely approximate the true number of receptors.

Cell Aggregation and Adhesion

Cell aggregates are formed when cells collide. Flow cytometers can define molecular assemblies as they relate to the behavior of two cell species (each associated with a specific adhesion molecule) in suspension (35) or to the behavior of one cell species and a related soluble ligand (23). Early in the technological development of these assays, collision was facilitated through the use of a stirring mechanism (a magnetic stir bar or a viscometer device) before sampling in the flow cytometer. Because magnetic stir bars create a poorly defined shear stress field for cells, the effect of shear on adhesion bonds in these early attempts was ill defined. Subsequent refinement of these methods has enabled flow cytometric analysis of cell mixtures subjected to a more uniform and quantifiable fluid shear environment generated in a cone-plate viscometer. Cell mixtures are sampled periodically from the viscometer into a formalin fixative solution for subsequent online flow cytometric analysis (76, 121). Experiments showed a remarkable potentiation of adhesion efficiency through the combined action of two

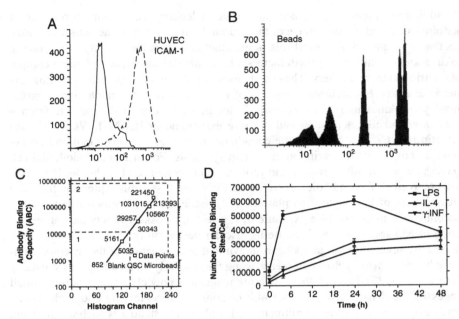

Figure 18.3. Schematic of site density determination by flow cytometric methods. (A) Fluorescence profiles of negative control ($-$), ICAM-1 expression at 4 h (- - -) and 24 h ($\cdots$) after lipopolysaccharide stimulation of endothelium. The fluorescence of the ICAM-1 staining is compared against the fluorescence of beads with fixed amounts of anti-ICAM-1 bound (B), and a standard curve is generated (C) as described in text. (D) Kinetics of ICAM-1 expression on endothelium in response to different cytokine stimulation.

sets of adhesion molecules and a progression of adhesion molecule use over time (48, 75, 77).

Alternatively, with the development of plug flow cytometry sample handling technology, an online analysis approach is also possible in which the cell mixtures are transported within seconds from the viscometer to the flow cytomer under computer control (32, 34; figure 18.4) Subsequently, technology was developed that attached viscometer type devices to a flow cytometer because of their ability to provide precise shear stress environments for long periods of time (figures 18.5 and 18.6).

To model collisions in a shear-stress environment, Smoluchowski's flocculation theory has been applied to obtain estimates of adhesion efficiency (48, 75–77, 121). The collisional model assumes the presence of a uniform linear shear field and a homogeneous dispersion of the interacting cells in the viscometer during the analysis interval. With this approach, it was shown that steady application of a threshold level of shear was sufficient to support homotypic neutrophil aggregation even in the absence of exogenous neutrophil activating chemotactic stimuli. However, temporal changes in radial cell concentration distributions in the viscometer that vary with cell or particle size and density and rotational velocity have been observed (B.S. Edwards and L.A. Sklar, unpublished data). This probably reflects the presence of secondary flow fields (104) and indicates that the collisional model may need to be refined to account for the effects of such cell concentration inhomogeneities. Indeed, this potential source of error

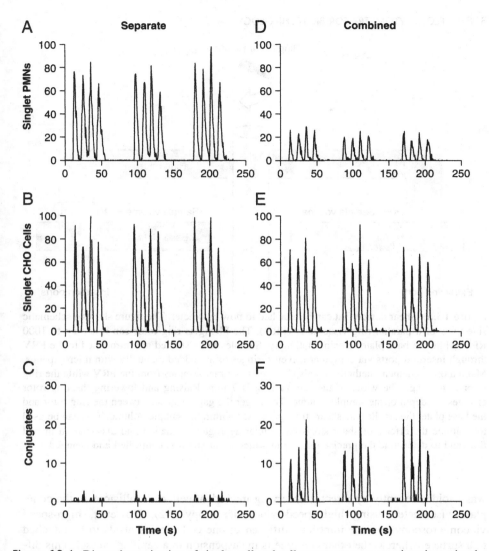

Figure 18.4. Direct determination of singlet cell and cell aggregate concentrations in a stirred cell suspension adhesion assay. Freshly prepared polymorphonuclear leukocytes (PMN) and Chinese hamster ovary cells (CHO) expressing human P-selectin were analyzed separately by plug flow cytometry to establish gating regions for detection of singlet PMNs (red fluorescence from Fura Red), singlet CHO cells (green fluorescence from GFP), and PMN-CHO cell conjugates (red and green cofluorescence). Twelve sample plugs of 5 μL each were analyzed within each gate to determine initial concentrations (M ± SD) of (A) singlet PMNs = 102,000 ± 6000 cells/mL and (B) singlet CHO cells = 102,000 ± 7000 cells/mL. (C) When PMNs were analyzed separately, the background concentration appearing in the conjugate gate was 1300 ± 400 particles/mL, which is comparable to the background observed when CHO cells were analyzed separately (not shown). PMNs and CHO cells were combined and stirred with a magnetic stirbar at 400 rpm for 5 min at 37°C. Twelve replicate sample plugs from the stirred suspension were then analyzed to determine the resulting concentrations of (D) singlet PMNs = 25,000 ± 4000 (E) singlet CHO cells = 66,000 ± 7000 cells/mL, and (F) conjugate particles = 13,000 ± 2000 particles/mL. [From (35), 2001, Copyright Wiley-Liss, Inc. Reprinted with permission of the publisher.]

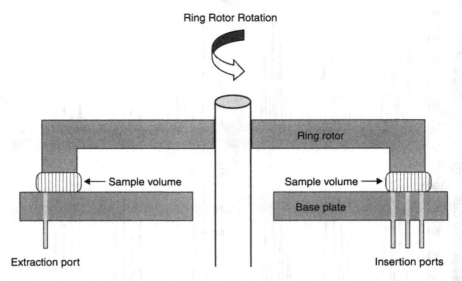

Ring Rotor Rotation

Figure 18.5. Shear device that can be attached to flow cytometer. The figure shows a schematic side view of a parallel ring viscometer (PRV). The Plexiglas ring rotor can rotate up to 1000 RPM while the base plate remains stationary. Samples to be studied are introduced to the PRV through insertion ports via computer-driven syringes or are added manually with micropipettes. Manual or mechanical methods can also be used to extract samples from the PRV while the ring rotor is moving. The width of the sample ring is 7 mm. Raising and lowering the ring rotor changes the depth of the sample volume. The larger the gap distance between the ring rotor and the base plate, the smaller the shear stress generated within the sample volume. This can be used to examine the effects of shear stress on cellular aggregate formation and dissociation in real time and to determine the precise manner in which shear stress is controlled and changed.

was avoided recently by directly observing individual cell–cell collisions in the cone-plate viscometer, using high-speed videomicroscopy (42). However, high-speed videomicroscopy has one important difference; one cell species needs to be attached to a surface, whereas the other cell type is in suspension in a capillary vessel. This difference may not affect the behavior of the adhesion bonds under stress once the bonds have been formed, but it will affect the probability of forming cell aggregates. More cell–cell collisions will occur in a cell suspension assay than in a parallel plate assay because cell suspension assays involve cells moving in three dimensions (83).

Another important concern in flow cytometric cell adhesion analysis is the constitution of the population of adherent cell conjugate particles. When conjugate particles are mostly doublets with one cell from each of the two input cell population, the calculation of adhesion is straightforward because the number of adherent cells of either cell population is equal to the number of observed cell conjugate particles. When the cell conjugate particles are heterogeneous aggregates containing more than one cell from each population, a simple count of conjugate particles may grossly underestimate the number of adherent cells of either input population. With a fluorescent dye that labels cells uniformly, it may be possible to discriminate the number of labeled cells in individual conjugate particles and to thereby more accurately determine the true num-

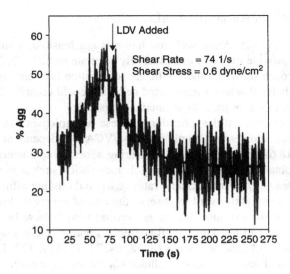

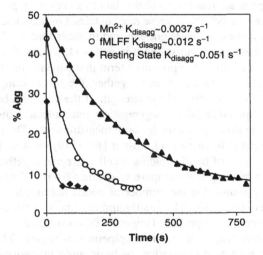

Figure 18.6. Cellular adhesion analysis in the parallel ring viscometer. (*A*) Measurement of cell disaggregation rate in U937 cells expressing VLA-4 and B78H1 transfected with VCAM-1. Solid curve is then fit to data. Kinetic data were obtained after mixing B78H1 cells (transfected with human VCAM-1) to U937 cells (expressing VLA-4) in a ratio of 3:1. Cells were added to a parallel ring viscometer, maintained at 37°C, and sheared at 74 s^{-1}. After 1 min of shearing, a VLA-4 blocking peptide containing the sequence LDV (leucine–aspartic acid–valine) was added to the sample volume in the parallel ring viscometer via a computer-driven syringe. During the insertion of cells and the addition of the blocking peptide, samples from the parallel ring viscometer were continuously removed through a port located in the lower nonrotating base plate, using a peristaltic pump. Cells removed from the parallel ring viscometer were mixed with 2% paraformaldehyde solution to prevent more aggregates from forming or dissociating while being transported to the Facscan. (*B*) Cell disaggregation rate in U937 cells transfected with formyl peptide receptor and B78H1 transfected with VCAM-1. B78H1/VCAM U937/VLA-4 cell aggregates were formed in suspension. After the addition of the blocking LDV peptide, cell disaggregation was followed. Three affinity states of VLA-4 integrin are shown: resting state, formyl peptide activated cells, and 1-mM Mn^{2+} activated cells. Data are plotted as percent of the aggregates versus time, following addition of the LDV blocking peptide.

ber of adherent cells (112). When such resolution is not feasible, a singlet depletion methodology may provide the most quantitatively accurate results (32, 35). In one application, this approach showed a more efficient recognition between eosinophils than between neutrophils for P-selectin transfected cells that could contribute to the preferential accumulation of eosinophils in asthmatic lungs (32).

As an example of successful application of these technologies, we have recently employed these technologies to examine VLA-4/VCAM-1-dependent cell adhesion (figures 18.5 and 18.6). The challenge of measuring adhesion characteristics of VLA-4 and VCAM-1 is obtaining molecular binding characteristics, such as number of bonds, forward kinetic rates (k_{on}), reverse kinetic rates (k_{off}), and binding affinity from cellular information. It is straightforward to observe the rate of aggregate breakup, using a flow cytometer to obtain a cellular k_{off} value. Several models have been proposed to describe the formation and breakup of cell aggregates and to relate this knowledge to the molecular qualities of adhesion bonds (8, 58, 64, 74, 75, 119, 122, 135). To obtain molecular properties of adhesion bonds, cellular k_{off} values are measured as a function of shear rates. The simplest approach is to fit the data to an equation postulated by Bell (8, 21). This equation relates cellular reverse rates obtained under shear stress to those rates one would expect if the cells were not exposed to shear forces. This equation assumes that the kinetic theory of bond strength in solids (136) describes cell adhesion bonds. This relationship is a simple exponential term that depends on the temperature, the number of bonds holding the aggregates together, the force acting on the aggregate, and a constant that is indicative of the strength of the adhesion bond.

We have recently compared cellular disaggregation and molecular dissociation rates for a soluble ligand, using flow cytometry for both measurements. The relationship between these rates has been defined by R.G. Posner [see (119)] as a multiplicative factor depending on the number of bonds holding a cell aggregate together. Recent studies on VLA-4/VCAM-1 binding that compare k_{off} values from cell aggregates to cell and ligand binding have found that the number of bonds holding the cell aggregates together is approximately two (137). The small number of bonds estimated to hold cell aggregates together is perhaps surprising. However, this result is consistent with estimates obtained using centrifugation and micropipette techniques (22, 64, 83, 108). Thus, flow cytometry can be used to measure the basic molecular adhesion bond characteristics such as number of bonds and reverse and forward rate constants. These receptor–ligand interactions can then be quantifiably related to cell aggregation and adhesion.

Receptor Affinity Changes

β_1 and β_2 integrins, unlike the selectins and mucins for which binding is constitutive, depend on conformational changes in the binding region, secondary to stimulation (19). The receptors must be activated, or turned "on" in response to stimulation, for binding to occur. In fact, one of the β_1 integrins, VLA-4, has been found to have several different affinity states, whereas other integrins such as LFA-1 and Mac-1 appear to have only two states (20, 23, 53). Several small molecules and peptidomimetics are available or are currently being developed as anti-integrin therapeutics, because aberrant cell adhesion has been implicated in the pathogenesis of several disease states, including inflammatory conditions, cancer, and coronary artery disease (25, 27, 50, 80).

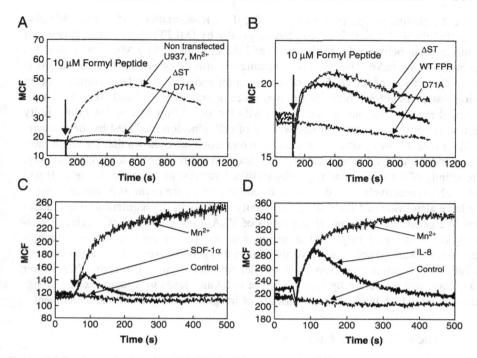

Figure 18.7. Response kinetics of LDV peptide binding to stimulation of CXCR4, CXCR2, wild-type FPR, and FPR mutants in U937 transfectants. Cell suspensions were incubated with 3 nM fluorescent VLA-4-specific peptide and stimulated with fMLFF (A and B), SDF-1α (C), and interleukin 8 (D). (A) Nontransfected U937, stimulated with Mn^{2+}, nondesensitizing ΔST, and nonactivating D71A ΔST. (B) Expanded scale for wild-type FPR, ΔST, D71A ΔST. (C) Rapid and transient response of CXCR4-transfected U937 cells to SDF-1α. (D) Response of CXCR2 receptor–transfected U937 cells to interleukin 8. Binding is shown as mean channel fluorescence versus time [From (23). Reprinted with permission of the publisher.]

Small molecules that are currently being developed or are available for $\alpha4\beta1$, $\alpha4\beta7$, $\alpha v\beta3$, $\alpha IIb\beta3$, and LFA-1 (27) and include drugs for asthma, inflammatory bowel disease, multiple sclerosis, rheumatoid arthritis, osteoporosis, and cancer. These integrin antagonists can also be used as reagents for studying affinity changes of receptors, in that they have very rapid "on" and "off" rates in comparison to monoclonal antibodies, which are slow. When these small molecules are used at a concentration that approximates their affinity constant for the high-affinity state, they will preferentially bind to receptors on cells in the high-affinity state. The major advantage of an assay of this type over conventional antibody techniques is that one can study the regulation of constitutively expressed receptors that alter their affinity states, such as VLA-4, rather than just study expression levels (23).

Using a fluorescein isothiocyanate–labeled VLA-4 specific peptidomimetic, we have recently developed a flow cytometry-based assay to study VLA-4 integrin affinity changes on cell adhesion (figure 18.7). Using this peptidomimetic, we have shown that in leukocytic cell lines, peripheral blood leukocytes, and leukemia cells, the affinity of the binding can be modulated rapidly by divalent cations as well as soluble ligands for different signaling receptors (formyl peptide, IL-5, IL-8, SDF-1, and others). We also found that the active state induced physiologically by fMLFF, IL-5, or IgE is one that is intermediate between the resting state and the one induced by Mn^{2+} or anti-α_4-integrin-activating mAb. The process of turning the affinity of the integrin on and off is tied to the activity of the chemoattractant receptor and appears to be accounted for by a single-step conversion between the two states, nominally "off" and "on." Taken together, these observations are consistent with the possibility that α_4-integrin affinity regulation contributes to the overall avidity of cell adhesion mediated by the integrin.

More recent flow cytometric studies from our group have shown that the small molecules and the native ligand VCAM-1 interact with the receptor in a comparable way. The affinity of the resting and physiologically activated states is approximately 10 μM and 1 μM, respectively. The dissociation rates constants are in the 10/s and 1/s ranges. We have also evaluated the role of the affinity changes in the regulation of cell adhesion. We generated different affinity states of VLA-4 integrin by physiologically activating cells through formyl peptide receptor or by divalent cations. Using a "conjugate" assay described above, we have also shown that in both cases, cell avidity, measured by cell disaggregation rate, changes in parallel with the affinity of integrin–VCAM pair (23). Thus, cell activation by chemokines or cytokines leads to a rapid inside-out signaling event, which upregulates the affinity of integrin, leading to cellular attachment.

Shear Stress and Adhesion

Endothelial cells respond to different types of biomechanical force: shear stress, cyclic strain, and hydrostatic pressure. The effects of these forces on cellular adhesion can be measured in a flow cytometer attached to shear devices. Cell exposure to shear stress in vitro leads to calcium signaling, phosphorylation of cytoplasmic proteins, changes or rearrangement of stress fibers, and gene expression including adhesion receptors including ICAM-1 (38, 93, 94, 109, 110, 123). Interestingly, VCAM-1 and E-selectin expression are not regulated by shear stress (72). In pulmonary endothelium, shear stress induces pronounced cortical cytoskeletal rearrangement, junctional protein tyrosine phosphorylation, and other effects (10). The properties of shear flow are also an important factor in cell binding. For example, endothelial cells subjected to disturbed laminar shear stress exhibit increased levels of nuclear-localized NF-kappaB, Egr-1, c-Jun, and c-Fos, compared with cells exposed to uniform laminar shear stress or maintained under static conditions (72). The effect of shear stress in vascular endothelium can be related to β_1 integrin. Anti-β_1-integrin-blocking mAb or TS2/16-activating mAb modulate the effect of shear stress on sterol regulatory element-binding protein 1 (63). Shear stress also can be a second signal (in addition to chemokine gradients across the endothelium) that is necessary for leukocytes' transendothelial migration. Shear provides mechanical signals coupled to G (i) protein signals at apical endothelial zones (24).

Emerging Applications

Flow cytometry has been at the forefront of biotechnology since its inception. It was originally established as an automated method for measuring optical or fluorescence characteristics of cells or particles in suspension. However, in recent history, flow cy-

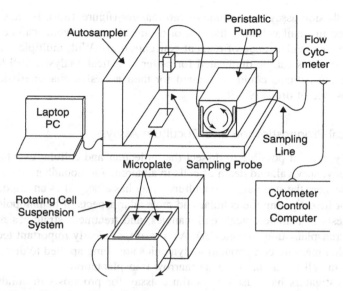

Figure 18.8. HyperCyt system including laptop computer, autosampler, probe, microwell plate, peristaltic tubing, peristaltic pump, interface junction, FACScan, G4 computer, and MARSS.

tometers have become increasingly important in clinical diagnostics, being used extensively in cell sorting (92), immunophenotyping of blood cells, and measuring DNA content to analyze cell cycle status (129). Several more recent developments in the clinical use of flow cytometry will be discussed.

Drug Development, High-Throughput Screening and HyperCyt

Modern drug discovery involves screening receptors or other cellular targets against millions of compounds. High-throughput flow cytometry has the potential for screening large numbers of compounds (36, 56, 78) from 96-well plates rapidly and would allow for end-point cell adhesion assays to be performed (56). One of these techniques, HyperCyt, enables sampling from microplate wells at rates of up to one sample per second (figure 18.8). HyperCyt works by picking up samples from multiwell plates through the use of a needle-like stainless steel probe attached to peristaltic tubing and directed from well to well by the arm of the autosampler while a peristaltic pump aspirates the sample and delivers it to the flow cytometer (56, 91, 92). The recent introduction of MARSS (micro assay rotational suspension system; 92) has made it possible to use HyperCyt for time-dependent cell adhesion assays. MARSS rotates its plate-holding platform by 360° at a rate of about four revolutions per minute. In doing so, every revolution inverts the plates and restores them to an upright position. Contributing to MARSS is the fact that the surface tension in the v-shaped wells (10 μL volume) prevents the spilling of contents when the plates are inverted. The cells fall from bottom to top when the plate is upside down, and then back to the bottom when the full rotation is completed. A further feature of this system is its ability to perform cell adhesion assays in a miniaturized manner (figure 18.9). Because the assay is performed in small volumes, the cost of chemicals and reagents can be reduced or performed over a wider range of reagent concentrations. With multiple wells sampled in series, samples can be duplicated for better statistical analysis. Cell lines and cells isolated from patients could be studied for their adhesive characteristics in the presence or absence of drugs (90, 91).

Clinical Prognostication and Coculture Assays

Flow cytometry make it possible to dissect the molecular and cellular events governing cell ontogeny, survival, and death of cells in adherent and nonadherent conditions. Over the last decade, the stromal cell coculture assay has emerged as an important and reliable tool for investigating the cellular and molecular aspects of hematopoiesis, disease pathogenesis in hematological malignancies, and treatment strategies related to bone marrow transplantation (see earlier). As a result, clinically important techniques are emerging that measure cell population dynamics are being applied to disease prognostication, stem cell research, and bone marrow transplantation.

Several investigators have used a coculture assay for prognosis in children with acute lymphoblastic leukemia (67, 68, 131). To improve the reliability of the coculture assay, Winter et al. (131–133) used biologically inert beads with well-characterized fluorescent properties to adjust for volumetric flow variations during acquisition (figure 18.10). The inclusion of fluorescent beads with the stromal cell assay significantly lowers the coefficient of variation as compared to samples analyzed without beads. To normalize for the volume of fluid that passed through the flow cytometer during a defined period of data acquisition, a predetermined number (5.0×10^5) of fluorescent Immunobeads were added to a fixed volume in each sample. This technique improves the reliability and precision of cell enumeration (132). With multiple, distinctly labeled mAbs, the survival of both normal and malignant cells may be studied simlutaneously. This application has lead to a prognostic indicator in several diseases, including those in which good prognostic markers did not previously exist (figure 18.11).

The stromal cell coculture assays have also been used for in vitro drug sensitivity assays. This work has involved the use of the stromal line HS-5 to study drug sensitivity in acute myeloid leukemia cells (41), that of normal donor samples to investigate disease resistance in myeloma and pre-B ALL cells (18, 28), and the effects of drug toxicity on hematopoietic progenitor cells (73). In these settings, sublethal doses of cytotoxic drugs have been added to the stromal cell coculture assay to assess for a

---➤

Figure 18.9. High-throughput cell adhesion assay. (*A*) A dot plot showing two cell populations capable of forming cell aggregates (U937 cells expressing VLA-4 and stained green, FL1, and CHO cells expressing VCAM-1 and stained red, FL2). (*B*, *C*, and *D*) Dot plots of time versus FL-2 showing the gated populations from plot (*A*) respectively. (*B*) Represents cells in gate 1 (CHO VCAM-1 singlets). (*C*) Represents cells in gate 3 (aggregates of CHO VCAM-1 and U937 cells. (*D*) Represents cells in gate 2 (U937 singlets). The first 10 wells sampled contained U937 cells alone, the next 10 wells sampled contained CHO VCAM-1 cells alone, and the last 10 wells sampled contained a mixture of U937 and CHO VCAM-1 cells. Each data cluster in (*B*), (*C*), and (*D*) represent cells from a separate well.

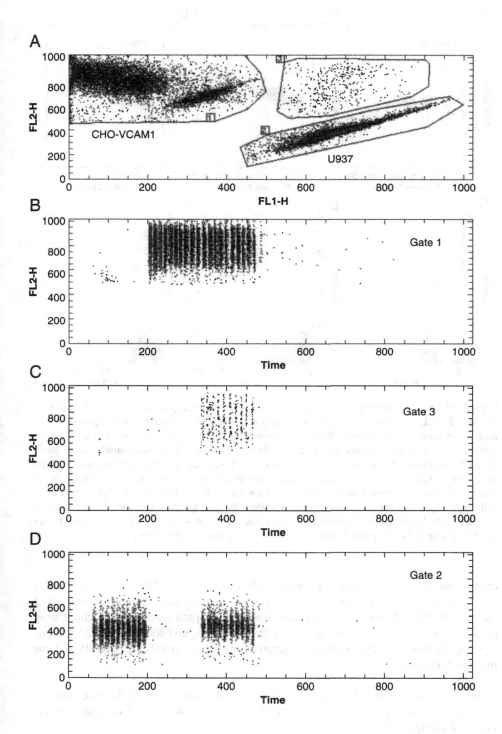

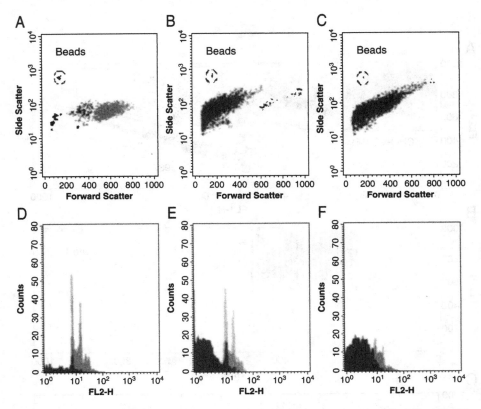

Figure 18.10. Representative histograms showing light-scatter properties of viable CD 5+, CD34+ T-ALL cells (green scattergram events) at 1 and 72 h, maintained in the presence and absence of bone marrow stroma. Fluorochrome-labeled beads (circled) are used to account for volume control in the analysis. (A) T-ALL cells harvested at 1 h of incubation showed more than 98% viability. (B) Enhanced recovery of viable T-ALL cells maintained on bone marrow stroma in comparison to (C) T-ALL cells maintained without stroma that shrink (blue scattergram events) in comparison to live cells (the forward scatter of live cells is higher than dead cells or debris). Representative histograms showing the Sub G_0 peak (apoptotic frequency) of leukemic cells harvested at 1 h (D) and in the presence (E) or absence of (F) bone marrow stroma at 72 h. [From (134). Reprinted with permission of Blackwell Publishing.]

disease resistance in the leukemic cell population of interest. Moreover, novel drugs are potentially screened for in vitro efficacy, using primary patient samples maintained on stromal cells. These coculture assays could be adapted for high-throughput drug screening in future therapeutic trials. In addition, high-speed sorting of homogeneous cell populations (5, 7, 29) promises to greatly enhance the viability and purity of bone marrow transplants.

As this chapter was being edited, we described the application of fluorescence resonance energy transfer in flow cytometry to the real-time analysis of intergrin conformational change (23a, 57a).

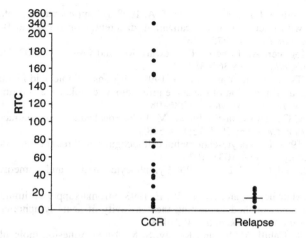

Figure 18.11. Stromal cell supported survival of T-ALL from patient samples. A dot indicates the ex vivo RTC (percentage cell recovery) of each patient sample studied. A high concordance rate* of 83% discriminates patients who are at risk for relapse from those who will remain in complete continuous remission (CCR). The median for each set is shown (−). Each sample was tested in duplicate. Coefficients of variation were less than 10%. [From (133). Reprinted with the permission of Wiley-Liss.]

References

1. Akman, H. O., Zhang, H., Siddiqui, M. A., Solomon, W., Smith, E. L., and Batuman, O. A. 2001. Response to hypoxia involves transforming growth factor-beta2 and Smad proteins in human endothelial cells. *Blood* 98:3324–3331.
2. Aplin, A. E., Howe, A., Alahari, S. K., and Juliano, R. L. 1998. Signal transduction and signal modulation by cell adhesion receptors: the role of integrins, cadherins, immunoglobulin-cell adhesion molecules, and selectins. *Pharmacol. Rev.* 50:197–263.
3. Ariel, A., Hershkoviz, R., Cahalon, L., Williams, D. E., Akiyama, S. K., Yamada, K. M., Chen, C., Alon, R., Lapidot, T., and Lider, O. 1997. Induction of T cell adhesion to extracellular matrix or endothelial cell ligands by soluble or matrix-bound interleukin-7. *Eur. J. Immunol.* 27:2562–2570.
4. Arroyo, A. G., Yang, J. T., Rayburn, H., and Hynes, R. O. 1996. Differential requirements for alpha4 integrins during fetal and adult hematopoiesis. *Cell* 85:997–1008.
5. Ashcroft, R. G. and Lopez, P. A. 2000. Commercial high speed machines open new opportunities in high throughput flow cytometry (HTFC). *J. Immunol. Methods* 243:13–24.
6. Bates, R. C., Lincz, L. F., and Burns, G. F. 1995. Involvement of integrins in cell survival. *Cancer Metastasis Rev.* 14:191–203.
7. Battye, F. L., Light, A., and Tarlinton, D. M. 2000. Single cell sorting and cloning. *J. Immunol. Methods* 243:25–32.
8. Bell, G. I. 1978. Models for the specific adhesion of cells to cells. *Science* 200:618–627.
9. Berke, G. 1985. Enumeration of lymphocyte-target cell conjugates by cytofluorometry. *Eur. J. Immunol.* 15:337–340.
10. Birukov, K. G., Birukova, A. A., Dudek, S. M., Verin, A. D., Crow, M. T., Zhan, X., DePaola, N., and Garcia, J. G. 2002. Shear stress-mediated cytoskeletal remodeling and cortactin translocation in pulmonary endothelial cells. *Am. J. Respir. Cell Mol. Biol.* 26: 453–464.
11. Blenc, A. M., Chigaev, A., Shuster, J. S., Sklar, L. A., and Larson, R. S. 2001. VLA-4 affinity correlates with the peripheral white blood cell count and DNA content in patients with B-ALL. *Leukemia* 17:21–4, 2003.

12. Bleul, C. C., Schultze, J. L., and Springer, T. A. 1998. B lymphocyte chemotaxis regulated in association with microanatomic localization, differentiation state, and B cell receptor engagement. *J. Exp. Med.* 187:753–762.

13. Bottino, D., Mogilner, A., Roberts, T., Stewart, M., and Oster, G. 2002. How nematode sperm crawl. *J. Cell Sci.* 115:367–384.

14. Brown, D. C., Tsuji, H., and Larson, R. S. 1999. All-trans retinoic acid regulates adhesion mechanism and transmigration of the acute promyelocytic leukaemia cell line NB-4 under physiologic flow. *Br. J. Haematol.* 107:86–98.

15. Burridge, K. and Chrzanowska-Wodnicka, M. 1996. Focal adhesions, contractility, and signaling. *Annu. Rev. Cell Dev. Biol.* 12:463–518.

16. Butcher, E. C. 1991. Leukocyte-endothelial cell recognition: three (or more) steps to specificity and diversity. *Cell* 67:1033–1036.

17. Butcher, E. C. and Picker, L. J. 1996. Lymphocyte homing and homeostasis. *Science* 272:60–66.

18. Campana, D., Manabe, A., and Evans, W. E. 1993. Stroma-supported immunocytometric assay (SIA): a novel method for testing the sensitivity of acute lymphoblastic leukemia cells to cytotoxic drugs. *Leukemia* 7:482–488.

19. Cavenagh, J. D., Cahill, M. R., and Kelsey, S. M. 1998. Adhesion molecules in clinical medicine. *Crit. Rev. Clin. Lab Sci.* 35:415–459.

20. Chen, L. L., Whitty, A., Lobb, R. R., Adams, S. P., and Pepinsky, R. B. 1999. Multiple activation states of integrin alpha4beta1 detected through their different affinities for a small molecule ligand. *J. Biol. Chem.* 274:13167–13175.

21. Chen, S. and Springer, T. A. 2001. Selectin receptor-ligand bonds: Formation limited by shear rate and dissociation governed by the Bell model. *Proc. Natl. Acad. Sci. USA* 98:950–955.

22. Chesla, S. E., Selvaraj, P., and Zhu, C. 1998. Measuring two-dimensional receptor-ligand binding kinetics by micropipette. *Biophys. J.* 75:1553–1572.

23. Chigaev, A., Blenc, A. M., Braaten, J. V., Kumaraswamy, N., Kepley, C. L., Andrews, R. P., Oliver, J. M., Edwards, B. S., Prossnitz, E. R., Larson, R. S., and Sklar, L. A. 2001. Real time analysis of the affinity regulation of alpha 4-integrin. The physiologically activated receptor is intermediate in affinity between resting and Mn(2+) or antibody activation. *J. Biol. Chem.* 276:48670–48678.

23a. Chigaev, A., Buranda, T., Dwyer, D.C, Prossnitz, E.R., Sklar, L.A. 2003. FRET detection of cellular alpha4-integrin conformational activation. *Biophysical Journal* 85:3951–3962.

24. Cinamon, G., Shinder, V., and Alon, R. 2001. Shear forces promote lymphocyte migration across vascular endothelium bearing apical chemokines. *Nat. Immunol.* 2:515–522.

25. Coutre, S. and Leung, L. 1995. Novel antithrombotic therapeutics targeted against platelet glycoprotein IIb/IIIa. *Annu. Rev. Med.* 46:257–265.

26. Critchley, D. R. 2000. Focal adhesions—the cytoskeletal connection. *Curr. Opin. Cell Biol.* 12:133–139.

27. Curley, G. P., Blum, H., and Humphries, M. J. 1999. Integrin antagonists. *Cell Mol. Life Sci.* 56:427–441.

28. Damiano, J. S., Cress, A. E., Hazlehurst, L. A., Shtil, A. A., and Dalton, W. S. 1999. Cell adhesion mediated drug resistance (CAM-DR): role of integrins and resistance to apoptosis in human myeloma cell lines. *Blood* 93:1658–1667.

29. Daugherty, P. S., Iverson, B. L., and Georgiou, G. 2000. Flow cytometric screening of cell-based libraries. *J. Immunol. Methods* 243:211–227.

30. de Fougerolles, A. R., Klickstein, L. B., and Springer, T. A. 1993. Cloning and expression of intercellular adhesion molecule 3 reveals strong homology to other immunoglobulin family counter-receptors for lymphocyte function-associated antigen 1. *J. Exp. Med.* 177:1187–1192.

31. Dong, C. and Lei, X. X. 2000. Biomechanics of cell rolling: shear flow, cell-surface adhesion, and cell deformability. *J. Biomech.* 33:35–43.

32. Edwards, B. S., Curry, M. S., Tsuji, H., Brown, D., Larson, R. S., and Sklar, L. A. 2000. Expression of P-selectin at low site density promotes selective attachment of eosinophils over neutrophils. *J. Immunol.* 165:404–410.

33. Edwards, B. S., Hoffman, R. R., and Curry, M. S. 1993. Calcium mobilization-associated and independent cytosolic acidification elicited in tandem with Na+/H+ exchanger activation in target cell-adherent human NK cells. *J. Immunol.* 150:4766–4776.

34. Edwards, B. S., Kuckuck, F., and Sklar, L. A. 1999. Plug flow cytometry: An automated coupling device for rapid sequential flow cytometric sample analysis. *Cytometry* 37:156–159.

35. Edwards, B. S., Kuckuck, F. W., Prossnitz, E. R., Okun, A., Ransom, J. T., and Sklar, L. A. 2001. Plug flow cytometry extends analytical capabilities in cell adhesion and receptor pharmacology. *Cytometry* 43:211–216.

36. Edwards, B. S., Kuckuck, F. W., Prossnitz, E. R., Ransom, J. T., and Sklar, L. A. 2001. HTPS flow cytometry: a novel platform for automated high throughput drug discovery and characterization. *J. Biomol. Screen.* 6:83–90.

37. Edwards, B. S., Nolla, H. A., and Hoffman, R. R. 1989. Relationship between target cell recognition and temporal fluctuations in intracellular Ca2+ of human NK cells. *J. Immunol.* 143:1058–1065.

38. Ferri, C., Desideri, G., Valenti, M., Bellini, C., Pasin, M., Santucci, A., and De Mattia, G. 1999. Early upregulation of endothelial adhesion molecules in obese hypertensive men. *Hypertension* 34:568–573.

39. Fors, B. P., Goodarzi, K., and von Andrian, U. H. 2001. L-selectin shedding is independent of its subsurface structures and topographic distribution. *J. Immunol.* 167:3642–3651.

40. Frenette, P. S., Mayadas, T. N., Rayburn, H., Hynes, R. O., and Wagner, D. D. 1996. Susceptibility to infection and altered hematopoiesis in mice deficient in both P- and E-selectins. *Cell* 84:563–574.

41. Garrido, S. M., Appelbaum, F. R., Willman, C. L., and Banker, D. E. 2001. Acute myeloid leukemia cells are protected from spontaneous and drug-induced apoptosis by direct contact with a human bone marrow stromal cell line (HS-5). *Exp. Hematol.* 29:448–457.

42. Goldsmith, H. L., Quinn, T. A., Drury, G., Spanos, C., McIntosh, F. A., and Simon, S. I. 2001. Dynamics of neutrophil aggregation in couette flow revealed by videomicroscopy: effect of shear rate on two-body collision efficiency and doublet lifetime. *Biophys. J.* 81:2020–2034.

43. Gonzalez-Amaro, R. and Sanchez-Madrid, F. 1999. Cell adhesion molecules: selectins and integrins. *Crit. Rev. Immunol.* 19:389–429.

44. Guyer, D. A., Moore, K. L., Lynam, E. B., Schammel, C. M., Rogelj, S., McEver, R. P., and Sklar, L. A. 1996. P-selectin glycoprotein ligand-1 (PSGL-1) is a ligand for L-selectin in neutrophil aggregation. *Blood* 88:2415–2421.

45. Hafezi-Moghadam, A., Thomas, K. L., Prorock, A. J., Huo, Y., and Ley, K. 2001. L-selectin shedding regulates leukocyte recruitment. *J. Exp. Med.* 193:863–872.

46. Hartwell, D. W. and Wagner, D. D. 1999. New discoveries with mice mutant in endothelial and platelet selectins. *Thromb. Haemost.* 82:850–857.

47. Helfrich, M. H., Aronson, D. C., Everts, V., Mieremet, R. H., Gerritsen, E. J., Eckhardt, P. G., Groot, C. G., and Scherft, J. P. 1991. Morphologic features of bone in human osteopetrosis. *Bone* 12:411–419.

48. Hentzen, E. R., Neelamegham, S., Kansas, G. S., Benanti, J. A., McIntire, L. V., Smith, C. W., and Simon, S. I. 2000. Sequential binding of CD11a/CD18 and CD11b/CD18 defines neutrophil capture and stable adhesion to intercellular adhesion molecule-1. *Blood* 95:911–920.

49. Howe, A., Aplin, A. E., Alahari, S. K., and Juliano, R. L. 1998. Integrin signaling and cell growth control. *Curr. Opin. Cell Biol.* 10:220–231.

50. Huang, Y. W., Baluna, R., and Vitetta, E. S. 1997. Adhesion molecules as targets for cancer therapy. *Histol. Histopathol.* 12:467–477.

51. Humphries, M. J. 2000. Integrin structure. *Biochem. Soc. Trans.* 28:311–339.

52. Hynes, R. O. 1992. Integrins: versatility, modulation, and signaling in cell adhesion. *Cell* 69:11–25.

53. Jakubowski, A., Rosa, M. D., Bixler, S., Lobb, R., and Burkly, L.C. 1995. Vascular cell adhesion molecule (VCAM)-Ig fusion protein defines distinct affinity states of the very late antigen-4 (VLA-4) receptor. *Cell Adhes. Commun.* 3:131–142.

54. Jockusch, B. M., Bubeck, P., Giehl, K., Kroemker, M., Moschner, J., Rothkegel, M., Rudiger, M., Schluter, K., Stanke, G., and Winkler, J. 1995. The molecular architecture of focal adhesions. *Annu. Rev. Cell Dev. Biol.* 11:379–416.

55. Kolanus, W. and Zeitlmann, L. 1998. Regulation of integrin function by inside-out signaling mechanisms. *Curr. Top. Microbiol. Immunol.* 231:33–49.

56. Kuckuck, F. W., Edwards, B. S., and Sklar, L. A. 2001. High throughput flow cytometry. *Cytometry* 44:83–90.

57. Larson, R. S. and Springer, T. A. 1990. Structure and function of leukocyte integrins. *Immunol. Rev.* 114:181–217.

57a. Larson, R.S., Davis, T., Bologa, C., Semenuk, G., Vijayan, S., Li, Y., Oprea, T., Chigaev, A., Wagner, C.R., and Sklar, L.A. In press. Dissociation of I domain and global conformational changes in LFA-1: Refinement of small molecule-I domain structure-activity relationships. *Biochemistry.*

58. Laurenzi, I. J. and Diamond, S. L. 1999. Monte Carlo simulation of the heterotypic aggregation kinetics of platelets and neutrophils. *Biophys. J.* 77:1733–1746.

59. Lawrence, M. B., Berg, E. L., Butcher, E. C., and Springer, T. A. 1995. Rolling of lymphocytes and neutrophils on peripheral node addressin and subsequent arrest on ICAM-1 in shear flow. *Eur. J. Immunol.* 25:1025–1031.

60. Lebow, L. T., Stewart, C. C., Perelson, A. S., and Bonavida, B. 1986. Analysis of lymphocyte-target conjugates by flow cytometry. I. Discrimination between killer and nonkiller lymphocytes bound to targets and sorting of conjugates containing one or multiple lymphocytes. *Nat. Immun. Cell Growth Regul.* 5:221–237.

61. Leeuwenberg, J. F., von Asmuth, E. J., Jeunhomme, T. M., and Buurman, W. A. 1990. IFN-gamma regulates the expression of the adhesion molecule ELAM-1 and IL-6 production by human endothelial cells in vitro. *J. Immunol.* 145:2110–2114.

62. Lewinsohn, D. M., Bargatze, R. F., and Butcher, E. C. 1987. Leukocyte-endothelial cell recognition: evidence of a common molecular mechanism shared by neutrophils, lymphocytes, and other leukocytes. *J. Immunol.* 138:4313–4321.

63. Liu, Y., Chen, B. P., Lu, M., Zhu, Y., Stemerman, M. B., Chien, S., and Shyy, J. Y. 2002. Shear stress activation of SREBP1 in endothelial cells is mediated by integrins. *Arterioscler. Thromb. Vasc. Biol.* 22:76–81.

64. Long, M., Goldsmith, H. L., Tees, D. F., and Zhu, C. 1999. Probabilistic modeling of shear-induced formation and breakage of doublets cross-linked by receptor-ligand bonds. *Biophys. J.* 76:1112–1128.

65. Luce, G. G., Sharrow, S. O., Shaw, S., Gallop, P.M. 1985. Enumeration of cytotoxic cell-target cell conjugates by flow cytometry using internal fluorescent stains. *Biotechniques* 3:270–272.

66. Luscinskas, F. W., Ding, H., Tan, P., Cumming, D., Tedder, T. F., and Gerritsen, M. E. 1996. L- and P-selectins, but not CD49d (VLA-4) integrins, mediate monocyte initial attachment to TNF-alpha-activated vascular endothelium under flow in vitro. *J. Immunol.* 157:326–335.

67. Manabe, A., Coustan-Smith, E., Behm, F. G., Raimondi, S. C., and Campana, D. 1992. Bone marrow-derived stromal cells prevent apoptotic cell death in B-lineage acute lymphoblastic leukemia. *Blood* 79:2370–2377.

68. Manabe, A., Murti, K. G., Coustan-Smith, E., Kumagai, M., Behm, F. G., Raimondi, S. C., and Campana, D. 1994. Adhesion-dependent survival of normal and leukemic human B lymphoblasts on bone marrow stromal cells. *Blood* 83:758–766.

69. Marui, N., Offermann, M. K., Swerlick, R., Kunsch, C., Rosen, C. A., Ahmad, M., Alexander, R. W., and Medford, R. M. 1993. Vascular cell adhesion molecule-1 (VCAM-1) gene transcription and expression are regulated through an antioxidant-sensitive mechanism in human vascular endothelial cells. *J. Clin. Invest.* 92:1866–1874.

70. Mogilner, A. and Oster, G. 1996. Cell motility driven by actin polymerization. *Biophys. J.* 71:3030–3045.

71. Myers, C. L., Wertheimer, S. J., Schembri-King, J., Parks, T., and Wallace, R. W. 1992. Induction of ICAM-1 by TNF-alpha, IL-1 beta, and LPS in human endothelial cells after downregulation of PKC. *Am. J. Physiol.* 263:C767–C772.

72. Nagel, T., Resnick, N., Dewey, C. F., Jr., and Gimbrone, M. A., Jr. 1999. Vascular en-

dothelial cells respond to spatial gradients in fluid shear stress by enhanced activation of transcription factors. *Arterioscler. Thromb. Vasc. Biol.* 19:1825–1834.

73. Naughton, B. A., Sibanda, B., Azar, L., and San Roman, J. 1992. Differential effects of drugs upon hematopoiesis can be assessed in long-term bone marrow cultures established on nylon screens. *Proc. Soc. Exp. Biol. Med.* 199:481–490.

74. Neelamegham, S., Munn, L. L., and Zygourakis, K. 1997. A model for the kinetics of homotypic cellular aggregation under static conditions. *Biophys. J.* 72:51–64.

75. Neelamegham, S., Taylor, A. D., Burns, A. R., Smith, C. W., and Simon, S. I. 1998. Hydrodynamic shear shows distinct roles for LFA-1 and Mac-1 in neutrophil adhesion to intercellular adhesion molecule-1. *Blood* 92:1626–1638.

76. Neelamegham, S., Taylor, A. D., Hellums, J. D., Dembo, M., Smith, C. W., and Simon, S. I. 1997. Modeling the reversible kinetics of neutrophil aggregation under hydrodynamic shear. *Biophys. J.*, 72:1527–1540.

77. Neelamegham, S., Taylor, A. D., Shankaran, H., Smith, C. W., and Simon, S. I. 2000. Shear and time-dependent changes in Mac-1, LFA-1, and ICAM-3 binding regulate neutrophil homotypic adhesion. *J. Immunol.* 164:3798–3805.

78. Nolan, J. P., Lauer, S., Prossnitz, E. R., and Sklar, L. A. 1999. Flow cytometry: a versatile tool for all phases of drug discovery. *Drug Discov. Today* 4:173–180.

79. Osborn, L., Hession, C., Tizard, R., Vassallo, C., Luhowskyj, S., Chi-Rosso, G., and Lobb, R. 1989. Direct expression cloning of vascular cell adhesion molecule 1, a cytokine-induced endothelial protein that binds to lymphocytes. *Cell* 59:1203–1211.

80. Paul, L. C. and Issekutz, T. B. 1997. *Adhesion Molecules in Health and Disease*. Marcel Dekker, New York.

81. Petering, H., Hochstetter, R., Kimmig, D., Smolarski, R., Kapp, A., and Elsner, J. 1998. Detection of MCP-4 in dermal fibroblasts and its activation of the respiratory burst in human eosinophils. *J. Immunol.* 160:555–558.

82. Pinsky, D. J., Naka, Y., Liao, H., Oz, M. C., Wagner, D. D., Mayadas, T. N., Johnson, R. C., Hynes, R. O., Heath, M., Lawson, C. A., and Stern, D. M. 1996. Hypoxia-induced exocytosis of endothelial cell Weibel-Palade bodies. A mechanism for rapid neutrophil recruitment after cardiac preservation. *J. Clin. Invest.* 97:493–500.

83. Piper, J. W., Swerlick, R. A., and Zhu, C. 1998. Determining force dependence of two-dimensional receptor-ligand binding affinity by centrifugation. *Biophys. J.* 74:492–513.

84. Postlethwaite, A. E., Holness, M. A., Katai, H., and Raghow, R. 1992. Human fibroblasts synthesize elevated levels of extracellular matrix proteins in response to interleukin 4. *J. Clin. Invest* 90:1479–1485.

85. Postlethwaite, A. E. and Kang, A. H. 1980. Characterization of guinea pig lymphocyte-derived chemotactic factor for fibroblasts. *J. Immunol.* 124:1462–1466.

86. Postlethwaite, A. E., Keski-Oja, J., Moses, H. L., and Kang, A. H. 1987. Stimulation of the chemotactic migration of human fibroblasts by transforming growth factor beta. *J. Exp. Med.* 165:251–256.

87. Postlethwaite, A. E. and Seyer, J. M. 1990. Stimulation of fibroblast chemotaxis by human recombinant tumor necrosis factor alpha (TNF-alpha) and a synthetic TNF-alpha 31-68 peptide. *J. Exp. Med.* 172:1749–1756.

88. Postlethwaite, A. E. and Seyer, J. M. 1991. Fibroblast chemotaxis induction by human recombinant interleukin-4. Identification by synthetic peptide analysis of two chemotactic domains residing in amino acid sequences 70-88 and 89-122. *J. Clin. Invest.* 87:2147–2152.

89. Prosper, F. and Verfaillie, C. M. 2001. Regulation of hematopoiesis through adhesion receptors. *J. Leukoc. Biol.* 69:307–316.

90. Ramirez, S. A. 2002. High throughput cell adhesion assay: effects of chemokines on VLA-4/VCAM-1 adhesion. Master's thesis. University of New Mexico, Albuquerque.

91. Ramirez, S. A., Aiken, C. M., Andrzejewski, B., Sklar, L. A., and Edwards, B.S. 2003. High-throughput flow cytometry: validation in microvolume bioassays. *Cytometry* 53A:55–65.

92. Ransom, J. T., Edwards, B. S., Kuckuck, F., Okun, A., Mattox, D. K., Prossnitz, E. R., and Sklar, L. A. 2000. Flow cytometry systems for drug discovery and development. *Proceeding of SPIE*, vol. 3921. Pages 90–100 in D.L. Farkas and R.C. Leif (eds.), *Optical Di-*

agnostics of Living Cells III. International Society of Optical Engineering/SPIE Publishing Services, Bellington, WA.

93. Resnick, N. and Gimbrone, M. A., Jr. 1995. Hemodynamic forces are complex regulators of endothelial gene expression. *FASEB J.* 9:874–882.

94. Resnick, N., Yahav, H., Khachigian, L. M., Collins, T., Anderson, K. R., Dewey, F. C., and Gimbrone, M. A., Jr. 1997. Endothelial gene regulation by laminar shear stress. *Adv. Exp. Med. Biol.* 430:155–164.

95. Reuss-Borst, M. A., Klein, G., Waller, H. D., and Muller, C. A. 1995. Differential expression of adhesion molecules in acute leukemia. *Leukemia* 9:869–874.

96. Robledo, M. M., Sanz-Rodriguez, F., Hidalgo, A., and Teixido, J. 1998. Differential use of very late antigen-4 and -5 integrins by hematopoietic precursors and myeloma cells to adhere to transforming growth factor-beta1-treated bone marrow stroma. *J. Biol. Chem.* 273:12056–12060.

97. Roecklein, B. A. and Torok-Storb, B. 1995. Functionally distinct human marrow stromal cell lines immortalized by transduction with the human papilloma virus E6/E7 genes. *Blood* 85:997–1005.

98. Rothlein, R. and Springer, T. A. 1986. The requirement for lymphocyte function-associated antigen 1 in homotypic leukocyte adhesion stimulated by phorbol ester. *J. Exp. Med.* 163:1132–1149.

99. Ruoslahti, E. 1997. Integrins as signaling molecules and targets for tumor therapy. *Kidney Int.* 51:1413–1417.

100. Sanchez-Garcia, J., Atkins, C., Pasvol, G., Wilkinson, R. J., and Colston, M. J. 1996. Antigen-driven shedding of L-selectin from human gamma delta T cells. *Immunology* 89:213–219.

101. Sastry, S. K. and Burridge, K. 2000. Focal adhesions: a nexus for intracellular signaling and cytoskeletal dynamics. *Exp. Cell Res.* 261:25–36.

102. Schwartz, M. A., Schaller, M. D., and Ginsberg, M. H. 1995. Integrins: emerging paradigms of signal transduction. *Annu. Rev. Cell Dev. Biol.* 11:549–599.

103. Scopes, J., Ismail, M., Marks, K. J., Rutherford, T. R., Draycott, G. S., Pocock, C., Gordon-Smith, E. C., and Gibson, F. M. 2001. Correction of stromal cell defect after bone marrow transplantation in aplastic anaemia. *Br. J. Haematol.* 115:642–652.

104. Sdougos, H. P., Bussolari S.R., and Dewey, C. F. 1984. Secondary flow and turbulence in a cone-and-plate device. *J. Fluid Mech.* 138:379–404.

105. Senior, R. M., Huang, J. S., Griffin, G. L., and Deuel, T. F. 1985. Dissociation of the chemotactic and mitogenic activities of platelet-derived growth factor by human neutrophil elastase. *J. Cell Biol.* 100:351–356.

106. Seppa, H., Grotendorst, G., Seppa, S., Schiffmann, E., and Martin, G. R. 1982. Platelet-derived growth factor in chemotactic for fibroblasts. *J. Cell Biol.* 92:584–588.

107. Seppa, H. E., Yamada, K. M., Seppa, S. T., Silver, M. H., Kleinman, H. K., and Schiffmann, E. 1981. The cell binding fragment of fibronectin is chemotactic for fibroblasts. *Cell Biol. Int. Rep.* 5:813–819.

108. Shao, J. Y. and Hochmuth, R. M. 1999. Mechanical anchoring strength of L-selectin, beta2 integrins, and CD45 to neutrophil cytoskeleton and membrane. *Biophys. J.* 77:587–596.

109. Shen, J., Luscinskas, F. W., Connolly, A., Dewey, C. F., Jr., and Gimbrone, M. A., Jr. 1992. Fluid shear stress modulates cytosolic free calcium in vascular endothelial cells. *Am. J. Physiol.* 262:C384–C390.

110. Shen, J., Luscinskas, F. W., Gimbrone, M. A., Jr., and Dewey, C. F., Jr. 1994. Fluid flow modulates vascular endothelial cytosolic calcium responses to adenine nucleotides. *Microcirculation* 1:67–78.

111. Simon, S. I., Chambers, J. D., Butcher, E., and Sklar, L. A. 1992. Neutrophil aggregation is beta 2-integrin- and L-selectin-dependent in blood and isolated cells. *J. Immunol.* 149:2765–2771.

112. Simon, S. I., Chambers, J. D., and Sklar, L. A. 1990. Flow cytometric analysis and modeling of cell-cell adhesive interactions: the neutrophil as a model. *J. Cell Biol.* 111:2747–2756.

113. Simon, S. I., Rochon, Y. P., Lynam, E. B., Smith, C. W., Anderson, D. C., and Sklar, L. A. 1993. Beta 2-integrin and L-selectin are obligatory receptors in neutrophil aggregation. *Blood* 82:1097–1106.
114. Springer, T. A. 1994. Traffic signals for lymphocyte recirculation and leukocyte emigration: the multistep paradigm. *Cell* 76:301–314.
115. St Pierre, Y., Hugo, P., Legault, D., Tremblay, P., and Potworowski, E. F. 1996. Modulation of integrin-mediated intercellular adhesion during the interaction of thymocytes with stromal cells expressing VLA-4 and LFA-1 ligands. *Eur. J. Immunol.* 26:2050–2055.
116. Staunton, D. E., Dustin, M. L., and Springer, T. A. 1989. Functional cloning of ICAM-2, a cell adhesion ligand for LFA-1 homologous to ICAM-1. *Nature* 339:61–64.
117. Staunton, D. E., Marlin, S. D., Stratowa, C., Dustin, M. L., and Springer, T. A. 1988. Primary structure of ICAM-1 demonstrates interaction between members of the immunoglobulin and integrin supergene families. *Cell* 52:925–933.
118. Storkus, W. J., Balber, A. E., and Dawson, J. R. 1986. Quantitation and sorting of vitally stained natural killer cell-target cell conjugates by dual beam flow cytometry. *Cytometry* 7:163–170.
119. Tandon, P. and Diamond, S. L. 1998. Kinetics of beta2-integrin and L-selectin bonding during neutrophil aggregation in shear flow. *Biophys. J.* 75:3163–3178.
120. Tauro, S., Hepburn, M. D., Peddie, C. M., Bowen, D. T., and Pippard, M. J. 2002. Functional disturbance of marrow stromal microenvironment in the myelodysplastic syndromes. *Leukemia* 16:785–790.
121. Taylor, A. D., Neelamegham, S., Hellums, J. D., Smith, C. W., and Simon, S. I. 1996. Molecular dynamics of the transition from L-selectin- to beta 2-integrin-dependent neutrophil adhesion under defined hydrodynamic shear. *Biophys. J.* 71:3488–3500.
122. Tees, D. F., Waugh, R. E., and Hammer, D. A. 2001. A microcantilever device to assess the effect of force on the lifetime of selectin-carbohydrate bonds. *Biophys. J.* 80:668–682.
123. Tseng, H., Peterson, T. E., and Berk, B. C. 1995. Fluid shear stress stimulates mitogen-activated protein kinase in endothelial cells. *Circ. Res.* 77:869–878.
124. Tsuji, T., Waga, I., Tezuka, K., Kamada, M., Yatsunami, K., and Kodama, H. 1998. Integrin beta2 (CD18)-mediated cell proliferation of HEL cells on a hematopoietic-supportive bone marrow stromal cell line, HESS-5 cells. *Blood* 91:1263–1271.
125. Uguccioni, M., Loetscher, P., Forssmann, U., Dewald, B., Li, H., Lima, S. H., Li, Y., Kreider, B., Garotta, G., Thelen, M., and Baggiolini, M. 1996. Monocyte chemotactic protein 4 (MCP-4), a novel structural and functional analogue of MCP-3 and eotaxin. *J. Exp. Med.* 183:2379–2384.
126. Vane, J. R., Anggard, E.E., and Botting, R. M. 1990. Regulatory functions of the vascular endothelium. *N. Engl. J. Med.* 323:27–36.
127. von Andrian, U. H. and MacKay, C. R. 2000. T-cell function and migration. Two sides of the same coin. *N. Engl. J. Med.* 343:1020–1034.
128. Wagner, D. D. 1993. The Weibel-Palade body: the storage granule for von Willebrand factor and P-selectin. *Thromb. Haemost.* 70:105–110.
129. Wedemeyer, N. and Potter, T. 2001. Flow cytometry: an "old" tool for novel applications in medical genetics. *Clin. Genet.* 60:1–8.
130. Welzenbach, K., Hommel, U. and Weitz-Schmidt, G. 2002. Small molecule inhibitors induce conformational changes in the I domain and the I-like domain of lymphocyte function-associated antigen-1. Molecular insights into integrin inhibition. *J. Biol. Chem.* 277:10590–10598.
131. Winter, S. S., Sweatman, J., Shuster, J.J., Link, M. P., Amylon, M. D., Pullen, J., Camitta, B. M., and Larson, R. S. 2002. Bone marrow stroma-supported culture of T-lineage acute lymphoblastic leukemic cells predicts treatment outcome in children: a Pediatric Oncology Group study. *Leukemia* 16:1121–1126.
132. Winter, S. S., Sweatman, J. J. and Larson, R. S. 2000. Improved quantification of cell survival on stromal monolayers by flow cytometric analyses. *Cytometry* 40:26–31.
133. Winter, S. S., Sweatman, J. J., Lawrence, M. B., Rhoades, T. H., Hart, A. L., and Larson, R. S. 2001. Enhanced T-lineage acute lymphoblastic leukaemia cell survival on

bone marrow stroma requires involvement of LFA-1 and ICAM-1. *Br. J. Haematol.* 115:862–871.

134. Zhao, L. C., Edgar, J. B., and Dailey, M. O. 2001. Characterization of the rapid proteolytic shedding of murine L-selectin. *Dev. Immunol.* 8:267–277.

135. Zhu, C., Bao, G., and Wang, N. 2000. Cell mechanics: mechanical response, cell adhesion, and molecular deformation. *Annu. Rev. Biomed. Eng.* 2:189–226.

136. Zhurkov, S. N. Kinetic Concept of the Strength of Solids. *Int. J. Fracture Mech.* 1, 311–323. 2002. 1-1-0065.

137. Zwartz, G. J., Chigaev, A., Foutz, T., Larson, R.S., and Sklar, L. A. 2004. Relationship between molecular and cellular dissociation rates for VLA-4/VCAM-1 interaction in the absence of shear. *Biophysical Journal.*86:1243–1252.

Index

Printed in the United States
By Bookmasters